Economic Complexity and Evolution

More information about this series at http://www.springer.com/series/11583

Stephan Dabbert • Iris Lewandowski •
Jochen Weiss • Andreas Pyka
Editors

Knowledge-Driven Developments in the Bioeconomy

Technological and Economic Perspectives

Springer

Editors
Stephan Dabbert
University of Hohenheim
Stuttgart, Germany

Iris Lewandowski
University of Hohenheim
Stuttgart, Germany

Jochen Weiss
University of Hohenheim
Stuttgart, Germany

Andreas Pyka
University of Hohenheim
Stuttgart, Germany

ISSN 2199-3173 ISSN 2199-3181 (electronic)
Economic Complexity and Evolution
ISBN 978-3-319-58373-0 ISBN 978-3-319-58374-7 (eBook)
DOI 10.1007/978-3-319-58374-7

Library of Congress Control Number: 2017947307

Printed on acid-free paper

This Springer imprint is published by Springer Nature
The registered company is Springer International Publishing AG
The registered company address is: Gewerbestrasse 11, 6330 Cham, Switzerland

Editorial

The transformation of our economic systems towards a knowledge-based bioeconomy has become an important vision at the beginning of the twenty-first century in order to strive for sustainability. New food and feed, energy from residual material, chemicals and plastics from plants and algae: bioeconomy opens new paths to new products, to new production processes and to a modern economy which no longer challenges the planetary boundaries. Its raw materials are bio-based, and it is supposed to overcome the heritage of the early industrial revolution in the nineteenth and twentieth century, namely the lock-in into a carbon-based economy. In a bioeconomy, resources come from plants, animals, or microorganisms. In production, it also uses biological processes for a more sustainable production that conserves energy and resources. Together with digitalization, robotics and changes in consumption behaviour, this transformation is supposed to fundamentally change the organization of the economy towards conserving the natural resources and at the same time performing cleaner production, decreasing energy usage and developing towards higher quality consumption—in short, towards sustainability.

To substantiate this vision, the bioeconomy needs a great deal of research, innovation and a fundamental change in the economy and society. Bioeconomy gives food security and adequate nutrition the highest priority. Climate change mitigation, health and species diversity are also important aspects. The University of Hohenheim has made bioeconomy the focus of its research and teaching. Unique in Germany, it has the comprehensive expertise necessary to cover the entire bio-based value chain. It covers every topic from plant and animal production to new technical processes to the necessary change processes in the economy and society. In this, agricultural, nutritional and food scientists work hand in hand with biologists and biotechnologists as well as economists and social scientists.

Obviously, the transformation to bioeconomy is not a national task. It can only succeed in its goals as a global project which exploits the synergies and commonalities among different geographical regions as well as socio-economic environments. Furthermore, by acknowledging the heterogeneity and variety of development paths, it becomes clear that there exists neither a unique nor an optimal strategy for a growing global bioeconomy. The University of Hohenheim has a long history of international cooperation in knowledge generation and diffusion which forms an important part of its bioeconomy focus. The Strategic Network Bio-based Economy (BECY) coordinated by the University of Hohenheim is a strategic network of six leading universities in the emerging research field of bio-based economy and related disciplines. Besides the University of Hohenheim, the members of this DAAD-funded network are the University of São Paulo (USP), São Paulo State University (UNESP, both Brazil), the Metropolitan Autonomous University (UAM, Mexico), the University of Guelph (UoG, Canada) and the University of Copenhagen (KU, Denmark). The aim of BECY is to set up sustainable cooperation in mutual and complementing areas of bio-based economy in order to strengthen excellence in the quality of research and teaching.

In October 2015, BECY organized a conference entitled "Strategies for Knowledge-Driven Developments in the Bioeconomy" where besides the BECY partners also external participants from Australia, Costa Rica, France, Italy, the Netherlands and Malaysia were invited to broaden the international cooperation and to provide a global perspective. During the three conference days, more than 50 participants exchanged their views on topics covering strategies for the bioeconomy implementation as well as biomass production, biomaterials, sustainable food processing and socio-economic approaches for the transformation towards a knowledge-based bioeconomy. The conference contributions are the basis for the book at hand. After the discussions at the conference, we started an editorial process to select the various chapters of this book which were written to document the state of the discussion and to outline potential future directions of collaborative research. Due to its global and interdisciplinary orientation, this book can be considered a manifesto for moving ahead in the implementation of the vision to transform our socio-economic systems towards a knowledge-based bioeconomy.

The first part of the volume, **Bioeconomy systems: Theoretical underpinnings**, approaches the knowledge-based bioeconomy from a systemic perspective. In his chapter "Transformation of Economic Systems: The Bio-economy Case", Andreas Pyka (University of Hohenheim, Germany) explains why the introduction of a knowledge-based bioeconomy will require a radical and qualitative transformation. From a modern innovation economic perspective, the establishment of a sustainable and inclusive bioeconomy will not be achieved by the substitution of one or more technologies, but will require a whole package of mutually dependent technologies, infrastructural developments, behavioural adaptations and institutional changes. The nature of such structural change and the necessary contribution of new knowledge are explored in the following chapter by Pier Paolo Saviotti (Utrecht University, Netherlands). In his contribution "Structural Change, Knowledge and the Bioeconomy", he demonstrates what changes in the economic structure (e.g. the composition of economic sectors) as well as in the knowledge structure (e.g. through the emergence of new fields of knowledge such as molecular biology) are currently observed and can be expected in the future. In his chapter "Some Thoughts about the Bio-economy as Intelligently Navigated Complex Adaptive System", Hugo de Vries (University of Montpellier, France) sets out to solve the observed paradox between the general upscaling and homogenization of production processes and the necessity to balance economic growth with social and environmental goals. By understanding the knowledge-based bioeconomy as a complex adaptive system, the author derives quite concrete principles to guide future bioeconomy research with the goal to improve the efficiency of innovations in terms of their contribution to the bioeconomy.

Part two, **Framing the Bioeconomy: Regional and National Approaches**, presents cases of bioeconomies observed around the world illustrating a diversity of implementation approaches of the bioeconomy from nation to nation. In their paper "Varieties of Knowledge-Based Bioeconomies", Sophie Urmetzer and Andreas Pyka (University of Hohenheim, Germany) portray national pathways of the European Union member states towards a knowledge-based bioeconomy. By compiling and analysing economic, environmental and social indicators of each member state, they show regional and national differences as well as similarities of national innovation systems, thus presenting a basis for policy learning and the improvement of innovation systems. The following paper, "International Bioeconomy Innovations in Central America" written by Mercedes Montero Vega and Olman Quiros Madrigal (University of Costa Rica), gives an overview of current bioeconomy policy and research in Latin America. According to the authors, the great bioeconomic potential of this species-rich and climatically favoured region should be exploited more effectively and sustainably in order to trigger economic growth while protecting the environment and improving the lives of the rural population.

The remaining contributions of this part present four cases of bioeconomy potential and strategies on the national level. The authors Alcides L. Leao, Ivana Cesarino (both São Paulo State University, Brazil), Suresh Narine (Trent University, Canada) and Mohini Sain (University of Toronto, Canada) examine the state of the art and the possible future of the Brazilian bioeconomy in their chapter "Potentials for the Brazilian Bioeconomy Innovation System". It is argued that due to a tendency towards deindustrialization and a lack of innovation incentives Brazil currently underachieves its enormous agricultural productivity. Just like in Brazil, agriculture makes up for quite a high share in the national economy of the Australian island state of Tasmania. In "Tasmania's Bioeconomy: Using All Capitals to Sustain Innovative and Entrepreneurial Agrifood Value Chains", Holger Meinke, Laurie Bonney, Katherine Evans (University of Tasmania, Australia) and Morgan Miles (University of Canterbury, New Zealand) take a critical look at the national policies to strengthen the country's agricultural sector. In order to tap Tasmania's full agricultural potential to move towards a knowledge-based bioeconomy, the government has to concentrate on investing in research, knowledge creation, marketing, value chain innovations and capability development instead of just improving the technical conditions for agriculture. The authors of "Agricultural Biomass Utilisation as a Key Driver for Malaysian Bioeconomy", Ismael Norli, Arfifin Fazilah (University of Sains, Malaysia) and I. Mohamad Pazli (Northern Skills Development Centre Sdn. Bhd. Penang, Malaysia), present the Malaysian government's quite detailed plans to foster bioeconomic research and development in this agriculturally strong country. Quite different from the previous examples, the Danish case is illustrated by K.E. Markedal, J.C. Sørensen and S. Sørensen (University of Copenhagen, Denmark) in their contribution "University-Industry Relationships in the Bioeconomy Innovation System of Denmark". In this highly developed and densely populated country, the authors expect potential for improving bioeconomic performance in extending product portfolios and optimizing value chains. The focus of the national bioeconomic development should thus be put on strengthening the collaboration among enterprises up- and downstream and between enterprises and academia for facilitating the creation, diffusion and absorption of new knowledge critical for bioeconomic applications and processes.

The last two parts of the volume deal with the more practical aspects of the bioeconomy: the challenge of producing the required raw material in a sustainable way and the question of optimal processes and supply chains. The third part, **Resources of the Bioeconomy: Sustainable Biomass Supply,** starts out with "Increasing Biomass Production to Sustain the Bioeconomy", in which the author Iris Lewandowski (University of Hohenheim, Germany) explores the question whether the global biomass production can meet the increasing demand on a sustainable basis. Demand-driven politics have succeeded in globally increasing the market for renewable resources, but the amounts of biomass and arable land that currently are sustainably managed are limited. This calls for strategies that increase productivity in sustainable agricultural intensification, including plant breeding, modern agricultural production technologies and empowerment of farmers. Once again, it becomes obvious that solutions cannot be found in higher-further-faster strategies but in more intelligent ways of handling and using the resources available. Brazil's sugarcane sector serves as a good example for the diversification of the uses of a formerly very specific resource. In their chapter "Importance of Sugarcane in Brazilian and World Bioeconomy", Reges Heinrichs (Sao Paulo State University, Brazil), Rafael Otto (University of Sao Paulo, Brazil), Aline Magalhães and Guilherme Constantino Meirelles (Sao Paulo State University, Brazil) present the multiple uses of all parts of the sugarcane plant within the bioeconomy for the production of fuel, electricity, feed and food and emphasize the role of Brazilian sugarcane for the domestic and global bioeconomy. Another important source of renewable energy from Brazil is eucalypt for woodchip production. The colleagues Saulo Philipe Sebastião Guerra, Guilherme Oguri, Izabel Cristina Takitane, Caterina Giulia Lembo, Maura Esperancini and Seiko Tsutsui (Sao Paulo State University, Brazil) have calculated the efficiency of different felling techniques and presented their results in "Economic Evaluation of Short Rotation Eucalyptus Plantation Harvesting System: A Case Study". Muhammad Arif, Muhammad Riaz, C. Joe Martin, Yarmilla Reinprecht (University of Guelph, Canada), Leonardo Simon (University of Waterloo, Canada), Bill Dean and K. Peter Pauls (University of Guelph, Canada) close this part with examining the sustainability of using plant fibres for bioproduct manufacturing in general and identifying corn-based composite characteristics. The authors conclude that using the fibre remaining from the corn harvest for the production of bio-filled thermoplastic composites is feasible and sustainable. For its use on a commercial scale, however, fibre quality will have to be standardized first.

The final part, **Bioeconomy Applications: Optimizing Processes and Management,** focuses on supply chains and technologies making up the bioeconomy. As many of the authors of this volume have pointed out, the path towards a sustainable bioeconomy is paved by new technologies and alternative materials, but must be accompanied by new combinations of known resources and technologies as well as by efficiency gains and organizational optimisation. Traditional supply chains and conventional production processes are thus challenged. For an overview of biomass-based supply chain models, Stephan Fichtner and Herbert Meyr (University of Hohenheim, Germany) have reviewed the recent literature on the long-term strategic planning of biomass-based supply chains. In their chapter "Strategic Supply Chain Planning in Biomass-Based Industries—A Literature Review of Quantitative Models", they reveal certain characteristics typical for biomass-based supply chains that must be taken into consideration in strategic planning: they usually comprise quite heterogeneous participants, they deal with high uncertainty in resource supply and transportation costs of raw material (biomass) are very high. The following chapter "Structuring the Planning Tasks in Biomass-Based Supply Chains" resumes exactly these challenging features and uncertainties of biomass-based supply chains and presents the relevant planning tasks for their management. The authors Hendrik Butemann and Katja Schimmelpfeng (University of Hohenheim, Germany) claim that the specific structure and the distinct stakeholders of a biomass-based supply chain require a different management approach and simply adapting established models would not work. The final two chapters present two specific cases of new technologies to avoid the emission of greenhouse gases and minimize waste. A low-tech example is showcased by Olman Quirós Madrigal and Lady Arias Fallas (University of Costa Rica), who have calculated the efficiency of the installation of a biodigester for the production of biogas and organic fertilizer from excessive cattle excreta accumulated by a Costa Rican farmers' cooperative in their chapter "The Use of Biomass for Energy Production and Organic Fertilizer for Mitigating Climate Change and Improving the Competitiveness of the Agricultural Enterprise: The Case of the UPAP in Puriscal, Costa Rica". The colleagues from Sao Paulo, Marisa Aparecida Bismara Regitano-d'Arce, Naiane Sangaletti-Gerhard and Larissa Braga Bueno-Borges (University of Sao Paulo, Brazil), have found a new method to optimize biodiesel production. In their chapter "Bioethanol as the Sole Solvent for Vegetable Oil Extraction and Biodiesel Production", they argue that oil extraction with ethanol does not require the usual oil refining steps prior to biodiesel production, making the process more economical and environmentally friendly. Furthermore, the by-products of this process are of such quality that they can be used by other industries.

The four parts of this volume illustrate the variety of research areas that are essential for putting the sustainable knowledge-based bioeconomy on track. We have to

- Further strengthen the theoretical base of bioeconomy systems by applying an global, inter- and transdisciplinary perspective

- Monitor and steer national, regional and global action and accomplishment towards this goal
- Grapple the ecology and biology of the central resources
- Keep searching for new materials, alternative technologies, novel configurations and processes as well as social innovations no one has even thought of before.

Stephan Dabbert
Iris Lewandowski
Jochen Weiss
Andreas Pyka

Contents

Part I
Bioeconomy Systems: Theoretical Underpinnings

Transformation of Economic Systems: The Bio-Economy Case

Andreas Pyka

Abstract To improve sustainability, the global economic system has to undergo severe transformation processes. This chapter deals with the possibility of an innovation-triggered transformation towards a knowledge-based bioeconomy, which is supposed to overcome the current lock-in into a fossil fuel-based CO_2-intensive production. To do this, a Neo-Schumpeterian view is applied that highlights the complex interplay in knowledge-generation and -diffusion processes between firms, consumers and government institutions. By applying the Neo-Schumpeterian approach it becomes obvious that innovation and economic growth are part of the solution and not part of the sustainability problem. The shift from quantitative growth—prevailing in textbook economics—to qualitative development—prevailing in Neo-Schumpeterian economics—makes the difference and affects all agents and institutions in an economic system, which needs to be designed as a *dedicated innovation system* supporting the transformation towards a knowledge-based bioeconomy.

1 Introduction

After more than 200 years of industrial production, large parts of the world population are richer than ever before. Simultaneously, past industrial production is closely linked with the exploitation of natural resources and the strong accumulation of environmentally harmful greenhouse gases, thereby endangering human survival. It is evident that things cannot continue as before. But how can future development be shaped without threatening our natural basis of life and contributing to a high and increased level of welfare at the same time? At the beginning of the twenty-first century, many economies all around the world place big hope in the so-called *knowledge-based bioeconomy*. Is this a possible way out? Can economic growth and development, widely the cause of the problem, also become part of the solution? The following contribution discusses the possibility of transforming the global

A. Pyka (✉)
Institute for Economics: Innovation Economics (520I), University of Hohenheim, Stuttgart, Germany
e-mail: a.pyka@uni-hohenheim.de

S. Dabbert et al. (eds.), *Knowledge-Driven Developments in the Bioeconomy*, Economic Complexity and Evolution, DOI 10.1007/978-3-319-58374-7_1

production system towards a knowledge-based bio-economy from the perspective of modern innovation economics.

Almost all economists agree that technological development substantially triggers quantitative growth in income per head. However, there is less consensus with respect to the qualitative characteristics of economic development: whereas mainstream-oriented parts of economics—often summarized under the heading neoclassical economics—focuses only on quantitative aspects and thus shows a short-term orientation, Neo-Schumpeterian economics focuses on qualitative aspects and thus on a change of fundamental economic structures over longer periods.

Generally, change can be either of an incremental type in terms of small improvements along well-known trajectories, or it can be more fundamental, leading to structural changes like the emergence of new and the disappearance of old industries. To simplify, we assume that incremental technological changes are based on existing technological solutions, whereas radical technological changes question major existing production processes. They might lead to massive changes of the global production system in the sense of "creative destruction" (Schumpeter 1943).

This chapter deals with a fundamental transformation of production systems: overcoming the lock-in situation of present production systems towards fossil fuels (Unruh 2000) and establishing a knowledge-based bio-economy at the same time (Pyka 2017; Pyka and Buchmann 2016). Without doubt this transformation process is radical, qualitative, and effective in the long-run only and it has to be considered under the Neo-Schumpeterian approach to innovation economics. It was already in his work *Business Cycles,* published in 1939, when Schumpeter revitalized *Kondratieff's Theory of long waves* in order to explain this process as a *regular process* in long-term economic development. His illustration of this change, which is characterized by its discontinuous nature, is famous: "*Add successively as many mail coaches as you please, you will never get a railway thereby*" (Schumpeter 1934, p. 64). Industrialization around the year 1800 represented the first long wave and was fueled by the steam engine and by cotton processing. Then, starting around the year 1850, the widespread availability of steel and the diffusion of railways constituted the second long wave that was again, at the beginning of the twentieth century, replaced by electrical technology and the chemical industry. In the middle of the previous century, the third long wave gained momentum by mass production and the automobile as well as the petrochemical industries. Thus, manufacturing activities focused on oil as a second fossil fuel apart from coal. Since the 1980s, one refers to the fifth long wave, which is reflected in the fast and ubiquitous diffusion and application of information and communication technology solutions.

Now, at the beginning of the twenty-first century, another paradigmatic change is in the air, being characterized, however, by one major difference to previous revolutions: whereas previous cycles were driven by technological bottlenecks and their overcoming, humans in the twenty-first century face the vital question of how to restore environmental sustainability of economic activities. The k*nowledge-based bio-economy* plays a key-role in this transformation process which, of course, like previous radical changes is characterized by fundamental uncertainty (Knight 1921).

Today, literature provides many alternative terms for the massive change shaking global production systems: Freeman (1991) and Dosi (1982) call them *techno-economic paradigm changes*, Sahal (1985) uses cartographic analogies and refers to *technological guideposts* that are pointing to *technological avenues*. All authors highlight the confrontation with profound changes economic systems are faced with over longer periods of time which question all established production approaches. Not a single technology is responsible for this phenomenon, but several complementary developments that include, apart from a package of mutually dependent technologies (e.g., combustion engine, petro chemistry, assembly line production), numerous infrastructural developments (e.g., road structure, filling station network), behavioral changes (e.g., suburbs and commuter flow, shopping malls outside the city centers) as well as institutional changes (e.g., spatial planning and commuter allowance, etc.). The old paradigm will not be replaced by the new one until all these elements interact.

The Neo-Schumpeterian approach provides us with crucial hints on the process of the forthcoming change. For this purpose, we introduce in the second section to the economic discussion of transformation processes and shortly outline the consideration of growth-pessimistic approaches that enjoy great popularity, such as *post-growth* or *de-growth* approaches. These are contrasted with the growth-optimistic approaches that cherish Schumpeter's intellectual heritage and rely on the creative forces of capitalistic economic systems to overcome the fundamental problems of the human society. Innovations are supported by the discovery and successful spread of new knowledge. Therefore, knowledge-based economies organize *innovation systems* composed of different actors which establish a creative environment for mutual learning and knowledge creation. That is what the third section of this chapter is about. No innovation would have ever been established if it had not attracted consumers' interest and if it had not been leveraged by their purchasing power. We will focus on these questions in section four. Knowledge-based societies consider new concepts in the sense of 'responsible innovation' that are decisive in bringing an entire economy on a new sustainable trajectory shaping growth and development. Section five deals with the massive economic impacts originating from these technological and knowledge-driven changes. It requires, besides technological change, also institutional change in a co-evolutionary fashion, if new sustainable technologies are to achieve the aspired transformation of the economic system.

2 Limits to Growth

The sustainability of a capitalistic organization of production, as it has been set up in western industrialized economies since the beginning of the industrial revolution at the end of the eighteenth century, has been questioned at the latest since 1972 when "The Limits to Growth" was published by the *Club of Rome* (Meadows et al. 1972). Since then, two fundamentally different solution strategies are being discussed within society: *conservation of resources by the abstinence from growth* on the one hand or *decoupling of growth and resource exploitation* on the other hand. The supporters of the first approach (Blewitt and Cunningham 2014; Kallis et al. 2014), summarized under the headings of "abstinence" and "downscaling", claim a renunciation of a way of life that is based on consumption and increasing deployment of resources. According to these approaches, market-oriented economic systems are not believed to manage endogenously a change towards sustainability. There are considerations that even call for a return to small-scale regional agriculture or subsistence economies, respectively. This is considered the only way to enable a sustainable and resource-friendly lifestyle and form of economic activity. To summarize, it is easy to see that these approaches are based on the neo-classical line of thought with the underlying assumption of stable economic structures and an understanding of economic growth as a sheer quantitative process.

The second approach, instead, is strongly characterized by the observation that innovations, market forces, structural change, and urban ways of life are both, part of the problem and part of the solution to the sustainability problem. This second approach is assigned to the Neo-Schumpeterian perspective with its qualitative perspective on economic development. Innovation-triggered development is characterized by both, a quantitative, i.e., income-increasing dimension and a qualitative, i.e., structure-changing dimension. In particular, at the end of the twentieth century and at the beginning of the twenty-first century, capitalist-oriented economies have demonstrated impressively their global power of change: in a short time more people are brought out of poverty (one of the 17 objectives of the UN's agenda 2030) by creative entrepreneurship in free markets than before by 50 years of development policies. Obviously, these developments have aggravated the resource problem and pollution to some extent; however, higher income economies move along the environmental Kuznets curve and organize cleaner production (Fagerberg et al. 2015). New creative solutions are able to reform our future economy in the sense of sustainability, thereby supporting the achievements of the UN's objectives towards a sustainable development and ensuring growth and development at the same time (Mazzucato and Perez 2015).

The leading idea of a knowledge-based economy is based on the notion that abstinence in the sense of economic down-scaling is neither the first nor the only solution. In principle, the opinion is shared—which includes both demand side and participatory elements—that, in accordance with the supporters of the first method, certain past patterns of production and consumption require urgent adjustments. Especially concepts resulting in a more intensive use of goods and therefore contributing to the economization of resources ('*sharing-economy*') are important. The same applies for closed-loop material cycles, recycling systems, and intelligent waste treatment. These concepts are perfectly applicable to triggering learning and behavioral changes on the demand side. However, the core idea consists of *supplying and demanding* new technological solutions within a comprehensive economic transformation process (Geels 2002), i.e., different goods and services are produced and demanded in different ways, which are characterized by sustainability. Realizing the technological possibilities of the bio-economy not only creates new investment opportunities but is also the prerequisite for a necessary socio-economic and cultural change. The consumers' acceptance of bio-based products and their demand are a conditio sine qua non for a successful change. Consequently, innovations, functioning markets, and changed consumer attitudes are complementing conditions for the creation of a sustainable production system.

Supporters of the Neo-Schumpeterian school (Dosi et al. 1988; Lundvall 1992, 1998; Nelson 1993) emphasize the systemic character of innovation processes in knowledge-intensive economic sectors. So-called innovation systems consist of different actors (companies, research institutions, political actors, consumers, etc.) and linkages between these actors (flows of goods, R&D cooperation, knowledge transfer relationships, user-producer-relationships, etc.). These linkages are required to ensure mutual learning and common knowledge development to solve complex innovation challenges. Such systems are characterized by their dynamic and co-evolutionary nature and are thus enormously complex, as both, actors and their knowledge and linkages and interactions between actors, may change over time.

Dosi (1982) takes this systemic conception as a starting point in defining technological paradigms as "[. . .] set of procedures, or a definition of the 'relevant' problems and of the specific knowledge related to their solution". Transferred to the knowledge-based bio-economy, the core idea is substitution, i.e., replacing carbon-based materials and energy with bio-based materials and energy. This can only be achieved by applying a variety of technological processes in the entire breadth and depth of the value-added chain. In this process the exploration of economic complementarities in terms of cross-fertilization of different knowledge fields matters. For example, to a large extent, digitalization allows for an extension of value chains by increasing the added value in new sustainable production sectors in a CO_2-neutral way (e.g., by electric mobility based on renewables, by development of smart grids, etc.). The concept of technological paradigms also illustrates that a paradigm shift is not possible at any time. A *window of opportunity* will only occasionally be opened and allow for a paradigm shift when several interconnected technologies are established and the creation of conducive demand-side and institutional conditions happens simultaneously. This, of course, holds for the emergence of a new bio-economic innovation system, too.

3 Innovation Systems and Knowledge

The theory of industrial life cycles, which emphasizes the strong dynamics in the emergence and decline of industries, gives a first hint on the meaning of the development of a dedicated innovation system supporting the transformation towards a knowledge-based bioeconomy. Typically, industrial development is divided into four stages: (i) a development phase (new knowledge creates prerequisites for innovation), (ii) an entrepreneurial and growth phase (many market entries of smaller innovative firms), (iii) a saturation phase and consolidation phase (formation of industrial standards, mergers and acquisitions as well as market exits), (iv) a downturn phase (oligopolistic competition in only less innovative industries) (Audretsch and Feldman 1996). Although the bio-economy does not represent a well-defined industrial sector, understanding the theory of industrial life cycles is of crucial importance to structure the transformation process towards the knowledge-based bio-economy. Without doubt, the bio-economy has to be characterized as cross-sectional. On the one hand, several new sectors will emerge, e.g., in the fields of bio-plastic, waste management, or bio-refineries. On the other hand, already existing sectors in the fields of vehicle construction, battery technology, pharmaceuticals, etc., will gain new momentum by the arrival of bio-economic approaches. Therefore, we argue that new sectors will emerge by establishing bio-economical technologies, and development dynamics of some already existing industries will receive new impetus at the same time. Adjustments of old and development of new institutions (e.g., in Germany the Renewable Energy Act, the Greenhouse Gas Emissions Trading Law, etc.), adjustments of consumer habits, and the emergence of new educational opportunities in terms of co-evolution will accompany these processes and establish the institutional, the industrial, and the consumer pillars of a dedicated innovation system.

The patterns and nature of new businesses in the Bioeconomy are thus strongly influenced by national institutions and organizations (Casper et al. 1999; Whitley 1999). Institutions are defined as 'a set of rules, formal or informal, that actors generally follow, whether for normative, cognitive, or material reasons'. 'Organizations are durable entities with formally recognized members, whose rules also contribute to the institutions of the political economy' (North 1990; Hall and Soskice 2001). In this interplay between organizations and institutions, the knowledge-base of an economy is created by the education and research system and represents one of the most important prerequisites for the transformation towards a bio-economical production system (Geels 2002). This automatically relates to a high level of uncertainty in particular concerning the required right future competences. In this complex process numerous individual knowledge fields are potentially relevant for the transformation and are already identified, e.g., synthetic chemistry, process engineering, genetic engineering, food technology, or informatics. It is decisive to understand the dynamics of these knowledge fields and the possibilities of their recombination with other knowledge fields and adequate actors in order to create an innovation system. In many cases, linkages of different knowledge fields ('*cross-fertilization*') are responsible for the emergence of extensive technological opportunities: for instance, a complete new industry, bioinformatics, has been initiated by the fusion of two so far unrelated knowledge fields: database technology and molecular biology. Consequently, because the link between different knowledge fields often implies true uncertainty, governmental innovation policies matter a lot. Knowledge about future potentials is essential for supporting research and innovation policies: the analysis of knowledge and network dynamics allows for the identification of development trajectories showing sectors requiring public attention and support concerning research and development in order to close existing knowledge gaps and build bridges between still unconnected knowledge domains (Burt 2004; Zaheer und Bell 2005).

4 Innovation in Knowledge-Based Societies

It has already been mentioned that also consumer knowledge plays an important role for the development and establishment of sustainable consumption patterns in a knowledge-based bio-economy (Geels 2002). Therefore, the analysis of the transformation process has to include the interaction of technological development, demand, and acceptance of innovative solutions as well as sociological variables. The latter include, e.g., education, age, income and gender. All are important explanatory factors determining attention and readiness to deal with bio-economic issues. A bio-economic innovation will only be successful when consumers accept it. The direction of the transformation process is, comparable to the importance of the policy realm, determined by consumers and their demand, i.e., an important question has to deal with consumers' openness to bio-economics and its products.

Finally, (real and virtual) social networks matter for the establishment of new consumption patterns. They can contribute significantly to a diffusion of consumers' behavioral patterns and values (Robertson et al. 1996; Valente 1996; Nyblom et al. 2003; Deffuant et al. 2005). Recent studies show that attitudes are substantial for the development of social relationships and that in turn, social relationships considerably influence behavior and attitudes. In the field of renewable energies, for example, the initiative of municipal utilities' customers has led in many cases to a 'green' orientation of regional power supply. In some cases, citizens' networks finally transformed to investment companies that are engaged in wind farms.

Critical issues are to be dealt with in democratic processes in order to be widely accepted. Not everything that is technically possible is also socially desirable. In the field of the bio-economy, this may, for instance, include the use of genetically modified organisms in agriculture. In fact, these organisms promise efficiency advantages with regard to the consumption of land and water etc., but their long-term health and environmental risks cannot be completely (as with any new technology) anticipated. Accordingly, technological developments require consumers' acceptance and attitude and thus depend on the level of education in an economy. This raises the question of a society's openness towards innovations that are fundamentally associated with uncertainty. The concept of *Responsible Innovation* summarizes the future-oriented organization of development and is currently discussed with a high priority by European policy makers and institutions. A comprehensive working definition has been developed by von Schomberg (2011). He describes responsible innovation as "a transparent, interactive process by which societal actors and innovators become mutually responsive to each other with a view to the (ethical) acceptability, sustainability and societal desirability of the innovation process and its marketable products (in order to allow a proper embedding of scientific and technological advances in our society)." This means that innovations are not exclusively evaluated by their economic efficiency, but different aspects (e.g., consumer protection or ecological aspects) also matter and are to be evaluated. Discussions on fossil fuels ('fuel vs. food') show that both, a pure economic and a one-dimensional ethical perspective is not sufficient. The quality of these discussions depends on the discussants' mutual understanding which in turn depends on the participants' level of knowledge.

Modern plant breeding and production of seeds are bio-economical fields of innovation in which issues of responsibility are frequently and controversially discussed. German consumers are skeptical about interference with the genome of food crops, but individual points of criticism remain unclear. New breeding techniques introduced, e.g., *Genome Editing,* enable scientists to selectively modify DNA strands of crop plants. These techniques are considered innovative as they may allow breeding of potentially efficient plants in fast and cheap ways. Species developed in this way hardly differ from those of conventional breeding. The Central Advisory Committee for Biological Safety does not classify these techniques as genetic engineering, especially because no new combinations of genetic material are made. As the Genetic Engineering Act does not explicitly address these techniques, legal clarification is still necessary as to whether these techniques are classified as genetic engineering at all. Dissemination potential and acceptance are influenced by this result. Here again, the necessity to include education and information policies becomes evident to support the transformation towards a knowledge-based bio-economy.

The concept of '*Social Innovation* '(e.g., Hanusch and Pyka 2013) emphasizes the importance of active citizenship in innovation. Thus, according to the understanding of the European Commission, this term includes innovations that are social, both in relation to their objective and their instruments. In particular, this includes innovations referring to the development and the application of new ideas (for products, services and models), covering at the same time social demand and creating new social relationships or collaborations. The whole society should benefit and contribute to generate new impetus for improvement. Social innovations can make a major contribution to rural development and promote economic resilience in these regions by strengthening cooperative behavior. Rural cooperatives (e.g., regional producer and marketing associations, winegrowers' cooperatives, tourism associations etc.) can help to develop regional competitiveness considering ecological and social aspects. As a consequence, within the framework of a bio-economy, rural regions that are notably affected by the already imminent demographic change and subsequent depopulation receive new opportunities for economic development.

5 The Economics of Change

The sections above illustrate that a transformation of the prevailing economic system towards a bio-based economy is an extremely complex process. Various different actors participating in different roles are contributing different pieces of knowledge. In this process, innovative adjustments in already existing industries as well as the emergence of new and the disappearance of mature industries can be observed simultaneously. In addition to the substitutive relations of new bio-based industries to traditional oil-based industries, there are numerous essential complementary relations giving further momentum for the transformation process. First and foremost there are the possibilities and application fields of digitalization. Digitalization allows to replace many oil-based products and energy-intensive services simply by bits and bytes. Simultaneously, digitalization offers a wide range of opportunities by coordinating decentralized and very detailed bio-economical technologies and processes such as energy production and distribution. This affects the composition of individual sectors where a coexistence of large diversified companies and small high-specialized technology companies is a likely solution. Finally, digitalization also offers consumer platforms to efficiently organize 'sharing economy'-approaches. Finally, successful knowledge generation and diffusion of relevant bio-economic knowledge depends on dynamic innovation networks (Pyka 2002) in which different actors jointly share and create new knowledge. The consumers, represented, for example, by consumer associations or politics, will play a key role in these innovation networks and will help to establish networks in early stages of technology development.

In a knowledge-based bio-economy, investments and economic growth still represent a crucial element for employment, international competitiveness and income generation. The bio-economy can make important contributions to accelerate investments by providing new investment opportunities generated fundamental innovations and thereby bringing currently available large quantities of liquidity to a productive use. This, in turn, accelerates the technological paradigm shift (Perez 2010).

The time path of the transformation process represents another critical component and has been explored only partially so far. On the one hand, it is high time to reduce carbon-based production methods. On the other hand, there will be frictions in the transformation process being caused for example by a lack of specialists and required competences. In this context, the so-called *sailing ship effects* (Howells 2002), frequently observed with radical innovations, could be made good use of. In the middle of the nineteenth century, when the existence of the established sailing ship technology was threatened by the arrival of new steam ships, shipbuilders—not having changed their technologies for many decades, if not centuries—began to innovate again. Due to the threat of innovative technologies, adjustment reactions in predecessor technologies can be observed with the aim to prevent the ancient technologies to be quickly replaced. Such adjustment reactions are, for example, fuel-efficient combustion engines and hybrid technologies as a reaction to the emergence of electric vehicles. These adjustments are advantageous since they pursue the same environmental objectives (e.g., inner-city fine dust and noise reduction, etc.) and thus provide more time to develop new technologies. Accordingly, the transformation process will for longer periods of time feature a co-existence of traditional and bio-based industries. Furthermore, it will be important to concurrently steer the relevant innovation processes in traditional technologies. This co-existence further increases complexity. At the same time, innovation policy is given room for maneuver and yet insufficiently developed technologies are prevented from being introduced prematurely which might cause promising approaches to fail.

Distributional effects of the transformation process are important for social acceptance. A bio-based economy on an industrial scale will largely represent a knowledge-based economy. Consequently, additional demand for high-skilled workers arises whereas opportunities for low-skilled workers decrease. This means a potential loss of jobs for less skilled workers in traditional industrial production. But apart from that, there will be demand for different goods and services whose compensation potential with regard to added value and employment is still unclear, though. Moreover, it remains open to what extent companies are prepared for this transformation into the bio-economy. Transformation processes will lead to a devaluation of competences so far responsible for economic success. How do established companies deal with the so-called 'not-invented-here-syndrome', overcome operational blindness and shape transformation processes actively in order to obtain added value at their established locations?

From this follows that distributional effects have an important regional dimension: does the bio-economy strengthen divergence processes between regions or does it help to achieve more convergence? The approach of creating networks in the sense of the so-called 'smart specialization principle' (Foray et al. 2009) connecting regional strengths along value-added chains in the best possible way, is promising but only sparsely implemented so far. Thus, in general, polarization tendencies leading to economic as well as political and cultural concentration of power and resulting in strong center-periphery structures can be avoided. But it still remains unclear, how strong and operational meaningful politically induced networks are in comparison to self-organized networks and how policy might exert influence. First findings indicate signs of a potential disintegration of the networks when political support is withdrawn (Green et al. 2013).

Transformation towards a knowledge-based bio-economic production system is supposed to terminate the existing negative relations between economic growth and environmental pollution, use of resources, climate change and energy consumption, and to promote a sustainable economy. The following questions are closely linked to the basic uncertainty of innovation and cannot be answered ex ante: 'which contributions are to be made by individual sectors?', 'what complex feedbacks for national and international competitiveness are to be expected?' and 'do so-called rebound effects possibly reduce or even overcompensate the positive effects of the transformation?' Institutional rules, such as a self-commitment of oil-producing countries to reduce their outputs due to the declining demand caused by bio-economics, are a way to reduce these uncertainties, at least partly. It remains necessary for all actors, companies, households and policy makers to refrain from optimization approaches and profit maximization in this transformation process. The complexity and uncertainty of this process requires the awareness of all actors to experimental behavior ('*trial-and-error*') which always also includes the possibility of failure.

6 Conclusions

Socio-economic systems have been exposed to permanent transformation processes since the industrial revolution. While development processes so far have been driven 'only' by result-oriented innovation processes, the character of the bio-economic transformation process is clearly concretized by society and politics. In the past, mainly bottlenecks caused by scientific-technological restrictions were overcome by vast technological revolutions, shifting the socio-economic system on new trajectories without giving direct instructions to the direction of the development process. At the beginning of the twenty-first century, however, the massive accumulation of greenhouse gases in the atmosphere since the beginning of the industrial revolution and the vulnerability of our present ecosystems reveal that global thresholds are almost surpassed. Thus, the level of freedom for future developments is restricted in order not to irreversibly damage natural conditions for human life and biodiversity. It is yet unclear whether this transformation process succeeds in the desired way and how it can be governed by political influence to achieve existential objectives of the global human society.

New technological developments alone are not enough to transform the socio-economic system. In a first step, they only create the necessary potential for radical changes affecting the economy as a whole. Converging trajectories and synergies that may finally introduce the paradigm shift necessarily require a broad social consensus on a specific use of these technologies. This means an initiation of a direction of development which connects investment decisions, innovations, and the tackling of basic uncertainty by politics (Pérez 2013). The 'green growth paradigm' based on bio-based technologies can be such a direction bringing together the potential of different technological developments and exploring their full potential. This requires political decisions supporting a new-orientation of research and innovation activities, exploitation of new energy sources, improvements in productivity of natural resources and new sustainable ways of living and producing (Pérez 2013). Moreover, in such a transformation process catching-up economies have to be provided with new opportunities for economic development without overstretching global natural resources and environment. Thus, a political and social direction is essential for a successful transformation process (Mazzucato and Perez 2015).

Examples include the development of new products within emerging bio-economic innovation systems. In this perspective, innovations require an interplay of actors along value added chains which might lead to the development of new industries. In the past, for example, the provision of cheap electricity led to the spread of fridges and freezers in private households which brought innovations in the fields of frozen food and packaging. Similarly, the creation of a '*Sharing Economy*' may lead to new digital coordination platforms and the creation of sustainable designs by product manufacturers in the bio-economy. *Planned obsolescence*, a phenomenon wasting resources and shortening product life cycles, would be eliminated this way and new sectors, for example, in the field of repair and maintenance services are initiated. Important determinants shaping long-term development are networks and clusters. They help to reduce uncertainty and support self-reinforcing effects. Furthermore, social changes and changing lifestyles are both, an expression and a driver of this transformation process (Mazzucato and Perez 2015).

Therefore, the role of governments is not only restricted to the correction of market failures. In fact, by ensuring investment safety and reducing risks and uncertainty, government instruments prepare the emergence and flourishing of new markets (Mowery et al. 2010). A crucial task for policies in the realm of innovation and entrepreneurship is the transition from invention to innovation, i.e., the expansion of bio-economical activities in a market. Correspondingly, a growth path based on bio-economics is more than a mere replacement of crude oil by renewable resources or renewable energies. It rather needs a *dedicated innovation system* creating synergies, knowledge transfer, and networks between manufacturers, suppliers, and consumers. It requires a comprehensive reorganization that includes the entire economy and renews production and consumption patterns in their present forms, which were shaped by previous transformation process within the oil-based paradigm.

The technological potential of a bio-economy is a necessary but insufficient condition for this transformation process. It also requires democratic consensus on the broad development and wide application of this technological potential. This includes the exploration of new trajectories and the fusion of new and existing technological trajectories. Markets in which innovations are profitable do not arise on their own but rather need feedback loops between political decisions, corporate strategies, and consumer preferences.

References

Audretsch, D., & Feldman, M. P. (1996). Innovative clusters and the industry life cycle. *Review of Industrial Organization, 11*(2), 253–273.

Blewitt, J., & Cunningham, R. (Eds.). (2014). *The post-growth project: How the end of economic growth could bring a fairer and happier society*. Green House, e-prints, Aston University.

Burt, R. S. (2004). Structural holes and good ideas. *American Journal of Sociology, 110*(2), 349–399.

Casper, S., Lehrer, M., & Soskice, D. (1999). Can high-technology industries prosper in Germany? Institutional frameworks and the evolution of the German software and biotechnology industries. *Industry and Innovation, 6*(1), 114–139.

Deffuant, G., Huet, S., & Amblard, F. (2005). An Individual-based model of innovation diffusion mixing social value and individual benefit. *American Journal of Sociology, 110*(4), 1041–1069.

Dosi, G. (1982). Technological paradigms and technological trajectories: A suggested interpretation of the determinants and directions of technical change. *Research Policy, 11*(3), 147–162.

Dosi, G., Freeman, C., Nelson, R., Silverberg, G., & Soete, L. (1988). *Technical change and economic theory*. London: Pinter.

Fagerberg, J., Laestadius, S., & Martin, B. R. (Eds.). (2015). *The triple challenge for Europe: Economic development, climate change, and governance*. Oxford: Oxford University Press.

Foray, D., David, P. A., & Hall, B. (2009). *Smart specialization – The concept* (Knowledge Economists Policy Brief, Vol. 9).

Freeman, C. (1991). Networks of innovators: A synthesis of research issues. *Research Policy, 20*(5), 499–514.
Geels, F. W. (2002). Technological transitions as evolutionary reconfiguration processes: A multi-level perspective and a case-study. *Research Policy, 31*(8), 1257–1274.
Green, L., Pyka, A., & Schön, B. (2013). A life-cycle based taxonomy of innovation networks – With focus on public-private collaboration. In F. Gallouj, L. Rubalcaba, & P. Windrum (Eds.), *Public-private innovation networks in services* (pp. 113–138). Cheltenham: Edward Elgar.
Hall, P. A., & Soskice, D. (2001). An introduction to varieties of capitalism. In P. A. Hall & D. Soskice (Eds.), *Varieties of capitalism: The institutional foundations of comparative advantage* (pp. 1–68). Oxford: Oxford University Press.
Hanusch, H., & Pyka, A. (2013). Social innovations in the perspective of Comprehensive Neo-Schumpeterian Economics. In C. Ruiz Viñals & C. Parra Rodríguez (Eds.), *Social innovation – New forms of organization in knowledge-based societies* (pp. 29–43). London: Routledge.
Howells, J. (2002). The response of old technology incumbents to technological competition – Does the sailing ship effect exist? *Journal of Management Studies, 39*(7), 887–906.
Kallis, G., Kerschner, C., & Martinez-Alier, J. (2014). The economics of degrowth. *Ecological Economics, 84*, 172–180. December 2012.
Knight, F. H. (1921). *Risk, uncertainty and profit*. Boston: Hart, Schaffner& Marx.
Lundvall, B.-A. (1992). *National innovation systems: Towards a theory of innovation and interactive learning*. London: Pinter.
Lundvall, B.-A. (1998). Why study national systems and national styles of innovation? *Technology Analysis and Strategic Management, 10*(4), 403–422.
Mazzucato, M., & Perez, C. (2015). Innovation as growth policy. In J. Fagerberg, S. Laestadius, & B. R. Martin (Eds.), *The triple challenge for Europe: Economic development, climate change, and governance* (p. 229). Oxford: Oxford University Press.
Meadows, D., Randers, J., & Behrens, W. W. (1972). *The limits to growth*. New York: Universe Books.
Mowery, D. C., Nelson, R. R., & Martin, B. R. (2010). Technology policy and global warming: Why new policy models are needed (or why putting new wine in old bottles won't work). *Research Policy, 39*, 1011–1023.
Nelson, R. R. (1993). *National innovation systems: A comparative analysis*. Oxford University: Press.
North, D. C. (1990). *Institutions, institutional change and economic performance*. Cambridge, MA: Cambridge University Press.
Nyblom, J., Borgatti, S., Roslakka, J., & Salo, M. A. (2003). Statistical analysis of network data – An application to diffusion of innovation. *Social Networks, 25*(2), 175–195.
Pérez, C. (2010). Technological revolutions and techno-economic paradigms. *Cambridge Journal of Economics, 34*(1), 185–202.
Pérez, C. (2013). Financial bubbles, crises and the role of government in unleashing golden ages. In A. Pyka & H.-P. Burghof (Eds.), *Innovation and finance* (pp. 11–25). Milton Park: Routledge.
Pyka, A. (2002). Innovation networks in economics – From the incentive-based to the knowledge-based approaches. *European Journal of Innovation Management, 5*(3), 152–163.
Pyka, A. (2017). Die Bioökonomie unter dem Blickwinkel der Innovationsökonomie. In J. Pietzsch (Ed.), *Bioökonomie für Einsteiger*. Berlin: Springer.
Pyka, A., & Buchmann, T. (2016). Die Transformation zur wissensbasierten Bioökonomie. In W. Burr & M. Stephan (Eds.), *Technologie, strategie und organisation* (pp. 333–361). Berlin: Springer.
Robertson, M., Swan, J., & Newell, S. (1996). The role of networks in the diffusion of technological innovation. *Journal of Management Studies, 33*, 333–360.
Sahal, D. (1985). Technological guideposts and innovation avenues. *Research Policy, 14*(2), 61–82.
Schumpeter, J. A. (1934). *The theory of economic development*. Harvard: President and Fellows of Harvard College.
Schumpeter, J. A. (1939). *Business cycles: A theoretical, historical and statistical analysis of the capitalist process*. New York: McGraw-Hill.

Schumpeter, J. A. (1943). *Capitalism, socialism and democracy*. New York: Harper.
Unruh, C. G. (2000). Undesrtanding Carbon lock-in. *Energy Policy, 28*, 817–830.
Valente, T. W. (1996). Social network thresholds in the diffusion of innovations. *Social Networks, 18*(1), 69–89.
Von Schomberg, R. (2011). Prospects for technology assessment in a framework of responsible research and innovation. In M. Dusseldorp & R. Beecroft (Eds.), *Technikfolgen abschätzen lehren: Bildungspotenziale transdisziplinärer Methode* (pp. 39–61). Wiesbaden: Springer.
Whitley, R. (1999). *Divergent capitalisms: The social structuring and change of business systems: The social structuring and change of business systems*. Oxford University: Press.
Zaheer, A., & Bell, G. G. (2005). Benefiting from network position: Firm capabilities, structural holes, and performance. *Strategic Management Journal, 26*(9), 809–825.

Structural Change, Knowledge and the Bioeconomy

Pier Paolo Saviotti

Abstract The emergence of the bioeconomy is presented here as a long term process entailing important structural changes and affecting both many of the technologies we use and the structure of the world economic system. The development of the bioeconomy will depend on the availability of fossil fuels, on the impact of human activities on the environment and on the progress of science and technology. The bioeconomy is going to be highly knowledge intensive and to rely on modern biotechnology. As all important innovations the bioeconomy will lead to creative destruction, giving rise to wealth creation and to the displacement of present economic activities. Suppliers of biological inputs are likely to benefit and those of fossil fuels to suffer. However, the balance of power will tend to favour countries and regions which are centres of knowledge creation rather than suppliers of natural resources.

1 Introduction

Starting from the industrial revolution our economic system used energy sources based on fossil inputs. While these energy sources allowed human beings to make an enormous economic progress they are not sustainable due to their excessive environmental impact and to the exhaustion of the required raw materials. The need to move towards a sustainable economic system is now generally perceived by citizens and decision makers. Amongst the changes required there are new modes of energy production and new types of industrial processes. In this context the vision of the bioeconomy has emerged as an important component of the construction of a sustainable human civilization. The term bioeconomy indicates an economic system in which most inputs to productive processes are of biological nature rather than of fossil origin. The following definition is an example:

- The Bioeconomy refers to the sustainable production and conversion of biomass into a range of food, health, fibre and industrial products and energy.

P.P. Saviotti (✉)
Innovation Studies Group, Faculty of Geosciences, Utrecht University, Utrecht, The Netherlands
e-mail: pier-paolo.saviotti@wanadoo.fr

S. Dabbert et al. (eds.), *Knowledge-Driven Developments in the Bioeconomy*,
Economic Complexity and Evolution, DOI 10.1007/978-3-319-58374-7_2

Renewable biomass encompasses any biological material (agriculture, forestry and animal-based including fish) as a product in itself or to be used as raw material.

The recent emergence of the concept of bioeconomy has been due to the combination of three conditions:

1. A shortage of raw materials, and in particular of those which are derived from oil, coal or gas.
2. The growing environmental impact of human activities.
3. The advances in science and technology, and in particular of those linked to modern biology.

(1) and (2) are constraints of the existence and seriousness of which most decision makers are increasingly convinced. They do not allow the continuation of the present mode of production based on fossil inputs in a number of sectors. However, if the transition to the bioeconomy is desirable, at present it is not yet economically feasible. Compared to present oil prices the cost of most biological analogues of fossil based processes is still too high. In order for the transition to the bioeconomy to occur for a wide range of processes the relative cost of biological processes will have to decrease substantially. This could happen for different reasons:

1. A considerable rise in oil prices.
2. If the natural disasters that one can clearly attribute to the environmental impact of human activities were to grow in frequency and size.
3. An increase in the efficiency of biologically based processes determined by the progress of science and technology.

Changes in any of these three conditions can in principle increase the relative efficiency of biological processes making them more competitive with fossil based ones.

The considerations in the previous paragraph could have been interpreted as implying that the transition to the bioeconomy will consist simply of the substitution of fossil processes by biological ones. Although in a limited number of cases existing fossil based processes can be easily converted to biological ones, more often biological processes could be very different from their fossil counterpart. In reality the transition to the bioeconomy can be expected to involve a high degree of structural change and to occur over a long period of time. Furthermore, it is likely to be heavily dependent on advances in knowledge. The following sections discuss the nature of the structural change to be expected and the potential contributions of knowledge.

2 The Bioeconomy

The use of non animal forms of energy was one of the most important factors leading to the industrial revolution. Coal, the most important source during the nineteenth century, was largely replaced by oil in the twentieth century. Our economic system is still heavily based on fossil fuels both for energy generation and for a large number of industrial activities. There is a growing awareness that oil reserves are decreasing and that the environmental impact of fossil based processes is unsustainable. The bioeconomy holds the promise to reduce the need for fossil inputs and the environmental impact of human activities in energy generation and in a wide range of industrial activities. However, the scale of the transformation from a fossil based to a bio based economy is huge. Even if we had available bio based processes that are cost competitive with respect to their fossil counterparts the investment required would be immense. Furthermore, event at present oil prices bio based processes are not economically efficient. Thus, a very large investment in R&D is required to make bio based processes more environment friendly and more economically efficient. This is not to imply that the bioeconomy can be conceived as the analogue of the fossil economy with the same processes using biological inputs and being more sustainable. We can expect new activities to emerge and the relationships between sectors to change sometimes even drastically. A good example can be given by the largely modified role that agriculture will need to play and by the modified relationships it will have with other sectors. Furthermore, in a classical case of creative destruction today's producers of fossil fuels are unlikely to be tomorrow's producers and exporters of biomass. Thus, the whole world economic system will need to undergo a profound reorganization.

To have an idea of the time horizon that can be expected for the transition to the bioeconomy we can compare it to previous energy transitions. First, we can notice that the bioeconomy is not new. Until the eighteenth century almost all food and energy were produced by a form of bioeconomy (see Fig. 1). The substitution of biological by manmade energy started at about the time of the industrial revolution with coal first and then with oil and gas. Then, it is clear that the time horizon of all

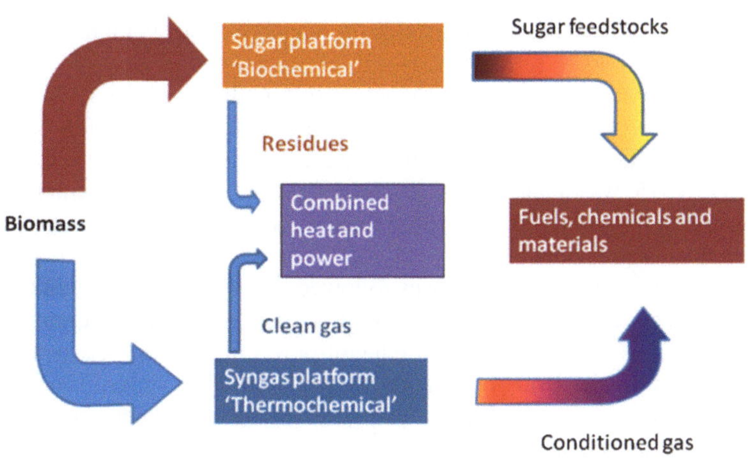

Fig. 1 The biorefinery, the equivalent of a petroleum refinery, can in principle transform any type of biomass into energy and chemical products

these transitions is of the order of one century. Thus, we can expect the transition to the bioeconomy to be spread over a large part of the twenty-first century.

At present biological processes play a limited role in our economic system. The two main fields already developed are biofuels and some bioproducts, such as chemicals, plastics etc.

A **biofuel** is a fuel that is produced through contemporary biological processes, such as agriculture and anaerobic digestion, rather than a fuel produced by geological processes such as those involved in the formation of fossil fuels, such as coal and petroleum, from prehistoric biological matter. ("What is biofuel? definition and meaning". BusinessDictionary.com. Retrieved 30 May 2015.) Biofuels can be derived directly from plants, or indirectly from agricultural, commercial, domestic, and/or industrial wastes (House of Lords 2014).

There are different generations of biofuels. First generation biofuels are made from the sugars and vegetable oils found in arable crops, which can be easily extracted using conventional technology. In comparison, second generation biofuels are made from lignocellulosic biomass or woody crops, agricultural residues or waste, (examples). Second generation biofuels are harder to extract because a series of physical and chemical treatments might be required to convert lignocellulosic biomass to liquid fuels suitable for transportation. However, they avoid to a greater extent competition with food production. Present production facilities produce mostly biofuels of the first generation. The second generation of biofuels is still under development.

Biofuels have some advantages as well as disadvantages. In principle, if well used they can reduce the environmental impact of human activities, for example by limiting global warming, and they can contribute to energy security by reducing imports of fossil fuels. However, they compete with food production. In spite of the barriers to their use biofuels have advantages in particular niches. They are unlikely to be competitive with solar or wind in the static generation of electrical energy, but they can have an advantage in some fields of transports (e.g. aviation, sea transport) or in the military field.

Brazil pioneered the industrial production of bioethanol in the 1970s when the first oil crisis resulted in gasoline shortages and raised awareness of the dangers of oil dependence. The National Alcohol Program ('Programa Nacional do Álcool'), launched in 1975, was a nationwide program financed by the government to phase out automobile fuels derived from fossil fuels, such as gasoline, in favor of ethanol produced from sugar cane.

The initial success in the 1970s, was followed by a slowdown in alcohol production as oil prices fell in the 1980s, leading to a crisis which was overcome only in the early 2000s, when confidence on ethanol-powered vehicles was restored with the introduction in the Brazilian market of flexible-fuel vehicles. The bioethanol industry in Brazil has become more competitive and no longer needs subsidies. In Brazil there are no general biorefineries but existing industrial plants produce bioethanol and/or sugar, and increasingly bioenergy by the combustion of wastes from bioethanol production.

Table 1 Distribution of world production of bioethanol in 2011 (millions of gallons)

Africa	Asia	Australia	Europe	North and Central America	South America
38.31	889.7	87.2	1167.64	14,401.34	5771.9

Table 2 Distribution of world production of bioethanol by country in 2011 (millions of gallons)

Brazil	Canada	China	European Union
5573.24	462.3	554.76	1199.31

Source: RFA, Accelerating Industry, 2016 Ethanol Industry Outlook

The USA is the main world producer of bioethanol. Together with Brazil they account for 90% of global production. However, differently from Brazil, in the USA bioethanol is derived mainly from corn and supplied together with other products, such as corn oil. In turn, corn oil can be used in various applications including biodiesel production, feed markets, and as the building blocks for a variety of industrial applications. The Renewable Fuels Association (RFA) estimates that U.S. ethanol producers supplied 175 million gallons of corn oil in 2011.

The world production of bioethanol in 2011 is shown in Tables 1 and 2. A study by Balat and Bala (2009) explored the potential world production of bioethanol for most countries in the world.

The other important biofuel is biodiesel. It is produced from oils or fats of different origin. Feedstocks for biodiesel include animal fats, vegetable oils, soy, rapeseed, jatropha, mahua, mustard, flax, sunflower, palm oil, hemp, field pennycress, *Pongamia pinnata* and algae. Biodiesel is the most common biofuel in Europe.

So far biofuels have been the most important product of the emerging bioeconomy but they are likely to be followed by a range of other products, thus giving rise to an industrial, or white biotechnology. This is the application of biotechnology for the processing and production of chemicals, materials and energy. It uses enzymes and micro-organisms to make products in sectors such as chemistry, food and feed, pulp and paper, textiles and energy (Europabio and ESAB 2010). Due to past investment patterns, which were manly concentrated in pharmaceuticals (red biotechnology) and agrochemicals (green biotechnology), other bioproducts are today still less important than biofuels. However, investment in the industrial applications of biotechnology (white) are expected to increase substantially in the near future.

In general the vision of the bioeconomy is centred around the biorefinery (Fig. 1). The biorefinery, the equivalent of a petroleum refinery, can in principle transform any type of biomass into energy and chemical products. In turn the chemical products produced by the biorefinery are inputs to many industrial sectors.

There are many possible types of biorefinery ranging from very general to very specialized ones. They can differ for the feedstock or the conversion techniques they use, or for the range of outputs they produce (Table 3).

Table 3 Typology of biorefineries

Biomass feedstock	Conversion techniques
i. Sugar/starch crops	i. Fermentation of sugar/starch crops
ii. Vegetable oil	ii. Fermentation of lignocellulosic biomass
iii. Lignocellulosic biomass	iii. Transesterification of triglycerides
iv. Jatropha oil	iv. Gasification: formation of syngas
v. Micro-algae	v. Fast pyrolysis
	vi. Fischer-Tropsch synthesis
	vii. Hydrogenation
	viii. Conversion of syngas to methane
	ix. Anaserobic digestion

At present there are very few general biorefineries capable of using many types of biomass feedstock and many types of conversion techniques. The future development of biorefineries is subject to many uncertainties. Their size, location and the type of feedstock they use can considerably affect the impact they have on the environment. Furthermore, future biorefineries could be general or specialized.

2.1 *Science, Technology and the Bioeconomy*

The bioeconomy is very old. The production of beer, yogurt, wine, bread etc. has been going on since the beginning of human history and can be considered the first generation of biotechnology. These processes underwent changes in productive efficiency during most of human history, but important advances started occurring in the nineteenth century when Pasteur contributed to significant advances in fermentation. The most path breaking changes in the underlying science and technology occurred during the twentieth century. The production of antibiotics (Aminov 2010) started after the second world war, following their discovery in the first half of the twentieth century, constituted the second generation of biotechnology. Then in the 1970s began the third generation of biotechnology, based on the combination of genetics with molecular biology. The origins of third generation biotechnology coincided with the birth of genetic engineering. There were two key events that have come to be seen as scientific breakthroughs beginning the era that would unite genetics with biotechnology. One was the 1953 discovery of the structure of DNA, by Watson and Crick, and the other was the 1973 discovery by Cohen and Boyer of a recombinant DNA technique by which a section of DNA was cut from the plasmid of an E. coli bacterium and transferred into the DNA of another.

The 1953 discovery of the structure of DNA gave a powerful impetus to the development of a new discipline called molecular biology, that is the study of the molecular underpinnings of the processes of replication, transcription, translation, and cell function. It is closely related to biochemistry and to genetics. This combination gave rise to genetic engineering, with which modern biotechnology is closely associated. This approach could, in principle, enable bacteria to adopt the genes and produce proteins of other organisms, including humans. Popularly referred to as "genetic engineering," it came to be defined as the basis of new biotechnology. Since the 1970s biotechnology has become one of the most rapidly advancing fields of science and technology.

Its progress has been in part due to the advances in ICT which gave rise to a hybrid discipline called bioinformatics (Hogeweg 2011). Bioinformatics is an interdisciplinary field that develops methods and software tools for understanding biological data. Bioinformatics combines computer science, statistics, mathematics and engineering to analyze and interpret Biological data. Bioinformatics has been used for in silico analyses of biological queries using mathematical and statistical techniques. The growing uses of bioinformatics represent a form of automation of research without which progress in biology after the 1970s would have been impossible. What was perhaps the most spectacular result of the research in biotechnology, the human genome project, would have taken much longer to complete had it not been for the contribution of bioinformatics.

Starting from the beginning of the twenty-first century the most important development in biotechnology has been the emergence of synthetic biology (Oldham et al. 2012). There is a debate about whether synthetic biology is a discontinuity or a continuation of the trend which began with recombinant DNA. Irrespective of the answer to this question, most definitions of synthetic biology tend to stress its engineering nature. Originally seen as a subset of biology, in recent years the role of electrical and chemical engineering has become more important.

- Synthetic Biology is (a) the design and construction of new biological parts, devices, and systems, and (b) the re-design of existing, natural biological systems for useful purposes. (syntheticbiology.org/).

Whatever its definition it is clear that synthetic biology is now the frontier of modern biotechnology. Globally, the first place of production of scientific publications in synthetic biology is the United States, which accounts for nearly 50% of the literature, followed by Europe and finally South-East Asia (China, Japan, South Korea). The rest of the world's production is relatively small and difficult to interpret, although there is a relatively large activity in Israel (58 articles) and Australia (50 articles). International collaborations in synthetic biology are numerous and globally balanced except for a major axis of collaboration between Europe and the United States. Finally, relatively high growth rates (50% for the most part) reveal a dynamic area, mainly in countries such as China (125%) or Australia (180%) (http://www.synbioproject.org/events/biosecurity).

The world distribution of patents in synthetic biology reveals an equivalent dominance of the USA but a much greater role of Japan and China (van Doren et al. 2013) (Fig. 2).

Industrial applications of biotechnology have been in existence since the 1970s, particularly in the pharmaceutical and agrochemical sectors. Even amongst the industrial applications of synthetic biology, right now there are relatively few

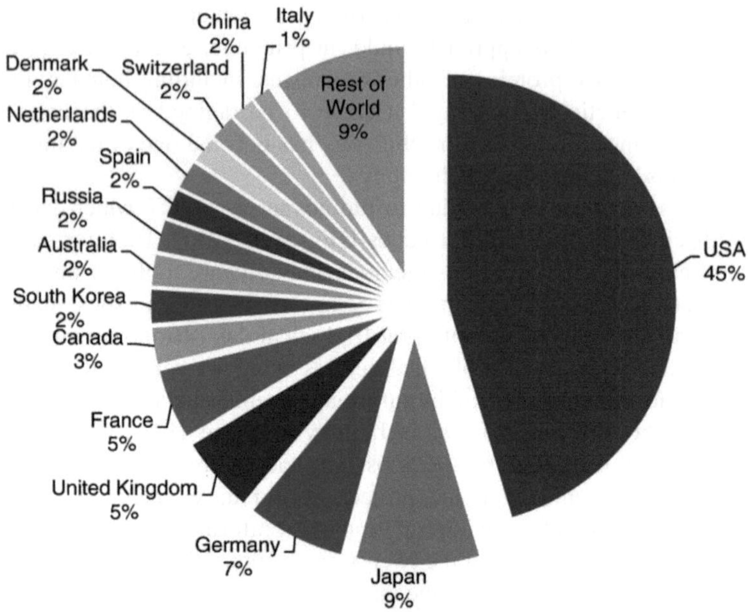

Fig. 2 World distribution of patents in synthetic biology

products on the market but several important applications are under development (BIO 2013; see Box 1). Since synthetic biology is highly dependent on progress in science and technology, its industrial developments are likely to depend on the technological capabilities of different countries, measured by their distance from the world science and technology frontier, and on the prevalent intellectual property rights (IPRs). These are going to be barriers to the adoption of new technologies used in the bioeconomy by less developed countries. Furthermore, since synthetic biology opens the possibility to combine biological parts to create complex macromolecules, cells or biological organisms, the availability of standardized parts is crucial. This requires a relaxed system of IPRs or, in the best case, an open source approach (Henkel and Maurer 2007). The use of standardized components has so far not been very common in biological research. However, there is already evidence that companies using stem cells prefer to start from a small number of standard lines for which they have shared experience (Maurer 2007). This approach is very similar to that of the electronics industry, where programs for a common operating system or game console are routinely used by competing companies. We can expect a similar approach to be adopted in synthetic biology. An important initiative in this direction is the project BioNet of the BioBricks Foundation, which aims at making standardized parts easily available (Biobricks).

Box 1 Products of Synthetic Biology

Cephalexin, a synthetic antibiotic: DSM, a Life Sciences and Materials Sciences company headquartered in the Netherlands, was one of the first companies to utilize synthetic biology, dramatically improving an existing process for commercial production of Cephalexin, a synthetic antibiotic.

Starting with a penicillin-producing microbial strain, DSM introduced and optimized two enzyme-encoding genes for a one-step direct fermentation of adipoyl-7-ADCA, which could then be converted into Cephalexin via two enzymatic steps.

The new process replaced a 13-step chemical process, resulting in significant cost and energy savings. DSM has gone on to build a business in antibiotics, vitamins, enzymes, organic acids, and performance materials within one of its emerging business areas called Biobased Products and Services.

Biotechnology innovation organization, https://www.bio.org/articles/current-uses-synthetic-biology

3 Economic Development and Structural Change

The advent of the bioeconomy can in principle reduce the environmental impact of human activities but under what conditions can it do so? and can it also provide the citizens of both industrialized and developing countries with the employment and income required? A solution which has been proposed is to eliminate growth and to move to a no growth economic system (Georgescu Roegen 1971; Daly 2007; Gowdy 1994). An even more extreme solution would be to go back to an economic system similar to those that existed before the industrial revolution. In this chapter we assume that (a) we do not necessarily want to eliminate growth but to change its content, and that (b) we want to preserve the present level of services but to reduce their environmental impact. Regarding the second of these assumptions we can observe that the products we purchase are not demanded for their material nature but for the services they supply (Lancaster 1966; Saviotti and Metcalfe 1984). Thus we need to preserve the level of services while reducing the environmental impact of their material production.

The bioeconomy represents the emergence of a new industrial structure at a level of aggregation superior to that of an industry and encompassing if not all sectors at least a wide range of them. In reality it is a reconversion of the economy similar to that which occurred in the twentieth century with the switch from coal to oil. As it was pointed out at the beginning, the main types of factors contributing to the emergence of the bioeconomy are a shortage of raw materials and energy, the environmental impact of human activities, and the progress of science and technology. Two other types of factors, energy security and geopolitical issues, are likely to affect the way the future bioeconomy will develop. The transition from our petroleum based economy to the bioeconomy will not consist uniquely of a simple choice of readymade technological alternatives available for purchase. On the contrary, it can be expected to require the creation of new knowledge and of very large investments. In particular, the transition will not occur only as the substitution of fossil based processes by equivalent bio-based ones. New processes will be required in conjunction with new forms of work and industrial organization, new supply chains, new institutions etc. Thus, the transition to the bioeconomy can be expected to be a very important example of structural change occurring at a high level of aggregation and involving changes at the national and international level.

If the exhaustion of natural resources and the impact on the environment are amongst the most important factors contributing to the emergence of the bioeconomy, the need to provide employment and income remain as important now as in all the previous phases of economic development. To understand how the bioeconomy can contribute to satisfy both environmental and socioeconomic requirements we now examine our present knowledge about structural change and economic development.

Recent research showed that there is a close relationship between structural change and economic development (Chenery 1960; Cornwall 1977; Metcalfe et al. 2006; Teixeira and Queirós 2016). Structural change can be defined as a change in the structure of the economic system, which is constituted by its components and by their interactions. Components can be either industrial sectors or less strictly economic activities such as education, R&D, labour markets, health care, tourism and leisure activities etc. All these components interact with links of variable intensity. For example, the health care system buys drugs from the pharmaceutical industry with governments or ministries as possible intermediaries. Alternatively, the tourism industry depends heavily on the presence of holidays and of disposable income. In general the emergence of economic activities producing new goods or services require the presence of appropriate institutions, a phenomenon called the co-evolution of technologies and institutions (Nelson 1994).

In the above definition of structural change components are mostly agents or organizations. In addition structural change can occur also in manmade artefacts (MMAs) and in knowledge. Structural change in knowledge occurs by the emergence of new fields qualitatively different from pre-existing ones. Examples of this type of structural change are given by the emergence of molecular biology, of biotechnology, of bioinformatics and of synthetic biology. In all types of structural change the creation of new components requires the presence of a discontinuity, which in turn implies that the new components are qualitatively different from pre-existing ones. Here we assume that molecular biology is qualitatively different from previous approaches to biology, and also that a discontinuity occurred at the emergence of bioinformatics. Whether the emergence of synthetic biology constitutes a discontinuity is not clear. One could say that it is simply an extension of the combination of recombinant DNA and of bioinformatics to the synthesis of new biological organisms, extension which is warranted by the enhanced reliability of techniques to 'cut and paste' DNA fragments, such as CRISPR Cas9 (Boglioli and Richard 2015). In this case as well as in the decision about whether two biological species or two industrial sectors are different, the distinction between qualitative and quantitative difference is not always straightforward. However, what matters is how different fields of knowledge are perceived and embodied in institutions and organizations. When pharmaceutical firms switched from organic chemistry to biotechnology they could not tell their researchers to transform overnight their knowledge base, but they had to hire researchers trained in different departments and faculties from the previous ones. In other words, even if the boundaries between two disciplines were not perfectly defined their inclusion in different institutions would enhance the presence of a discontinuity. This would then be reflected in barriers to the adoption of a qualitatively different knowledge base by a firm or by an organization. Thus, for a firm to hire researchers with the same knowledge base already used or with a completely different one involves different organizational problems. For example, researchers in a very new discipline may be in scarce supply. Furthermore, it is difficult to integrate researchers whose knowledge base has a very large cognitive distance with respect to that presently used by the firm (Krafft et al. 2014).

Structural change in knowledge can lead to structural change in industrial sectors, either by leading to the emergence of new sectors or by introducing important modifications into existing ones. For example, synthetic materials or consumer electronics owed their origin to important scientific advances while pharmaceuticals or telecommunications were heavily modified by advances in biology and in information technology. In modern industrialized countries R&D is closely integrated with production and contributes to the production of innovations.

The structure of the economic systems of today's industrialized countries changed enormously since the beginning of the industrial revolution by the emergence of new sectors and of new institutions and by the considerable expansion of existing ones. Examples of these changes are the institutionalization of industrial R&D and the diffusion of education. Of particular importance for this chapter are the following three trajectories which oriented development in the directions of increasing differentiation an of increasing output quality (Saviotti and Pyka 2013):

Trajectory 1. The efficiency of productive processes increases during the course of economic development. Here efficiency must be understood as the ratio of the inputs used to the output produced, when the type of output remains *constant*.

Trajectory 2. The output variety of the economic system increases in the course of time. Here such variety is measured by the number of distinguishable sectors, where a sector is defined as the set of firms producing a common although highly differentiated output.

Trajectory 3. The output quality and internal differentiation of existing sectors increases in the course of time after their creation. This means that if during the period of observation the type of output changes, what we will observe is a combination of growing productive efficiency and of quality change.

These three trajectories are not independent. A continuous increase in productive efficiency, if not accompanied by the emergence of new sectors and by their internal differentiation and rising quality, could have led the economic system to a bottleneck in which all demanded output would have been produced by a declining proportion of the labour force. According to (Pasinetti 1981, 1993) such a bottleneck could have been determined by the imbalance between continuously increasing productive efficiency and saturating demand. Even if demand did not exactly saturate with growing income, Saviotti and Pyka (2015) have shown that both the emergence of new sectors and their increasing quality and internal differentiation provided additional scope for further growth and allowed its continuation in the long run. Furthermore, both the emergence of new sectors and their growing quality and internal differentiation can compensate for the diminishing capability of incumbent and maturing sectors to create employment. Recently Hidalgo et al. (2007; Hidalgo and Hausmann 2009) showed that economic systems not only need to become more differentiated but also need to start producing more complex products, a finding that coincides with a mechanism proposed by Saviotti and Pyka (2013) who hypothesized that continued development also required the production of increasingly higher quality of goods and services. Of course, such trends towards goods and services of higher complexity or of higher quality require the creation of new knowledge.

The previous three trajectories have both a positive and a normative function since they describe what happened and indicate possible strategies of economic development. Thus, a development strategy based on export variety and technology would follow naturally from trajectories T2 and T3. The most successful developers of the second half the twentieth century, including Japan, South Korea, Taiwan and more recently China, followed exactly this strategy.

The previous considerations are related to the way the bioeconomy is likely to develop within each country. Two other factors, energy security and geopolitical issues, are likely to affect the development of the bioeconomy at the international level. Petroleum reserves and the demand for petroleum are distributed very asymmetrically around the world. This has given rise to very large trade flows from petroleum rich to petroleum using countries (IEA 2014). When these trade flows started in many cases the former were less developed countries while the latter were industrialized countries. In the period after the second world war many petroleum exporters became very rich, although these riches did not lead to a harmonious and well balanced socioeconomic development. This apparent contradiction between the growth of the GDP per capita and the poor performance in several important development dimensions (education, income distribution, health care etc.) gave rise to the literature on the so called Curse of Natural Resources (Auty 1997, 2001a, b; Gylfason 2001; Sachs and Warner 2001). Typical examples of this curse are countries such as Nigeria, Saudi Arabia, Angola and Venezuela. The economic models of these countries are seriously threatened by the expected scarcity of petroleum. Of course, not all petroleum exporting countries are affected by this curse. It now seems that the curse affects predominantly countries which when they started exploiting natural resources had very poorly developed technologies and institutions (Lederman and Maloney 2007; Wright and Czelusta 2004, 2007).

The world distribution of the production of biomass is less asymmetrical than that of petroleum. Although some countries are suitable for the cultivation of particular crops or plants, the range of possible feedstocks is so wide that hardly any country will be unable to use bio-based processes with local inputs. Nevertheless, some countries, such as Brazil, Argentina and Australia are very well endowed with the conditions required to produce biomass and could become important exporters of it (Biopact 2008). However, as the Curse of Natural Resources demonstrated, it is not necessarily the countries that are more abundantly endowed with the critical natural resources of a period that profit the most from them but the countries that have the relevant technologies and knowledge. Given that economic systems are becoming more and more knowledge intensive, we can expect competitive advantage in the bioeconomy to lie more with technological leaders than with resource abundant countries.

The petroleum based economy had created a world structure including flows of trade and of knowledge. The world structure of the emerging bioeconomy is likely to be considerably different, at least for what concerns the suppliers of biomass. To the extent that the capabilities to produce biomass are less asymmetrically distributed than those to produce petroleum we can expect the negotiating power of biomass suppliers to be reduced. Synthetic biology has the potential to create new plants which can produce types of biomass in countries where now they cannot be harvested, further reducing the asymmetry of biomass production (Wellhausen and Mukunda 2009; Wield et al. 2013). As a consequence, we can expect that, although some biomass producing countries could derive economic advantages from the transition to the bioeconomy, the overall economic and political advantage is likely to be skewed towards knowledge producers than towards biomass producers. In particular, the biomass producing countries that have weak technologies and institutions are likely to fall into a new curse of natural resources. Of course, these conclusions could be considerably affected if there were some cooperative agreements between knowledge producers and biomass producers.

The emergence of the bioeconomy will not only provide variable advantages for knowledge producers and biomass producers but it will also destroy the economic model of oil producing countries. Of course, the decline of the fossil based economy will not happen overnight and in principle that leaves oil producing countries the time to adapt. Yet, the scope of the changes required and the creative destruction in the world economy is likely to be so great that there will be both winners and losers. To avoid these potential conflicts new types of international agreements, such as joint cooperative investment by knowledge producers and biomass producers will be required.

4 Conclusions

The transition to the bioeconomy will occur because our present economic system is too dependent on the use of fossil fuels and it is not sustainable. If well developed, the bioeconomy could reduce the environmental impact of human activities. It could contribute positively to growth and employment, helping the world economic system to overcome its present crisis. However, its timing and the path that it will follow are very difficult to predict accurately and they will depend on factors such as (i) the scarcity of fossil resources, (ii) the evidence of environmental impact, (iii) the progress of science and technology. Furthermore, even if the three previous factors were to provide very favourable conditions, the scale of the investment required is such that the transition to the bioeconomy is likely to occupy a large part of the twenty-first century. It will involve a high degree of structural change occurring at different levels:

(1) At the national level, new industrial sectors could emerge and old sectors decline, the content of existing sectors could change dramatically and new inter-sector interactions could be created. As an example, We only need to think about the role of a changed agriculture and about its relationships with other sectors.
(2) At the geopolitical level, we can expect a decline of fossil fuel producers and a rise of biomass producers. While the latter should expect some economic benefits, such benefits will not necessarily be forthcoming unless favourable endowments of natural resources are accompanied by adequate technologies and institutions. We can expect the balance of advantages from the transition to be skewed more towards knowledge providers than towards biomass suppliers. In countries at a very low level of economic development it could possibly lead to a different type of curse of natural resources. Even amongst the countries at a relatively high level of economic development there could be possible reorganizations depending on their distance from the scientific and technological frontier and on the presence of knowledge discontinuities. Given the above, it could create both windows of opportunity and bottlenecks. It will involve a non negligible extent of creative destruction and it could give rise to new geopolitical conflicts. To avoid these potential conflicts new types of international cooperative agreements will be required.

References

Aminov, R. I. (2010). A brief history of the antibiotic era: Lessons learned and challenges for the future. *Frontiers in Microbiology Antimicrobials, Resistance and Chemotherapy, 1*, Article 134, 6.

Auty, R. M. (1997). Natural resources, the state and development strategy. *Journal of International Development, 9*, 651–663.

Auty, R. M. (2001a). The political economy of resource-driven growth. *European Economic Review, 46*, 839–846.

Auty, R. M. (Ed.). (2001b). *Resource abundance and economic development*. Oxford: Oxford University Press.

Balat, M., & Bala, H. (2009). Recent trends in global production and utilization of bio-ethanol fuel. *Applied Energy, 86*, 2273–2282.

BIO, Biotechnology Industry Organization. (2013). *Current uses of synthetic biology for renewable chemicals, pharmaceuticals, and biofuels*. 1201 Maryland Ave. SW, Suite 900, Washington, DC 20024.

BioBricks Foundation. https://biobricks.org/bionet/

Biopact. (2008). *Expected world trade flows, biomass, towards a green energy pact between Europe and Africa*. http://global.mongabay.com/news/bioenergy/2008/01/coals-deep-trouble-makes-biomass-highly.html

Boglioli, E., & Richard, M. (2015, September). *Rewriting the book of life: A new era in precision gene editing* (Working Paper). The Boston Consulting Group.

Chenery, H. B. (1960). Patterns of industrial growth. *American Economic Review, 50*, 624–654.

Cornwall, J. (1977). *Modern capitalism, its growth and transformation*. London: Martin Robertson.

Daly, H. A. (2007). *Ecological economics and sustainable development*. Cheltenham: Edward Elgar.

EuropaBio and ESAB. (2010). *Industrial or white biotechnology, A driver of sustainable growth in Europe* (Working document from to be used as input for the Industrial Biotechnology section of the European Technology Platform for Sustainable Chemistry).

Georgescu Roegen, N. (1971). *The entropy law and the economic process*. Cambridge, MA: Harvard University Press.

Gowdy, J. (1994). *Coevolutionary economics: The economy, society and the environment*. Dordrecht: Kluwer Academic Publishers.

Gylfason, T. (2001). Natural resources, education, and economic development. *European Economic Review, 45*, 847–859.

Halff, A., & IEA, International Energy Agency. (2014, July 14). *Changing global petroleum product trade flows*. EIA Energy Conference, Washington, DC.

Henkel, J., & Maurer, S. M. (2007). The economics of synthetic biology. *Molecular Systems Biology, 3*, 117. doi:10.1038/msb4100161.

Hidalgo, C. A., & Hausmann, R. (2009). The building blocks of economic complexity. *PNAS, 106*(26), 10575.

Hidalgo, C. A., Klinger, B., Barabasi, A.-L., & Hausmann, R. (2007). The product space conditions the development of nations. *Science, 317*, 482–487.

Hogeweg, P. (2011). The roots of bioinformatics in theoretical biology. *PLoS Computational Biology, 7*(3), e1002021. doi:10.1371/journal.pcbi.1002021.

House of Lords. (2014, March 6). *Waste or resource? Stimulating a bioeconomy*. Science and Technology Select Committee, 3rd Report of Session 2013–14.

Krafft, J., Quatraro, F., & Saviotti, P. P. (2014). The dynamics of knowledge-intensive sectors' knowledge base: Evidence from biotechnology and telecommunications. *Industry and Innovation*. doi:10.1080/13662716.2014.919762.

Lancaster, K. J. (1966). A new approach to consumer theory. *Journal of Political Economy, 14*, 133–156.

Lederman, D., & Maloney, W. F. (Eds.). (2007). *Natural resources: Neither curse nor destiny*. The World Bank, Stanford University Press. ISBN-13: 978-0-8213-6545-8, ISBN-10: 0-8213-6545-2.

Maurer, S. (2007). Open source biology: Finding a niche (or maybe several). *UMKC Law Review, 76*(2).

Metcalfe, J. S., Foster, J., & Ramlogan, R. (2006). Adaptive economic growth. *Cambridge Journal of Economics, 30*, 7–32.

Nelson, R. R. (1994). The co-evolution of technology, industrial structure, and supporting institutions. *Industrial and Corporate Change, 3*(1), 47–63.

Oldham, P., Hall, S., & Burton, G. (2012). Synthetic biology: Mapping the scientific landscape. *PLoS ONE, 7*(4), e34368. doi:10.1371/journal.pone.0034368.

Pasinetti, L. L. (1981). *Structural change and economic growth*. Cambridge: Cambridge University Press.

Pasinetti, L. L. (1993). *Structural economic dynamics*. Cambridge: Cambridge University Press.

Renewable Fuels Association. *2016 ethanol industry outlook*. www.ethanolrfa.org/wp.../

Sachs, J. D., & Warner, A. M. (2001). The curse of national resources. *European Economic Review, 45*, 827–838.

Saviotti, P. P., & Metcalfe, J. S. (1984). A theoretical approach to the construction of technological output indicators. *Research Policy, 13*, 141–151.

Saviotti, P. P., & Pyka, A. (2013). From necessities to imaginary worlds: Structural change, product quality and economic development. *Technological Forecasting and Social Change, 80*, 1499–1512.

Saviotti, P. P., & Pyka, A. (2015). Innovation, structural change and demand evolution: Does demand saturate? *Journal of Evolutionary Economics*. doi:10.1007/s00191-015-0428-2.

Teixeira, A. A. C., & Queirós, A. S. S. (2016). Economic growth, human capital and structural change: A dynamic panel data analysis. *Research Policy, 45*, 1636–1648.

van Doren, D., Koenigstein, S., & Reis, T. (2013). The development of synthetic biology: A patent analysis. *Systems and Synthetic Biology, 7*, 209–220. doi:10.1007/s11693-013-9121-7.

Wellhausen, R., & Mukunda, G. (2009). Aspects of the political economy of development and synthetic biology. *Systems and Synthetic Biology, 3*(1–4), 115–123.

Wield, D., Hanlin, R., Mittra, J., & Smith, J. (2013). Twenty-first century bioeconomy: Global challenges of biological knowledge for health and agriculture. *Science and Public Policy, 40*, 17–24.

Wright, G., & Czelusta, J. (2004). The myth of the resource curse. *Challenge, 47*(2), 6–38.

Wright, G., & Czelusta, J. (2007). Resource-based growth past and present. In D. Lederman & W. Maloney (Eds.), *Neither curse nor destiny: Natural resources and development*. Palo Alto, CA and Geneva: Stanford University Press and World Bank Publication.

Some Thoughts About the Bio-economy as Intelligently Navigated Complex Adaptive Systems

Hugo de Vries

Abstract Observations of major curves for greenhouse gas emissions, energy usage, world population, obesity (and health related diseases), national deficits, etc. are showing *exponential* growth. This is based on current economic forces favouring upscaling, higher throughputs and homogenization of mass production processes for food, (bio-)materials, molecules, energy, but also for services like tourism. On the other hand, biology—especially ecology—and sociology demonstrate that complex, viable ecosystems require (bio- and cultural-)diversity, differentiation and dynamic equilibria between species. Concomitant patterns are sinusoidal allowing dynamics and life at the edge of order and chaos both locally and globally. Consequently, the '**bio**-economy' faces a *paradox*. First thoughts about a conceptual approach for the bio-economy are presented that relate the economic evolving patterns with ecological dynamic equilibria. It is based on an iterative process of defining and monitoring images of a viable planet, exploiting complex adaptive systems (CAS) with continuously adapted rules and interventions. The emerging properties of the system *fuel* innovations. Those are all steered collectively by a Bio-economy Council such that these are disruptive in nature and cross-sector-oriented in order to re-direct the exponential curves towards sinusoidal patterns; this approach is introduced as intelligently navigated complex adaptive systems (INCAS).

1 Introduction

If one critically observes major indicators for our global society over time, one recognizes exponential growth curves (see Fig. 1; UNISDR 2016; Zabel 2009; FAO 2013; WHO 2016; EPA 2016). This holds for the emission of greenhouse gases, world population and mass productions of food, oil usage, obesity and health related diseases, public deficits of major states, plastic debris in the oceans, urbanization, to name a few. The *worldometers* (2016) website clearly demonstrates how indicators

H. de Vries (✉)
UMR IATE, Cirad, INRA, Montpellier SupAgro and University of Montpellier, INRA, F-34060 Montpellier, France
e-mail: hugo.de-vries@inra.fr

S. Dabbert et al. (eds.), *Knowledge-Driven Developments in the Bioeconomy*, Economic Complexity and Evolution, DOI 10.1007/978-3-319-58374-7_3

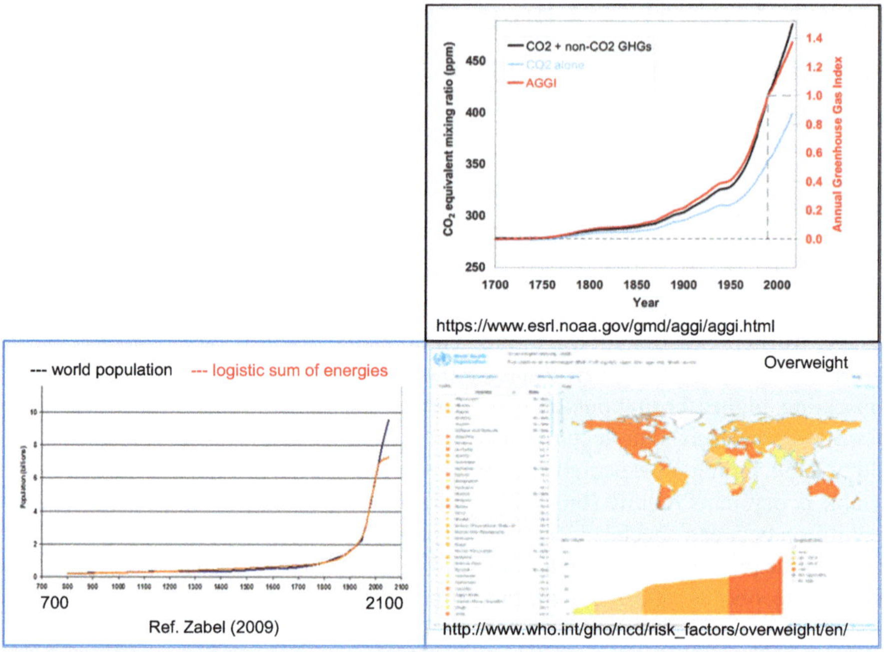

Fig. 1 Exponential and non-linear growth patterns of key societal and environmental indicators

are changing. Even if one questions the accuracy of figures one cannot misinterpret the overall patterns, e.g. as recently stated and visualized at the Climate conference in Paris, France (COP21 2015). Those patterns reveal systems being out of control, tending towards chaotic regimes. In contrast, viable ecosystems show patterns of self-organization.

In the past decades, the bio-economy concept has been formulated (SCAR 2011; Levidow et al. 2013; SCAR 2015). Those statements have led to new public-private actions, subventions, new R&D projects, first innovations, new curricula, etc. Those developments have shown that the thinking about the bio-economy has become a reality, that the underlying trends and challenges are recognized, even though that its adoption is highly country dependent (as demonstrated in this book by the various authors). Numerous countries are still struggling with the definition of an appropriate bio-economy agenda (Bioökonomierat Germany 2015). Also an internationally agreed biomass sustainability governance framework is now needed. Bosch et al. (2015) state: '*A starting point could be the total factor productivity (TFP) metric used to measure agricultural sustainability. The TFP reflects the rate of transformation of inputs (capital, labour, materials, energy and services) into outputs (biomass stock). A cost is attributed to each and to the negative social and economic impacts*'. Fortunately, a change is potentially feasible thanks to a number of sustainable solutions for e.g. biotechnology and synthetic biology (Carlson 2007), bio-refinery (Barakat et al. 2013), process engineering (example 'cold plasma' by Mastwijk and Nierop-Groot 2010), and new cooperation forms (Porter 2000; Porter and Kramer 2011).

Without doubt, those 'bottom-up' initiatives will change the curves by reducing slope rates. The question is whether it will be sufficient in order to maintain a viable planet. Do we know what a viable planet looks like? We don't need permanently horizontal curves, i.e. zero growth, with the risk of ending in a frozen, rigid, state; on the contrary, a viable planet would benefit from sinusoidal-like patterns that allow dynamics and systems to evolve, however within boundary layers defined by rules and incentives. Even though the boundary layer between the orderly and chaotic state has a certain bandwidth due to numerous external and internal conditions, the bandwidth is not unlimited (see Fig. 2). Within this boundary layer, all living creatures have to respectfully play their games, challenge the rules in time but accept the upper and lower limits, being revolutionary more creative, etc. Today, incremental innovations showing 10% improvements may motivate us, however, this is largely insufficient, since we are currently using on average two planets for our daily living conditions (Le Monde 2015).

Here, we face a *twenty-first century paradox* in the bio-based economy: the efficiency thinking or economy of scale concepts drives us towards in general upscaling and homogenization of production processes (food, biomaterials, biomolecules, eco-related services, ...), however, the environmental and social context forces us towards biodiversity, product differentiation, balanced income distribution, respecting cultural identities, etc. It is here, where agro-ecology and bio-economy concepts meet (Georgescu-Roegen 1971; Jordan et al. 2007).

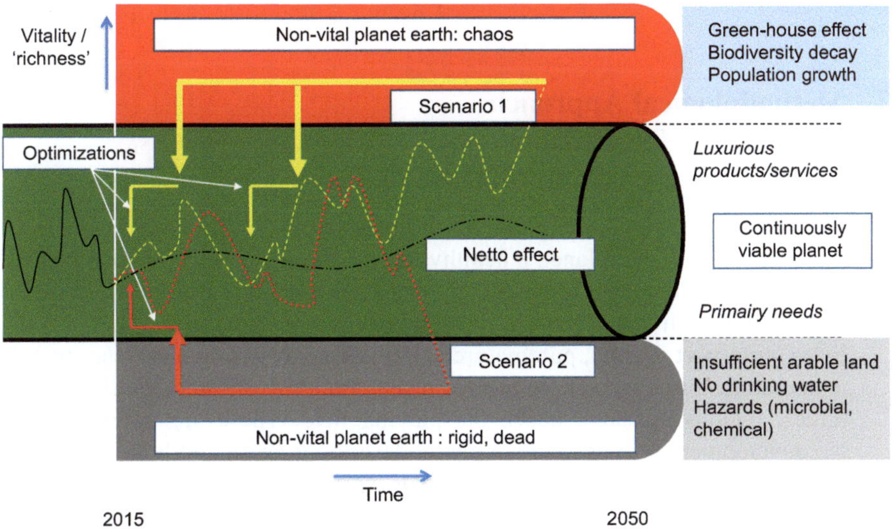

Fig. 2 The boundaries of a viable planet

Would it thus be possible to come up with an appropriate systemic approach and to either utilize existing methodologies or to develop new ones for the most complex system that we know, i.e. 'the viable planet', the only sustainable bio-economy? Utilizing the numerous existing methodologies save time especially if those can be tested locally and globally; but is this possible? Which parameters and rules should then be taking into account for all players in all kind of (new) configurations? What does it mean for the understanding and exploitation of the complex agro-resources? Which simplicities are possible—e.g. in the number of sustainable valorisation pathways of our renewable resources—without drastically changing our insights? Which radical innovations—technological, operational, and organizational—are adequate and really effective (European Commission 2015a)? Can we monitor their impacts in order to appropriately re-adjust our interventions? Could we do this in an iterative manner?

It is like playing a game with the viable planet as only acceptable outcome, however with the notion that our role is diverse. We are both players and observers facing an enormous diversity of agro-resources, socio-economic and environmental criteria as well as the adjusters of rules and incentives; hence the game is evolutionary in its nature but steered in practice. Thus, it is hard to see the—cascade or snowball—consequences of our interventions leading to either a dynamically, adaptive system, or chaos or rigidity. There is at least hope: as quite some authors have shown, complex adaptive systems have the advantage that the number of attractors—'dynamic' end-states—is huge but not unlimited due to power laws as previously described in literature as Pareto distributions (Kauffman 1993; Andriani and McKelvey 2011); thus, we can play an enormous amount of games providing freedom, flexibility and creativity. With this in mind we have started our challenging journey.

2 Methodological Approach

The methodological approach is divided into four inter-related steps:

(a) The starting point is sharing the notion that the only reasonable outcomes of the game are a viable planet, a healthy and well-appreciated bio-economy. We thus should build images of these outcomes based on imagination and aesthetics including basic living conditions in a prospective way (preferred food, healthy living, transport means), pure nature and biodiversity maintenance, acceptable temperature shifts, etc. These images reveal a mosaic of interacting territories (cities, regions, oceans), hence the playing field. These are translated into the development of a descriptive model describing green and blue landscapes and the overall viable planet, with a balanced focus on qualitative (aesthetics, opinions, emotions, perceptions, ...) and quantitative approaches [hard figures for production, biodiversity, temperature changes, population growth (like e.g. Fischer et al. 2012)].

Next, a symbolic observatory concept should be developed for the qualitative and quantitative approaches that monitor the boundaries of such a planet including all of its numerous interacting territories over time in a non-intrusive manner. An excellent base is the Bio-economy observatory (European Commission 2015b). It monitors the evolution of the bio-economy, like economic growth patterns and viable ecosystems and societies. The here proposed 'observatory' also include biological-environmental, (socio-)economic and sustainable technological knowledge to devise metrics (Lu et al. 2015) in order to link observations with potential bio-economy strategies and impacts of innovations utilizing knowledge engineering tools (decision support tools, argumentation models, fuzzy logics; Perrot et al. 2016) and artificial intelligence due to the enormous amount of data.

(b) The bio-economy strategies are proposed to be based on complex system theories that capture the playing field, the heterogeneity of agents, their interactions, the boundary layers, and the images and observations above. Those strategies should reveal self-organization, resilience and new emerging properties. The overall bio-economy system needs to have the adequate time for finding its own attractors due to its high level complexity and underlying less than astronomic number of interactions. It is divided into sub-systems following scalability rules (fractals or others) and inter-connections between those; the inter-connections should be incorporated in models in order to include consequences of inter-dependencies, non-linearity and feedback loops. It is of primary importance that rules and interventions are precisely defined and re-adjusted to maintain a focus on a viable planet organizing itself within clear boundaries. The emerging properties are the results of the interactions between different players, their assets (products, services, ...), the dimensions and shapes of the local or global playing fields, the interfaces between those, the timings, the internal and external restraints, etc. Game theory thinking should be included (Neumann and Morgenstern 1944; Holland 1998), in an evolutionary manner. Hereby, the selected interventions may lead to losses in the beginning but gains at the end, and vice versa; consequently, human behaviour should accept the principle of passing deep valleys before reaching higher peaks as well as accepting that those peaks are potentially sub-optimal. Applying evolutionary game theories for the bio-economy allows understanding the consequences of human interventions and innovations at well-suited time scales and weighing most relevant strategies for a viable planet.

(c) For the products of the bio-economy, a complex system approach would be beneficial in order to unravel the complexity of the structure-function relationships of agro-resources upon all imposed (environmental, engineering, ...) conditions. Two theories from physics are exploited, namely thermodynamics dealing with bulk properties and their dynamics upon changing external conditions (macro scale) and electromagnetism dealing with intra-molecular forces (micro-scale). A potentially new defined area is electro-thermodynamics that covers the meso-scale level; it concerns changing electron energies and electron densities as function of temperatures and pressures (like a plasma;

currently, receiving quite some research attention). It is interesting to research whether those theories provide a sound basis for complex systems. Targeted processing principles should then address macro, meso and micro levels in order to be highly effective.

(d) The notion 'that pathways revealing improvements of typically 10% are not sufficient to alter the course of exponential into sinusoidal patterns' leads to the fourth action. The viable landscape models, the non-intrusive and on-line observatory and the complex system approach for the bio-economy and its utilization of agro-resources should thus be integrated such that an innovation agenda can be drawn that evoke real breakthroughs (from 10% towards a factor of 2–10); in the complex system landscape it would mean to select attractors that are at least 2–10 times deeper than existing ones. This agenda will then be tested via a case-based approach to reveal the real improvement steps in the various domains (technology, organization, communication, ...). The outcomes are again monitored by our observatory in order to re-adjust interventions and landscapes; hence, a real iterative process.

3 Discussion About INCAS as a Conceptual Proposition Integrating Viable Planet Models, a Complex Adaptive Systems Approach and a Pragmatic Innovation Agenda

3.1 Descriptive Landscape Model and an Observatory for a Viable Planet and Local Systems

Brainstorm sessions and scenario discussions have led to numerous images of a viable planet (Agrimonde Terra 2011). Its division into a mosaic of territories (or sub-cultures and societies) makes the image more transparent; one can distinguish between the blue and green environments and rural areas, cities, agglomerations, etc. Here, we propose to translate the image into a generic model that comprises models for ecology (nature-oriented), agro-ecology, industrial ecology, value chains, and urban planning. Those models are fed with qualitative (aesthetics, opinions, emotions, perceptions, ...) and quantitative data (hard figures for biodiversity, temperature changes, population growth, added value, loss rates). The mosaic model imperatively includes the rules for the inter-connections between the territories, thus revealing the contours of the overall evolution picture. Those rules are not straightforward due to inter-correlations (*patchwork quilt theory,* Kauffman 1995), like: a growing population in a single territory impacts its neighbours in three dimensions. If more land is given back to nature, it impacts agro-ecology, food production, urban environments etc. Nevertheless, such an approach is a prerequisite for understanding of the overall image.

An *observatory* for monitoring a viable planet incorporates numerous functionalities. In order to reduce those, we have created the image of a bird (see Fig. 3;

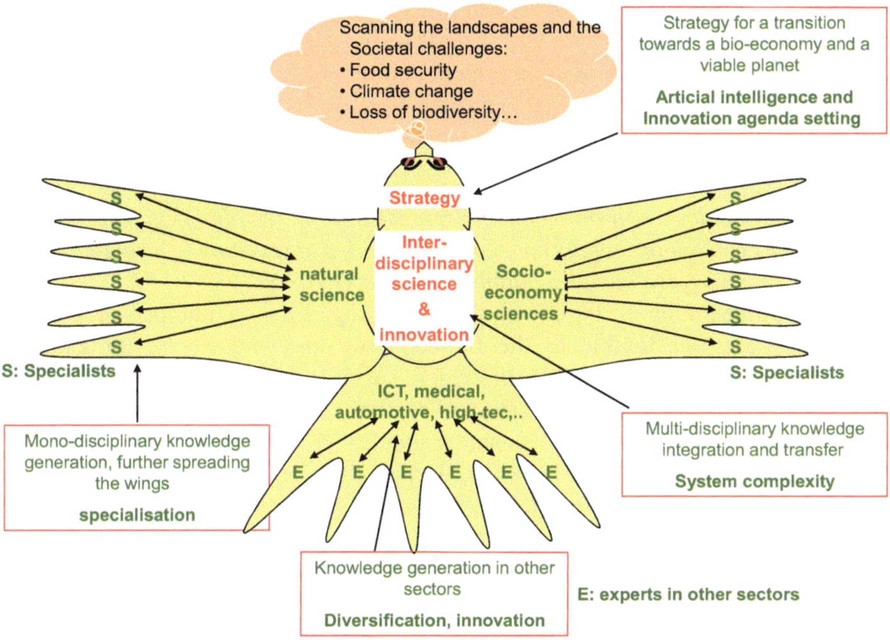

Fig. 3 The bird as dynamic and responsive observatory

modified image of Schepers 2010). The bird is able to observe—non-intrusively and on-line—global changes for different landscapes while changing its trajectory. The different images are combined for analysis and the development of scenarios. At the same time the bird metabolism integrates data and knowledge of different domains, which all of them are also separately deepened through new findings (widening of the span of the wings). The integration takes place in an artificial intelligent network that is coupled with incoming external signals. The network is capable of treating a large number of data. An example is the development of adequate decision support tools and argumentation models (Buche et al. 2013). Finally, the bird allows monitoring the impact of innovations listed below and re-adjusting interventions such that the main ambition of ending up and staying in a viable planet scenario is reached.

3.2 Complex System Approach for the Bio-economy

After having designed models for potential viable landscapes and observing the consequences of our behaviour on those landscapes, the next step concerns the development of a complex system approach for the bio-economy that describes the evolution of those landscapes locally and globally as mentioned above. The boundaries of the concept are determined by ecological and socio-economic conditions such as biodiversity and natural resilience, cultural diversity, food and nutritional security, sufficient employment and income. Clear boundaries allow its players to produce, to exchange, to consume, to recycle, etc.—in short 'to play'—in a dynamic manner with a lot of freedom, however, also within limits in order to avoid either chaos or rigidity.

Hypothesis *Ecology and bio-economy represent both complex systems; thus, the bio-economy can adopt ecology concepts and vice versa resulting in a single generic 'intelligently navigated complex adaptive systems (INCAS)'.*

The argumentation is based on the following. First, a number of insights in ecology are transferable to the bio-economy and vice versa, for example co-evolution, evolutionary stable strategies, survival of the fittest, biodiverse or heterogeneous agents, self-organization of systems, autocatalysis, emergence, non-linear interactions, butterfly effects and scalability (Kauffman 1993; Carbonara et al. 2010; Andrecut and Kauffman 2010). Hence, these are the major characteristics of Complex Adaptive Systems (CAS), being well positioned in what is called the melting zone (see Fig. 4).

Second, the notion and consequences of changing boundary conditions (restraints, voluntary and involuntary butterfly effects) for evolution are similar; this becomes even more apparent for agro-ecology and bio-economy. Some examples are (i) climate change impacts both local agro-ecology schemes as well as the eco-pyramid for efficient agro-resource usage, (ii) CO_2 increase leads to both adaptation of species and developing new carbon markets and soil-enriched production systems (iii) rising water levels impact local eco-systems and novel production locations (both positively and negatively), (iv) governmental interventions provoke changing markets, nature reserves and agro-ecological production sites and (v) economic crises reveal resilience of most fittest species and stakeholders.

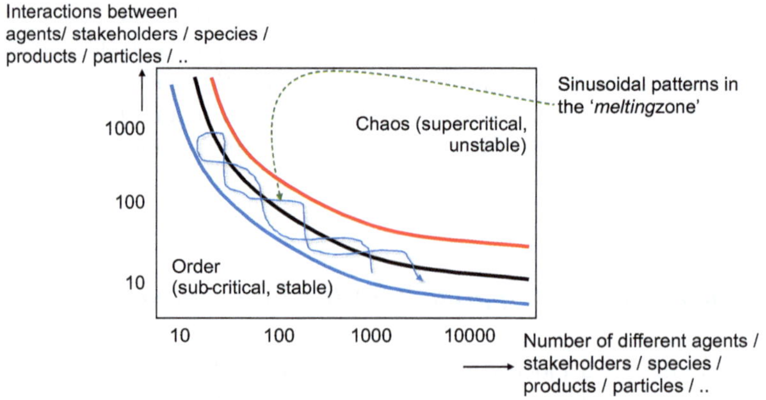

Fig. 4 The adapted thermodynamic 'order-chaos' scheme including the melting zone as home-base for complex adaptive systems

Third, even if the degree of heterogeneity is high and the interdependencies are largely distributed leading to extreme variance as stated in power law science (Andriani and McKelvey 2011), co-evolution and the interactions of heterogeneous agents show similarities in the origin of new species, bio-based product diversification and emerging eco-systems and markets; hence thinking about eco-pyramids—from added value for pharmaceuticals (high values, low volumes) via cosmetics, ... down to heat (low value, large volume)—show similar characteristics as the biological pyramid for species.

Fourth, thinking in sub-optimal instead of most optimal solutions allows reflecting on dynamic equilibria between e.g. different markets for agro-resources on one hand and different micro-ecosystems for species on the other. Herein, local bio-diverse production systems can exist in parallel to sustainable global value chains. However, it would mean a new dynamic equilibrium between those. If only global value chains and strategies of multi-nationals focusing on reducing number of high-volume products (potentially enriched with functional constituents/ingredients) will be adopted—the predominant breeze since the Second World War—a stable situation will not be reached; thus no sinusoidal patterns for all stakeholders.

Fifth, resilience is crucial in ecology and agro-ecology as well as in bio-economy in order to understand survival rates for species and marketed products. Resilience requires diversity; this underlines the importance of the remarks above.

Sixth, scalability plays both a role in (agro-)ecology as well as in economy. Micro-systems may add up arbitrarily or systematically revealing either fractal or highly dispersed patterns breaking scaling rules; numerous examples have been given for example for chemical and biological processes and deserve more attention (Perrot et al. 2016).

Finally, one may argue that, especially for behaviour and decision making—allowing to guide complex systems and coach stakeholders in a direction that corresponds with beneficial outcomes—the similarities between agro-ecology and bio-economy are sparse, however, also here scientific insights reveal interesting similarities as one can learn from Wilson (1990). These similarities deserve more attention looking to the diversity of players and their behaviour as individuals and within collectives.

A global bio-economical council may change the settings such that new balances arise between individuals and small collectives as well as between huge collectives and multinationals. Also the role of residents in the bio-economy is crucial since the introduction of ICT and personalized manufacturing like 3D-printing. The profession of some farmers is moving too towards entrepreneurship focusing on agro-production, bio- energy and providing tourism services at the same time. Even more, new industrial ecology (Frosch and Gallopoulos 1989) initiatives locally pop-up, like Agripark and Kalundborg, and other crossing-sector eco-innovation parks (BAFU 2014) challenging the balances mentioned above. It is here that solely a CAS approach is insufficient; intelligently navigated complex adaptive systems (INCAS) are more appropriate when being steered by a Bio-economical Council.

In all cases, large-scale high throughput mono-systems should be re-equilibrated with small-scale parallel-biodiverse production systems. Ecological studies have clearly revealed that a certain level of bio-diversity is necessary for survival, evolution and socio-economic development; the latter is recently presented in an assessment study by IPBES (2016) which shows that a growing number of pollinator species worldwide are being driven toward extinction by diverse pressures, many of them human-made, threatening millions of livelihoods and hundreds of billions of dollar worth of food supplies; however, measures can be taken to change these patterns. If INCAS thinking makes sense in linking (agro-)ecology and bio-economy, reflexions about sustainability and evolutionary games, governance and interventions, diversity in species and stakeholders as well as their interactions, etc. are getting their common conceptual frame. This allows sound decision making, targeted interventions and transparently revealing impacts.

3.3 Complex System Approach for Transformation of Bio-matter

Hypothesis *our understanding of 'bio-matter' should change in order to fully and intelligently make use of its constituents (molecules and functional fractions), properties and functionalities, caloric value, etc. as well as to design the best targeted, lowest energy and water consuming, processing methods for changing its structures into the most applicable end-user products.*

Unfortunately, bio-matter is often still a *black box* and approaches are 'trial & error' driven, leading to fractional—often wealthy—information about its structure, its constituents (and/or some of the 2200 additives as e.g. only in food) and the design of an enormous variety of food and materials. The statement that bio-matter should still be considered as black boxes is based on numerous observations and references. For food, Mezzenga et al. (2005) stated 'Foods make up some of the most complex examples of soft condensed matter (SCM) with which we interact daily. Their complexity arises from several factors: the intricacy of components, the different aggregation states in which foods are encountered, and the multitude of relevant characteristic time and length scales. Because foodstuffs are governed by the rules of SCM physics but with all the complications related to real systems, the experimental and theoretical approaches of SCM physics have deepened our comprehension of their nature and behaviour, but many questions remain.'

Other references reveal similar remarks and numerous research questions are posed today. Some examples are (INRA 2016), do we understand the formation of the heterogeneous protein networks and morphologies, as function of humidity and temperature during the production phase? What does that mean for e.g. efficient isolation of proteins? Why are we not able, after decennia of research, to mimic the natural rubber properties by synthetically produced rubber? How do all constituents contribute to these unique properties? Do we understand (ir-)reversible nuclei and agglomerate formation of powders? What is the impact of single grains on stability and which micro or macro avalanches are evoked? Which processes are underlying oxidative stress and adaptation mechanisms in 'living' materials? Is it simply multi-stage coupled electron transfer? What about electron densities? How and why do micro-organisms act under stress situations? Are assemblies formed to increase survival rates for organisms at their core? Which signal-transduction pathways are established? At what moment could microbial processes becoming 'sterilized' auto-catalytic processes at which reactions are steered by the system itself?

Those questions require a more than basic understanding of bio-matter (including microflora) at microscopic, mesoscopic and macroscopic level. It deals with constituents (proteins, polysaccharides, lipids, micronutrients, water, micro-organisms, ...), their make-up in a well-defined environment (e.g. grains, tissues, ...), their fluxes, their interactions (agglomeration, network forming capacity, ...), their catalytic actions, their phase-changes, their surface tensions, their plasticity, their reversibility and/or irreversibility potential, etc. A biologist may use other words like their adaptability, their self-organising capacity, their attractors, their enzymatic strength or autocatalytic behaviour, and so on. It is at the crossing border of physics, chemistry, biology, mathematics, process engineering and computer science that we may redefine bio-matter and inherent dynamics.

Thus a more generic approach and frame for food & biomaterials sciences (Wrangham 2009) is needed to understand and visualize process-structure-function relationships (Fischer and Windhab 2011). The thermodynamic approach adopted for other complex systems (biology, economics, ...) offers a starting point. In particular, thermodynamic approaches based on statistical physics and mechanics provides a general framework for relating the microstructure of individual atoms and molecules to the bulk properties (Prigogine and Stengers 1985). A system (like a food/bio-matter matrix full of constituents) could be either frozen in a very stable state or in a highly chaotic state or in the melting zone (see Fig. 4). The position of a 'system' in the scheme depends on the variety of its constituents (ingredients, persons, etc.), the interactions between those and the external (extreme) conditions. A position at the edge of order and chaos is most favourable because it allows adapting or responding to changing external conditions like processing conditions, while either maintaining its structure-properties-functions (for soft condensed matter) or showing clear phase changes or micro and macro-avalanches (Bak et al. 1988). As previously discussed, also here emergent properties (foaming, gelling, ...) are said being the result of self-organisation of the system as a whole; however, not with endless degree of freedom because the boundary conditions form the frame and restraints. The INCAS concept seems well suitable to acquire deep insights in bio-matter via appropriate interventions.

It is here, where experiments and numerical modelling met in order to deepen our knowledge of bio-matter. The numerical simulations, strengthened by the rapid increase in computational resources (GPU, HPC ...), can be used to investigate the microscopic origins of complex phase transitions in structural properties of soft matter and disordered systems such as colloids, polymers, emulsions, granular materials ... or cellular assemblies. Discrete numerical modelling (Multibody Contact Dynamics, Lattice Boltzmann, Monte Carlo Molecular Modelling ...) of interactions between large number of particles and under various boundary and environmental conditions gives new insight in complexity, order-disorder transition, chaos ... as emerging from the local scale of material heterogeneities and physical-chemical constituents (Augier et al. 2010). Furthermore, the association with stochastic approaches and statistical homogenization allows overcoming the lack of information regarding the geometry of particles or the initial state and gives a theoretical framework for upscaling properties in a constitutive thermodynamically-based equation of state.

The deepened insight in bio-matter also allows developing targeted intervention/processing methods focused on exactly modifying what is needed. If one likes to inactivate spores, it may be sufficient to destabilize a membrane-bound protein via the opening of a Sulphur bridge in an electrical field chamber. Hence, a thermodynamic intervention at macro-level (for example 'heating at 120 °C') should be replaced by an electromagnetic intervention at micro-level. If one likes to align micro-fibres in a lignocellulosic matrix, a single electromagnetic pre-treatment followed by a uni-directional shear force may be adequate; if correctly combined it concerns a meso-level intervention. It is here that electromagnetism dealing with intra-molecular forces (micro-scale) meet thermodynamics at macro-scale; this may be called electro-thermodynamic interventions at meso-scale; a promising area for future research and innovation.

3.4 Innovation Agenda

The emerging properties of INCAS as described above provide some key principles for the innovation agenda, which are potentially 2–10 times more efficient than today. This could be a principle that (i) avoids unnecessary exploitation of resources, (ii) efficiently transforms and uses agro-resources either via rapid iterative pathways or via snowball effects and (iii) valorises co-products and waste streams.

(i) *Avoiding eco-unfriendly- and over-exploitation of agro resources if other solutions exist, via:*

- The transition from products towards services (Herman et al. 1989; Schepers 2010), thus from materialistic towards non-materialistic assets.
- Lifetime optimization, dietary changes favouring low density—high satiety food above over-consumption and restorative economies building longer lasting technologies (WRAP 2016).
- Unnecessary breakdowns of constituents in multiple production stages, e.g. the replacement of meat proteins by plant or insect proteins (Aiking et al. 2006; CVT AllEnvi 2015).
- Waterless systems in processing such as dry bio-refineries (Barakat et al. 2013), closed water recycling systems, and producing salt tolerant species, halophytes, in coastal areas.
- Novel biotechnology and synthetic biology pathways (Carlson 2007).

- Energy production via solar energy capturing systems and not using biomass from an entropy point of view except after having valorised all its constituents in the eco-pyramid (Michel 2012).
- New breeding strategies for entire plant usage since often major plant parts are not valorized (Hofte 2013, 2015; IEA 2009; Poyry 2011).
- Local bio-refineries at the farm (Sanders 2015) avoiding unnecessary transport of water, re-utilizing co-products locally, avoiding empty returns, combining loads and avoiding car-shopping.

(ii) *Efficient transformation and usage of agro-resources via:*

- New ICT driven processes, like virtual design of bio-products to avoid unnecessary trials and unforeseen losses, domotics and eventually 3D printing (Lipson and Kurman 2013) if co-products are well taken into account, for guiding eco-efficient handling and consumption.
- Eco-efficient storage concepts based on feedback systems to maintain product quality.
- Water reduction systems utilizing high precision water-droplet systems, alternatives for cleaning and disinfection (Voigt et al. 2012), energy efficient desalting of sea water (MIT 2016).
- Energy reduction based on process intensification and control (Stankiewicz and Moulijn 2003), grinding up-to the level of functional fractions for both food and non-food applications (thus not down to molecular levels; Abecassis et al. 2014), volumetric heating replacing surface heating (de Vries and Matser 2011; Nutripulse 2013), biochemical synthesis (Jensen et al. 2012), auto-catalytic systems and pathways for efficient functionalization of molecules, more efficient transport (multi-modal, energy-intelligent containers).
- Novel biomaterials (Colonna 2015) and material reduction, counteracting the accumulation of packaging materials in oceans (UNEP 2011), no re-packaging, in package-treatments, etc.

(iii) *Valorisation of co-products and losses via:*

- Novel processing schemes for co-product and waste valorisation since one deals with enormous volumes (Poyry 2011).
- Introducing anaerobic digestion (Cazier et al. 2015) especially at farm level, however, after having extracted and valorised all other functional fractions.
- Adapted logistic and sustainable production schemes for co-product and waste recycling in rural and urban areas like aquaponics systems in order to create employment locally, recover useful constituents for agriculture (Magid et al. 2006) and to avoid accumulation of toxins.
- Development of fully new integrated, cross-sector, business (Porter and Kramer 2011) and public cooperation concepts such as industrial ecology, agriparks and eco-innovation parks in order to reach multiple sub-optimal solutions in terms of economic, social, environmental and aesthetic perspectives within a single concept. It includes closed circles for (all)

renewable resources, multiple markets locally and globally, new public-private partnerships, local embeddedness, novel branding strategies, close connections to consumers and job creation; it is applicable in diverse landscapes and (sub-)urban environments. Some business examples are: Kalundborg Industrial Symbiosis (Christensen 2006), Agripark Holland, Pomacle France; Shanghai Modern Agricultural Parks, Agrotechnology Park Singapore (Smeets 2011; Boix et al. 2012; BAFU 2014).

4 Conclusions

In the past decades, we have luxuriously been provided by fossil fuels. They have allowed playing an enormous variety of games to challenge and please us. The number of games has been large—by some considered as endless—however it's becoming more and more apparent that the number is not unlimited; renewable and recycled resources as well as solar energy input and the earth capacity have their limits. We may question 'how large?' That depends on our creativity and passion to play as well as to the new boundary conditions and their understandings. Examples of the latter are e.g. the limited available fossil fuel resources and the continuous rise in atmospheric CO_2 concentrations due to our daily requests for more energy. Sea levels rise and weather conditions may change more rapidly and extremely; consequently, this will restrict us to deliberately choose where and how to live, what to produce, consume and utilize. Even more, we are not solely facing environmental and economic constraints, but also social and ethical issues. Unequal global balances in 'access to and availability of food and healthy diets' is a key and persistent example of the latter; this leads to food insecurity, malnutrition and stunting for a major part of the growing (demands of the) world population (FAO 2015).

In this paper, five major questions have been addressed:

i. Can we create images of a viable planet and design observatories that make sense in terms of accurate information provision and steering?
ii. Which systemic approach could be developed for the bio-economy taking into account (agro-) ecology concepts?
iii. Which common concept should be developed for unravelling the complexity of bio-matter and their efficient transformations into useful end-products?
iv. What innovation agenda makes sense taking into account the necessity to have real breakthrough innovations reducing the sustainability impact with a factor of 2–10?
v. Based on the answers to the previous questions, are we able to introduce a suitable concept?

The first question reveals that a viable planet requires ecological and socio-economic boundary conditions and permits self-organization in such a way that outcomes of activities show sinusoidal patterns with amplitudes well between the boundaries (see Fig. 2). This allows freedom to play numerous games, however, not endlessly. It also provides us with numerous emergent properties (products, services) fulfilling the human sense for curiosity.

The second question deals with observatories (see Fig. 3) allowing non-intrusive and on-line monitoring to reveal dynamics of numerous, evolutionary and inter-connected, landscapes. They capture diverse information flows and show impacts of potential innovations, utilizing an intelligent artificial system to link observations with preferable landscapes and provide recommendations.

The third question addresses a concept that links bio-economy with ecology, and in particular agro-ecology, both at small (local agro-ecological production) as well as at large scales (globalized agro-production in an overall sustainable manner). The concept is based on complex adaptive systems incorporating both economic and ecological characteristics (biodiverse products and services, biodiversity, interacting ecological and economic agents, etc. (see Fig. 4). The concept results in a better understanding of the variety of options in the bio-economy and provides the basis for interventions. Thus, intelligently navigated complex adaptive systems (INCAS) seem to be most logical. The approach tackles the existing paradox between economics and ecology and provides an alternative approach. It also touches upon the circular economy concept (Mc Arthur foundation 2014); however, closed circles neither exist in ecology nor in (socio-)economy. Sinusoidal-like patterns resemble spirals, allowing the existence of a wider range of dynamic end-states, and thus evolutionary processes. It should be noted that self-organization within complex adaptive systems reduces the number of potential landscapes and 'limits the number of attractors'. It is in this phase that interventions are most likely contra productive; the system needs a while finding its own attractors before navigated actions are undertaken. An example is the living specie itself; if the genome consists of 100,000 genes and they present 2 states 'on' or 'off'; the potential number of end-states is $2^{100,000}$; however, the number of cell types is more or less the square root of number of genes (315), hence a very well self-organized 'system', most likely do to the limited number of grouping of 'cooperating' genes. This power law can be well mimicked by Boolean network and N–K models (Kauffman 1995) and exploited for other systems.

Intelligently navigated complex adaptive systems (INCAS) have also been discussed as powerful approach to unravel the complexity and richness of (new) nature resources in order to develop targeted engineering interventions. This approach is principally different from the approach starting with basic building blocks and, step-wise, assembling more complex molecules. Now, the starting point is the understanding and exploitation of the complexity and diversity of (novel) bio-matter, thus the interactions between constituents in all dynamic stages along the agri-food-bio-product chain. Numerous new functionalities and products pop up according to self-organizing principles, that may better fulfil the demands of tomorrow, reduce losses, create—local and global—value for all plant parts by using intelligently new sustainable (targeted, small-scale) biotransformation methodologies.

Consequently, an appropriate breakthrough innovation agenda could be defined focusing on (i) avoiding (over-)exploitation of resources, (ii) more efficient usage of resources and (iii) valorising waste and co-products. The first series of breakthrough innovations are organizational in nature focusing on alternative strategies in the bio-based economy and adapted behaviour. The second series of innovations are technological, either at micro- (electromagnetism), meso (interacting at the complex adaptive product matrix level, where thermodynamics and electromagnetism meet) or macro level (thermodynamics). The third series follows the eco-pyramid for value creation for co- and waste streams, with highest value smallest volumes at the top (pharma) and lowest value largest volumes at the bottom (heat generation). The final innovation agenda will contain a mixture of innovation items, based on the local opportunities for most efficient pathways.

The combination of above mentioned approaches leads to a bio-economy conceptual approach reasonably utilizing its resources, combining aesthetics, non-intrusive monitoring, artificial intelligence, complex adaptive systems, real innovations and steering by re-adjusted interventions by a Bio-economy Council (see Fig. 5). The here introduced concept is INCAS for intelligently navigated complex adaptive systems.

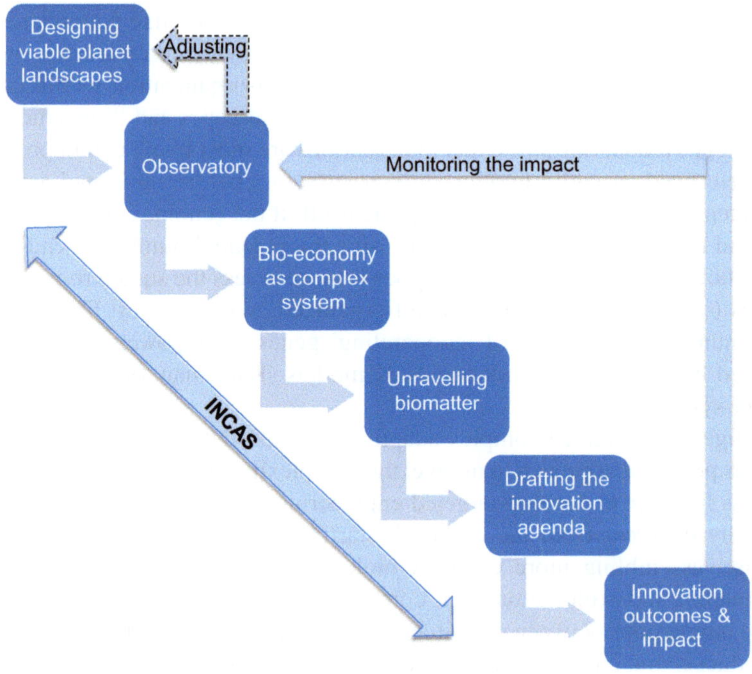

Fig. 5 The bio-economy as 'intelligently navigated complex adaptive systems (INCAS)'

5 Future Perspectives

In order to support a transition towards a viable bio-economy, the following actions are recommended to start with. The first four are best included in an in-depth foresight study with a limited number of scenarios and recommendations for policy and decision makers. The fifth would allow including all stakeholders in this transition.

- Better *understanding complex adaptive systems* both for the entire bio-economy and the linkage with agro-ecology concepts as well as for the multiple bio-matter matrices.
- Clarify the choice of technology and its Technology Readiness Level in the bio-economy and identify ways to *improve substantially their effectiveness* using economic (added value, competitiveness, clustering, ...), environmental (energy, water, greenhouse gases, biodiversity, loss, mitigation ...), and social (employment, food security, cultural values, ethics, urban habitats ...) criteria.
- Identify, test and evaluate public and private levers for re-equilibrating *self-organized, competitive, localized, systems with empowerment strategies* utilizing well-defined constraints, rules, incentives and decision making processes, thus exploiting *INCAS*.
- A science program focused on *industrial ecology concepts linked to local agro-ecology concepts and global value chains*; herein the unravelling of bio-matter black-boxes, both in the pre-harvest and post-harvest phases, for selecting best end-product solutions, plays again a key role. Other topics are the cycles of nutriments, minerals and 'dead' organic matter in comparison to those of material fluxes and waste recycling, the meaning and understanding of symbioses, reversed thinking (natural feedbacks and market-driven actions) and resilience towards external stress.
- An '*evolutionary game for a viable planet*' should be designed, including numerous sub-games for creating sustainable solutions. An example is given here. The earth is the playground. The field partitions are agro-ecological production fields, living areas, nature reserves and transport-lines. The pieces are a number of renewable resources and technologies, water (drinking, process) and energy sources (solar, wind, biomass). The finally obtained bio-based products and services also serve as new pieces after having incorporated their transaction costs. The duration is split into 4 years. The winning outcomes are the number of overall best fits—sub-optimal solutions—for the needs of current and five future generations; the outcomes are spread out over the various basic needs. The rules are defined by the government: (i) increased local biodiversity of 10% per decade, (ii) 15% dematerialization as compared to the current situations, (iii) maximized loss of 5% of an integral renewable resource in the initial production step, (iv) full usage of all losses and co-products within five follow-up steps, (v) 5 times higher energy and water efficiencies as compared to given references, (vi) acceptance rate of 90% by culturally diverse consumers for end-products and services. The government is the referee and definer of rules, the knowledge institutions

are the advisory board members and the NGOs are objectively auditing. The global citizens are the players bringing collective intelligence to provide the best, appreciated, solutions both locally and globally.

Acknowledgements The author likes to warmly thank his colleagues of the Joint Research Centre IATE, of the CEPIA-Inra division and of the Food Technology Centre of Wageningen UR as well as the partners in the European Project NovelQ. Special thanks are attributed to Dr. Monique Axelos, Inra, and Prof. Stéphane Guilbert, Montpellier SupAgro for introducing me in bio-based society research in France.

References

Abecassis, J., de Vries, H., & Rouau, X. (2014). New perspective for bio-refining cereals. *Biofuels, Bioproducts and Biorefinery, 8*, 462–474.

Agrimonde Terra. (2011). http://www.foresight-platform.eu/wp-content/uploads/2011/10/EFP-Brief-No.-196_Agrimonde.pdf

Aiking, H., de Boer, J., & Vereijken, J. (Eds.). (2006). *Sustainable protein production and consumption: Pigs or peas? Environment and Policy* (Vol. 45, 245 pp). Heidelberg: Springer. ISBN 1-4020-4062-8.

Andrecut, M., & Kauffman, S. A. (2010). Phase transition in a class of nonlinear random networks. *Physical Review E, 82*(2).

Andriani, P., & McKelvey, B. (2011). From skew distributions to power-law science. In P. Allen, S. Maguire, & B. McKelvey (Eds.), *The Sage handbook complexity and management*. London: Sage. ISBN 978-1-84787-569-3.

Augier, F., Idoux, F., & Delenne, J. Y. (2010). Numerical simulations of transfer and transport properties inside packed beds of spherical particles. *Chemical Engineering Science, 65*(3), 1055–1064.

BAFU. (2014). http://www.bafu.admin.ch/publikationen/publikation/01756/index.html?lang=en

Bak, P., Tang, C., & Wiesenfeld, K. (1988). Self-organized criticality. *Physical Review A, 38*(1), 364–374.

Barakat, A., de Vries, H., & Rouau, X. (2013). Dry fractionation process as an important step in current and future lignocellulose bio-refineries: A review. *Bioresource Technology, 134*, 362–373.

Bioökonomierat Germany. (2015). http://www.biooekonomierat.de/fileadmin/Publikationen/berichte/Bioeconomy-Policy_Part-II.pdf

Boix, M., Montastruc, L., Pibouleau, L., Azarro-Pantel, C., & Domenech, S. (2012). Industrial water management by multi-objective optimization: From individual to collective solution through eco-industrial parks. *Journal of Cleaner Production, 22*(1), 85–97.

Bosch, R., van de Pol, M., & Philp, J. (2015). Policy: Define biomass sustainability. *Nature, 523*, 526–527.

Buche, P., Croitoru, M., Fortin, J., Mosse, P., Tamani, N., & Thomopoulos, R. (2013). Argumentation and negociation for biodegradable food packaging selection. *Revue d'Intelligence Artificielle, 27*(4/5), 515–537.

Carbonara, N., Giannoccaro, I., & McKelvey, B. (2010). Making geographical clusters more successful: Complexity-based policies. *E:CO Issue, 12*(2010), 21–45.

Carlson, R. (2007). Laying the foundations for a bio-economy. *Systems and Synthetic Biology, 1*(3), 109–117.

Cazier, E., Trably, E., Steyer, J.-P., & Escudié, R. (2015). Biomass hydrolysis inhibition at high hydrogen partial pressure in solid-state anaerobic digestion. *Bioresource Technology, 190*, 106–113.

Christensen, J. (2006, November 27–December 1). *The history of the industrial symbiosis at Kalundborg, Danemark* (49 pp). Scientific Workshop 'Frontiers of Research in Industrial Ecology', University of Lausanne.

Colonna, P. (2015). http://www.3bcar.fr/images/Mat%C3%A9riaux_Biosourc%C3%A9s_-_Forum_Recherche_Industrie_2015_-_Web.pdf

COP21. (2015). http://www.cop21.gouv.fr/comprendre/10-photos-pour-voir-que-le-monde-change/

CVT AllEnvi. (2015). *Protéines végétales & Alimentation.*

De Vries, H.S.M., & Matser, A. (2011). *NovelQ.* http://cordis.europa.eu/publication/rcn/13339_en.html

EPA. (2016). http://www3.epa.gov/climatechange/science/causes.html

European Commission. (2015a). *The junction of health, environment and the bioeconomy: Foresight and implications for European.* RTD-PUBLICATIONS@ec.europa.eu.

European Commission. (2015b). *Bioeconomy observatory.* https://biobs.jrc.ec.europa.eu/

FAO. (2013). *Food wastage footprint, impact on natural resources* (Summary report). Rome: FAO.

FAO. (2015). *The state of food insecurity in the world 2014.*

Fischer, G., Nachtergaele, F. O., Prieler, S., Teixeira, E., Tóth, G., van Velthuizen, H., Verelst, L., & Wiberg, D. (2012). *Global Agro-Ecological Zones (GAEZ v3.0).* Laxenburg and Rome: IIASA and FAO.

Fischer, P., & Windhab, E. J. (2011). Rheology of food materials. *Current Opinion in Colloid and Interface Science, 16*, 36–40.

Frosch, R. A., & Gallopoulos, N. E. (1989). Strategies for manufacturing. *Scientific American, 261*, 144–152.

Georgescu-Roegen, N. (1971). *The entropy law and the economic process.* Cambridge, MA: Harvard University Press.

Herman, R., Ardekani, S. A., & Ausubel, J. H. (1989). Dematerialization. In J. H. Ausubel & H. E. Sladovich (Eds.), *Technology and Environment* (pp. 50–69). Washington, DC: National Academy Press.

Hofte, H. (2013). L'amélioration des cultures dédiées à la production de biomasse lignocellulosique. *Innovations Agronomiques, 26*(2013), 17–23.

Hofte, H. (2015). http://www.biomassforthefuture.org/

Holland, J. (1998). *Emergence: From chaos to order.* Redwood City, CA: Addison-Wesley.

IEA. (2009). http://www.globalbioenergy.org/uploads/media/0908_IEA_Bioenergy_-_Bioenergy_%E2%80%93_A_sustainable_and_reliable_energy_source_ExSum.pdf

INRA. (2016). http://umr-iate.cirad.fr/en/research-topics and http://www.cepia.inra.fr/Toutes-les-actualites/actu-recherches-et-innovations-2015

IPBES. (2016). http://www.ipbes.net/article/press-release-pollinators-vital-our-food-supply-under-threat

Jensen, K., Jensen, P. E., & Møller, B. L. (2012). Light-driven chemical synthesis. *Trends in Plant Science, 17*(2), 60–63.

Jordan, N., Boody, G., Broussard, W. N., Glover, J. D., Keeney, D., McCown, B. H., McIsaac, G., Muller, M., Murray, H., Neal, J., Pansing, C., Turner, R. E., Warner, K., & Wyse, D. (2007). Environment – Sustainable development of the agricultural bio-economy. *Science, 316*(5831), 1570–1571.

Kauffman, S. (1993). *The origins of order: Self-organization and selection in evolution.* New York: Oxford University Press.

Kauffman, S. (1995). *At home in the universe, The search for the laws of self-organization and complexity.* Oxford: Oxford University Press.

Le Monde. (2015). *Global Footprint Network.* http://www.lemonde.fr/planete/article/2015/08/13/a-partir-de-ce-soir-la-terre-vit-sur-ses-reserves_4722826_3244.html#rRWbxTxz27wkLpBe.99

Levidow, L., Birch, K., & Papaioannou, T. (2013). Divergent paradigms of European agro-food innovation: The Knowledge-Based Bio-Economy (KBBE) as an R&D Agenda. *Science, Technology, and Human Values, 38*(1), 94–125.

Lipson, H., & Kurman, M. (2013). *Fabricated: The new world of 3D printing*. Indianapolis: Wiley. ISBN 978-1-118-35063-8.

Lu, Y., Nakicenovic, N., Visbeck, M., & Stevance, A.-S. (2015). Policy: Five priorities for the UN sustainable development goals. *Nature, 520*, 432–433.

Magid, J., Eilersen, A. M., Wrisberg, S., & Henze, M. (2006). Possibilities and barriers for recirculation of nutrients and organic matter from urban to rural areas: A technical theoretical framework applied to the medium-sized town Hillerod, Denmark. *Ecological Engineering, 28*, 44–54.

Mastwijk, H. C., & Nierop-Groot, M. N. (2010). Use of cold plasma in food processing. In D. R. Heldman, D. G. Hoover, & M. B. Wheeler (Eds.), *Encyclopedia of biotechnology in agriculture and food* (pp. 174–177). New York: Taylor & Francis. ISBN 9780849350276.

Mc Arthur Foundation. (2014). *Towards the circular economy: Accelerating the scale-up across global supply chains*. Prepared in collaboration with the Ellen MacArthur Foundation and McKinsey and Company. Geneva: World Economic Forum.

Mezzenga, R., Schurtenberger, P., Burbidge, A., & Michel, M. (2005). Understanding foods as soft materials. *Nature Materials, 4*, 729.

Michel, H. (2012). The nonsense of biofuels. *Angewandte Chemie International Edition, 51*, 2516–2518.

MIT. (2016). https://www.technologyreview.com/s/534996/megascale-desalination/

Neumann, J. V., & Morgenstern, O. (1944). *Theory of games and economic behavior*. Princeton, NJ: Princeton University Press.

Nutripulse. (2013). http://www.innovation-xl.com/uploads/Brochure2013/Nutri-Pulse_e-Cooker_2013.pdf

Perrot, N., De Vries, H., Lutton, E., van Mil, H. G. J., Donner, M., Tonda, A., Martin, S., Alvarez, I., Bourgine, P., van der Linden, E., & Axelos, M. (2016). Some remarks on computational approaches towards sustainable complex agri-food systems. *Trends in Food Science and Technology, 48*, 88–101.

Porter, M. E. (2000). Location, competition, and economic development: Local clusters in a global economy. *Economic Development Quarterly, 14*, 15–34.

Porter, M., & Kramer, M. (2011). Creating shared value. *Harvard Business Review, 89*(1/2), 62–77.

Poyry. (2011, June 29). *Biomass imports to Europe and global availability*. VGB-Forschungsstiftung.

Prigogine, I., & Stengers, I. (1985). *Order out of chaos: Man's new dialogue with nature*. London: Flamingo.

Sanders, J. (2015). Personal communication, Wageningen UR, The Netherlands.

SCAR. (2011). *3rd SCAR foresight exercise: Sustainable food consumption and production in a resource-constrained world*. Brussels: Standing Committee on Agricultural Research, Foresight Expert Group. http://ec.europa.eu/research/agriculture/scar/pdf/scar_feg3_final_report_01_02_2011.pdf

SCAR. (2015). *Sustainable agriculture, forestry and fisheries, in the bioeconomy – A challenge for Europe. 4th SCAR Foresight Exercise*. Brussels: Standing Committee on Agricultural Research, Foresight Expert Grouped. KOVACS B. ISBN 978-92-79-47538-2.

Schepers, H. (2010). Personal communication, Wageningen UR, The Netherlands.

Smeets, P. J. A. M. (2011). *Expedition agroparks. Research by design into sustainable development and agriculture in the network society* (320 pp). Wageningen: Wageningen Academic Publishers. ISBN: 978-90-8686-163-7.

Stankiewicz, A., & Moulijn, J. A. (2003). *Re-engineering the chemical processing plant: Process intensification*. New York: CRC Press.

UNEP. (2011). *Plastic debris in the Ocean*. UNEP Year Book.

UNISDR. (2016). https://www.unisdr.org/we/inform/disaster-statistics

Voigt, E., Jaeger, H., & Knorr, D. (2012). *Securing safe water supplies: Comparison of applicable technologies, EFFoST Critical Reviews 1*. Oxford: Academic Press. ISBN: 978-0-12-405886-62-405886-6.

WHO. (2016). http://www.who.int/gho/ncd/risk_factors/overweight/en/

Wilson, E. O. (1990). *The ants*. Cambridge: Harvard University Press. ISBN 0-674-04075-9.
Worldometers. (2016). http://www.worldometers.info/
Wrangham, R. (2009). *Catching fire: How cooking made us human* (320 pp). New York: Basic Books.
WRAP. (2016). http://www.wrap.org.uk/sites/files/wrap/Final%20Report%20EVA128_SEI%20%281%29%20JB%20SC%20JB3.pdf
Zabel, G. (2009). London School of Economics, MSc Demography/Energy Economics. Originally published by *Energy Bulletin*. http://www2.energybulletin.net/node/48677

Part II
Framing the Bioeconomy: Regional and National Approaches

Varieties of Knowledge-Based Bioeconomies

Sophie Urmetzer and Andreas Pyka

Abstract Governments around the world seek for strategies to overcome the reliance on fossil resources and provide solutions for the most challenging contemporary global issues: food shortage, depletion of natural resources, environmental degradation and climate change. A very recent and widely diffused proposition is to transform economic systems into bio-based economies, which are based on new ways of intelligent and efficient use of biological resources and processes. If taken seriously, such endeavour calls for the creation and diffusion of new knowledge as basis for innovation and behavioural change on various levels and therefore often is referred to as knowledge-based bioeconomy. In the current debate, the requirement for innovation is mostly seen in the advance of the biotechnology sector. However, in order to fulfil the requirement of sustainability, which implicitly is connected with the bio-based economy, the transformation towards a bioeconomy requires a fundamental socio-economic transition and must comprise changes in technology as well as in markets, user practices, policy, culture and institutions. To illustrate a nation's capability for this transition, we refer to the concept of national innovation systems in its broad approach. With the help of an indicator-based multivariate analysis we detect similarities and dissimilarities of different national systems within the European Union as basis for a transition towards a knowledge-based bioeconomy. The analysis allows to compare the different strategies and to identify bottlenecks as well as success factors and promising approaches in order to design policy instruments to foster this imperative transformation.

1 Introduction

Based on the European Commission's (2012b) understanding of the bioeconomy and the OECD's (1996) definition of a knowledge-based economy combined with considerations taken from Schmid et al. (2012) we define the knowledge-based bioeconomy as:

S. Urmetzer • A. Pyka (✉)
Institute for Economics: Innovation Economics (520I), University of Hohenheim, Stuttgart, Germany
e-mail: a.pyka@uni-hohenheim.de; sophie.urmetzer@uni-hohenheim.de

S. Dabbert et al. (eds.), *Knowledge-Driven Developments in the Bioeconomy*, Economic Complexity and Evolution, DOI 10.1007/978-3-319-58374-7_4

> An economy that is based on the production and dissemination of new knowledge about renewable biological resources and their potential to be sustainably converted into food, feed, bio-based products and bioenergy with the aim to overcome the wastefulness of production and consumption in its full dependency on fossil resources.

Formally, there is a global agreement for the imperativeness of such transformation of our current economic systems explicitly highlighted in policy agendas and strategies for the bioeconomy on global (OECD 2009), regional (EC 2012a), national (US Government 2012; BMELV 2013) and sub-national (MWK 2013) levels. However, the responses of the specific national systems to the above mentioned global challenges and their capability to innovate by developing new policy strategies and institutional reforms vary. In their reflection on innovation systems in the learning economy, Lundvall et al. (2002) conjecture that "some national systems may, for historical reasons, be better prepared to cope up with the new context than others" (p. 234). Without doubt this conjecture holds for the transformation towards a knowledge-based bioeconomy as well.

To examine the various national conditions for innovation towards a knowledge-based bioeconomy within the 28 member states of the European Union (EU), we empirically analyse and compare the specific national innovation systems. The Sect. 2 illustrates the theoretical foundation of our analysis. Section 3 describes the analytical approach by specifying the variables (indicators) measured and the methodology deployed. This is followed by the presentation of the resulting clusters and some carefully deduced implications of these results in Sect. 4. The concluding Sect. 5 closes with an outlook and some critical remarks.

2 Theoretical Background

The comparison of different political economies has occupied scholars and political actors for many years. Politicians and scientists seek to understand how differences in the organisation of national systems are responsible for certain economic outcomes and why there is more than one model that delivers economic success (Hall and Gingerich 2009; Hall and Soskice 2001). Different theoretical frameworks have been applied for comparative analyses between nations. One of them is the concept of national innovation systems (NIS) (Lundvall 1992; Patel and Pavitt 1994). It emerged during the 1990s and illustrates the underlying structure and processes of the interdependent evolution of technologies, industries and institutions in an economy. The basic assumption of the broad approach of NIS is that those institutions that directly promote the acquisition and diffusion of new knowledge are embedded in a specific socio-economic system (Lundvall 1992). Within this system, "political and cultural influences as well as economic policies help to determine the scale, direction and relative success of innovation" (Freeman 2002, p. 194). The NIS framework has been the basis for many theoretical as well as empirical studies and was subject to refinement and further elaboration in many ways during the last two decades. Such comparative studies have been undergone at the level of national economies (e.g. the comparison of the Danish and the Swedish innovation systems undertaken by Edquist and Lundvall 1993), sectors (Malerba 2005) or individual parameters (e.g. the future-orientation of innovation systems of Central and Eastern European countries analysed by Hanusch et al. 2010). Some of these undertakings have not only served for describing dissimilarities between systems, but also to uncover cross-national similarities in the structure and innovation performance (Balzat and Pyka 2006). By identifying the extent and the areas of such structural similarities within empirically determined groups or clusters of national innovation systems, such research can have an impact on the efficiency of mutual learning processes for policy planning. Just as the differences between path-dependent NIS prohibit a one-fits-all political solution for a common problem, structural similarities in certain fields allow for sectorial policy learning across national borders (Lundvall and Tomlinson 2002). This holds especially for the countries of the EU which are embraced by common European institutions and share certain cultural characteristics.

The special challenge of examining the national systems regarding their capability to move towards a knowledge-based bioeconomy is emerging from the overarching and yet quite specific nature of the bioeconomy. While a mere analysis of innovativeness within e.g. the biotechnology sector or the agricultural sector would not allow for drawing conclusions on the state of the bioeconomy in a country, the examination of the entire economic system of a nation would fall short of the specific requirements for a development towards the bioeconomy. The connection between the concept of NIS and the bioeconomy has recently been endeavoured by Roberto Eposti (2012). He proposes the creation of an EU-wide knowledge and innovation system for bioeconomy (KISB) with the aim to overcome the sectorial boundaries, improve agricultural innovations, acknowledge the heterogeneity of involved actors and adapt the EU research policy to the emerging structures of the bioeconomy. This proposal entails important challenges of the transition towards the bioeconomy (namely transdisciplinarity, innovativeness, governance and policy convergence) and describes a process still "largely in progress, incomplete and country-specific" (Eposti 2012, p. 253). We will step back and try to assess the grounds for such concept by a comparative analysis of the underlying national systems on a broad empirical basis.

One of the latest adaptations of the NIS approach for innovations towards higher resource productivity and lower environmental impact has been coined by Stamm et al. (2009) and further refined by Altenburg and Pegels (2012): the sustainability-oriented innovation systems (SoIS). They are defined as to comprise the network of those institutions that foster innovations "to reduce environmental impacts and resource intensity to a level commensurate with the earth's carrying capacity" (Altenburg and Pegels 2012, p. 10, based on Freeman 1987). Many of the implications that comprise such SoIS also apply to our endeavour undertaken in this study and have found their expression in the identification of measured indicators. Another implication of NIS for bioeconomy is expected to be the impact of public attitudes towards the environment, technological progress and the consumers' willingness to change[1] (USDA 2012).

3 Analytical Approach

The factors that shape a national system's capacity to adopt a knowledge-based bioeconomy are unknown and highly complex. The varieties of historical, geographical, political and socio-economic conditions across the European countries (i.e. the "given factors") as well as the multitude of potential expressions of a well-functioning knowledge-based bioeconomy (i.e. the "desirables") mark the grounds of our analysis. We can at best try to approximate reality by subjectively defining relevant indicators for measurement. In a theoretical comparative analysis of the evolution of bio-industrial complexes as building block of an emerging bioeconomy in five different OECD countries, Mats Benner and Hans Löfgren (2007) focussed on the degree of state intervention as characteristic difference between countries. Since the scope of our understanding of the knowledge-based bioeconomy goes beyond the emergence of bio-industrial complexes (see our definition above), we extend the frame of analysis by including indicators that describe the "relevant institutions and economic structures affecting the rate and direction of technological change [in the field of bioeconomy (the authors)] in the society" (Edquist and Lundvall 1993).

3.1 Indicators

Indicators for monitoring innovation towards the bioeconomy have not yet been defined. For the purpose of this study, the indicators proposed by the OECD to monitor green growth (OECD 2011) as well as the goals defined within the European Bioeconomy Strategy (EC 2012a, b) and the implications of SoIS (Altenburg and Pegels 2012) have served as a basis for an eclectic identification of relevant units of measurement.

[1] "Public attitudes toward and understanding of biobased products are important for the growth of the bioeconomy for at least two reasons. First, the government's commitment and ability to financially support the growth of the bioeconomy relies on a willing public. Second, public attitudes toward and understanding of biobased products will influence the demand for these products, which ultimately will determine the future of the bioeconomy. Measuring public attitude could be used as a leading indicator." (USDA 2012 p. 49)

The following six categories of data for the empirical assessment of the potential to introduce the bioeconomy are covered by our selection:

1. The **environmental and resource productivity** of production and consumption: Indicating an economy's ability to minimize non-renewable resource consumption per unit of output (i.e. decoupling production from non-renewable resources).
2. The base of relevant **scientific, applied and public knowledge**: Indicating a nation's potential to tackle future challenges in the field of the bioeconomy with the help of education on different levels. The European Commission (2012a, b) states that innovation in bioeconomic sectors requires a workforce that has the right mix of skills including experienced workers with new qualifications and professionals for interdisciplinary tasks who understand "the economic and societal impact of their activities, fostering cross-talk between sectors" and across society. At the same time, public understanding about the ethical, environmental, health and safety implications of the bioeconomy affects the acceptance and the economic success of new products and processes (EC 2009; USDA 2012).
3. **Policy responses and bio-economic opportunities**: Indicating a nation's potential and will to innovate and proceed in technological and institutional terms. This becomes evident by assessing activities that foster innovation in general and specifically in environmental science and technology (Global Innovation Index, R&D expenditures, research personnel etc.). In addition, these indicators shall measure political efforts and social acceptance to support a move towards a resource-efficient and environmentally-friendly economy.
4. The **natural asset base**: Indicating an economy's capability to maintain the quantity of their natural assets. This measure takes account of the fact that naturally re-growing resources are not infinite and must be sustainably managed.
5. The environmental dimension of **quality of life**: Indicating the social well-being in terms of access to an intact environment (including clean air, intact nature etc.). The desired increase in utilisation of biological resources must not be achieved at the expense of a loss in environmental quality—an asset hardly measurable in economic terms and to be kept separate from the natural asset base measured quantitatively (indicator group no. 4).
6. General **socio-economic structure**: Indicating the socio-economic context in which the different economies act. Even among the EU member states, structural and socio-economic differences exist that may influence the overall performance of their development towards the bioeconomy, including differing attitudes of the population.

To examine the disposition of the 28 nations to move towards a knowledge-based bioeconomy, indicators for each of the above introduced categories have been identified, amounting to a total of 47 variables (see Annex 1).

3.2 Methodology

Cluster analysis is a multivariate statistical technique which is used to group objects (in this case: countries) based on the characteristics they possess (Hair et al. 2010). In the context of this chapter, the grouping emerges from the specific national values for each of the indicators identified to characterise the individual NIS with respect to their capacity for a transformation towards a knowledge-based bioeconomy. Maximum homogeneity within a cluster and maximum heterogeneity between the clusters allow for better handling and easier interpretation of the large amount of data. However, the main advantage of such classification is to reveal relationships among the observed innovation systems that are hard to detect on the basis of the individual national data (Hair et al. 2010). At the same time, our analysis indicates the degree and the areas of structural similarities between the countries, which are analysed and can potentially provide guidance to the improvement of mutual learning (Lundvall and Tomlinson 2002 in Balzat and Pyka 2006). Clusters are formed on a global data level (i.e. comprising all 47 variables) as well as on each of the six category levels introduced above.

We determine the coherence of a cluster and the diversity between clusters by calculating the distance values between the countries based on their measured characteristics. Of the various methods to calculate such distances, we apply the Euclidian distance. To measure similarity between clusters, we use the average-linkage method, since this procedure measures the averages of clusters and is therefore only to a small extent affected by extreme values. Furthermore, its main aspiration is to produce clusters with small within-group variation rather than seeking to form necessarily equally sized clusters. Because of the differing scales and magnitudes of the variables, original data has been standardised by converting the variables to standard scores (also known as Z-scores) before clustering the countries.

Since the number of clusters is not known beforehand, we apply an agglomerative hierarchical clustering process. The rationale behind this approach is to repetitively merge similar objects in groups and then similar groups together in bigger groups until you reach the maximum amount of heterogeneity between the groups while at the same time remaining at an acceptable level of homogeneity within each group. There is no strict method to determine the optimal number of clusters. One way to achieve a suitable number of clusters is to plot the heterogeneity coefficient against the number of steps taken along the agglomerative clustering process (Ketchen and Shook 1996). The step, within which the rising line of the graph suddenly steepens, i.e. where heterogeneity within the cluster strongly increases, is considered one-step "too far" (see red arrow in Fig. 1). The number of recommended clusters c is calculated as:

$$c = n - f$$

Fig. 1 Heterogeneity coefficient plotted against the orders of steps along the agglomerative clustering process. Note: the case displayed here would call for $c = n - f = 27 - 17 = 10$ clusters

where n is the number of cases and f is the order of fusion step along the agglomerative clustering process, which produces the sudden increase in heterogeneity.

The clusters emerging from this calculated number of clusters are thus formed by countries that are relatively similar in terms of the measured indicators. We will present the results of the cluster analysis in the following section and carefully interpret them thereafter, not without considering possible biases originating from the indicators chosen, imperfect data and uncertain causalities.

3.3 *Interpretation*

Comparison and benchmarking of different national innovation systems is difficult and must be undertaken with care. As suggested by Lundvall and Tomlinson (2000), it should not focus too much on the comparison of quantitative data, but on the efficiency of a system in achieving the goal in question. Only this way, the results of our analysis will be able to stimulate reflection and support learning among the countries examined. Quantitative comparisons will therefore be restricted to structurally similar countries, i.e. within detected clusters and to the illustration of differences between clusters regarding indicator values explicitly describing the efficiency of a system towards a bioeconomy transformation (e.g. CO_2 emission per capita).

How efficient are the European NIS in achieving the goal of a knowledge-based bioeconomy? The paths towards an economy "based on the production and dissemination of new knowledge about renewable biological resources and their potential to be sustainably converted into food, feed, bio-based products and bioenergy with the aim to overcome the wastefulness of production and consumption in its full dependency on fossil resources" (see above) are expected to be manifold and hard to measure and to compare. Well aware of the shortcomings of the underlying measurements, including the restrictions in data access and indicator relevance, statistical imperfection of the method of cluster analysis as well as the general uncertainty and path dependency of strategies towards a less wasteful and sustainable way of production and consumption, we take a chance to offer some interpretations and derive some implications of the results after they have been presented in the following.

4 Results

When calculating the distances of all variables across the European Union in a global analysis, seven groups of countries with similar structures can be identified[2] (Fig. 2). The similarities partly show a geographical distribution with a Northern cluster (Sweden and Finland), a North Sea cluster (Denmark, Ireland, Netherlands and United Kingdom), a central cluster (Austria, Belgium, France, Germany and Slovenia) and a South-and-East cluster (Bulgaria, Croatia, Cyprus, Czech Republic, Greece, Hungary, Italy, Latvia, Lithuania, Poland, Portugal, Romania, Slovakia and Spain). The remaining countries form single-country clusters indicating a special profile in the indicators measured (Estonia, Luxembourg and Malta respectively). The North cluster countries Finland and Sweden are wealthy nations with outstanding environmental quality and a very strong focus on education and training (strong knowledge base). They range among the most innovative countries in the EU and rely on a wealth of natural assets. This is accompanied by a high environmental awareness of the population. The countries forming the North Sea cluster are also relatively wealthy, build upon a very strong knowledge base and show high innovative capacity. Their natural assets are scarcer than in the North, but the environmental quality of life is above average. A high proportion of the countries' surface is used agriculturally and forest is scarce. The agricultural sector (including forestry, hunting and fishing) does not contribute much to the total domestic value added. The countries belonging to the central cluster form an average throughout the complete set of variables.

The extensive South-and-East cluster consists of the least wealthy countries of the EU with the largest agricultural sectors, comparably low innovation activity and a small proportion of employment in science and technology. The governments are

[2]The clusters emerging from the global analysis and the analyses according to the different categories are presented in different shades on maps (Figs. 2, 3, 4, 5, 6, 7, and 8). In addition, the clusters are presented numerically in a table (Annex 2).

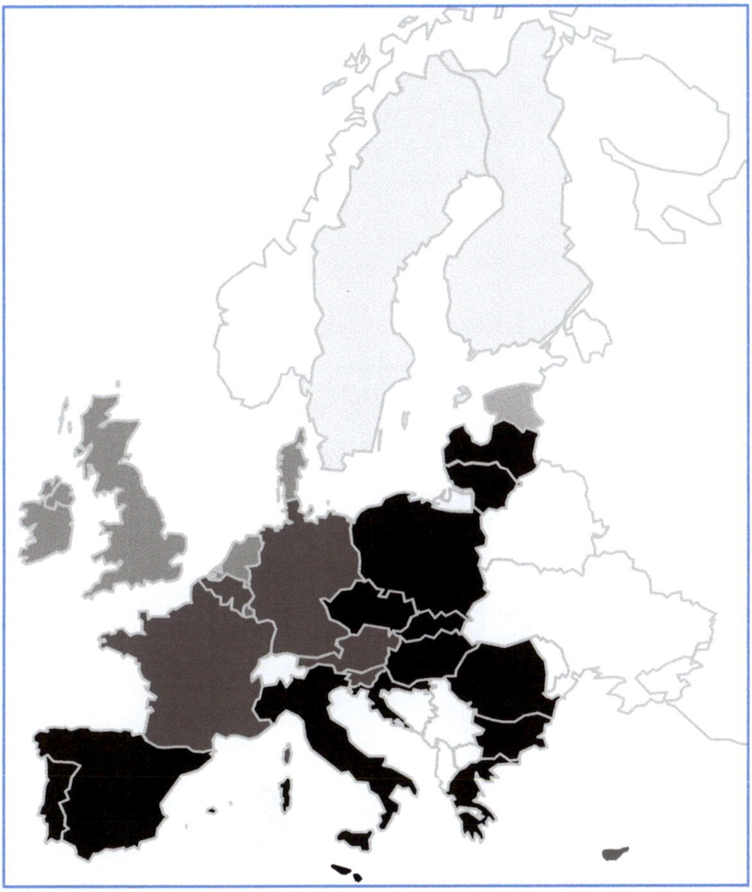

Fig. 2 The seven clusters according to the global analysis

less dedicated to education and training and the population is not so much concerned about the environment which is partially very healthy and partially heavily polluted. However, because of the relatively low income per head and the correlation to economic activities their overall CO_2 emissions per capita are relatively low.

Estonia's innovation system is very different from the other EU countries' and thus forming its own cluster. As a country with the seventh lowest GDP in the EU and a very carbon intensive economy, it possesses a remarkably strong knowledge base and the highest number of biotechnology patents per million inhabitants. The nation with a strong natural asset base and relatively unpolluted environment is home to an optimistic and environmentally aware population. Luxembourg is probably the most exceptional country in the EU. With the highest GDP and largest proportion of researchers in the active population it emits the most CO_2 per capita. It has a population with a very high environmental awareness and very little trust

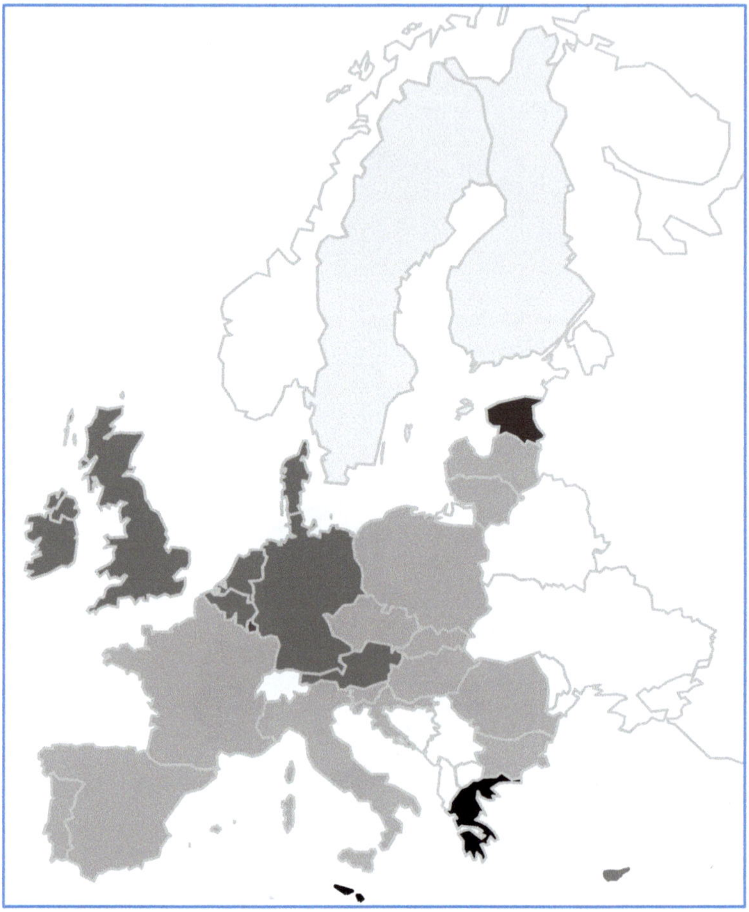

Fig. 3 The seven clusters according to environmental and resource productivity indicators

in science and technology. The third one-country cluster is formed by Malta. The Mediterranean island of medium economic wealth and very limited natural space and resources produces at a highly resource efficient, but carbon intensive scale. The high governmental expenditures on education have not shown any effect on the knowledge base of the country. Environmental awareness is very high, pollution partially very heavy.

4.1 Environmental and Resource Productivity

Since the category of environmental and resource productivity obviously encompasses those indicators that are most directly connected to the achievement of

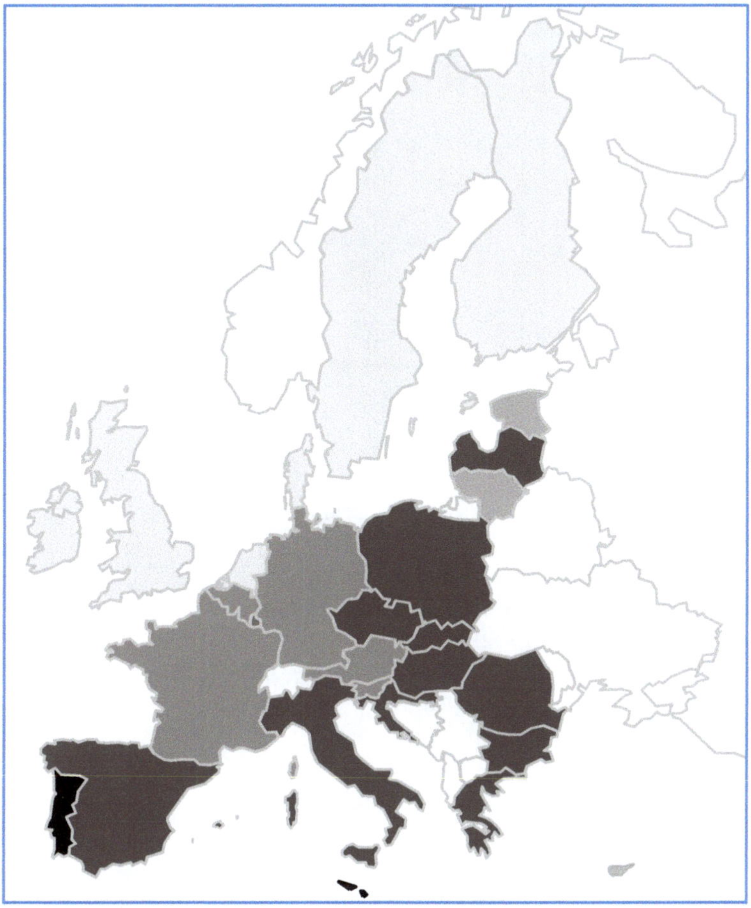

Fig. 4 The six clusters according to knowledge base indicators

the goals formulated in our definition of a knowledge-based bioeconomy, the similarities and dissimilarities of NIS in this category deserve special consideration. Here, again seven clusters emerge, however in a somewhat different distribution than in the global analysis (Fig. 3).

While the North cluster of Finland and Sweden remains to form a group, parts of the central cluster (except France and Slovenia) and Cyprus merge with the North Sea cluster. Both clusters produce a relatively high amount of CO_2 emissions. The North cluster is less dependent on fossil energy and excels in the share of renewable energy in gross final energy consumption, whereas the central/North Sea/Cyprus cluster proves to be more energy and resource efficient. The largest cluster comprises Bulgaria, Croatia, Czech Republic, France, Hungary, Italy, Latvia, Lithuania, Poland, Portugal, Romania, Slovakia, Slovenia and Spain. These countries produce on average less CO_2 and waste, have a low recycling rate

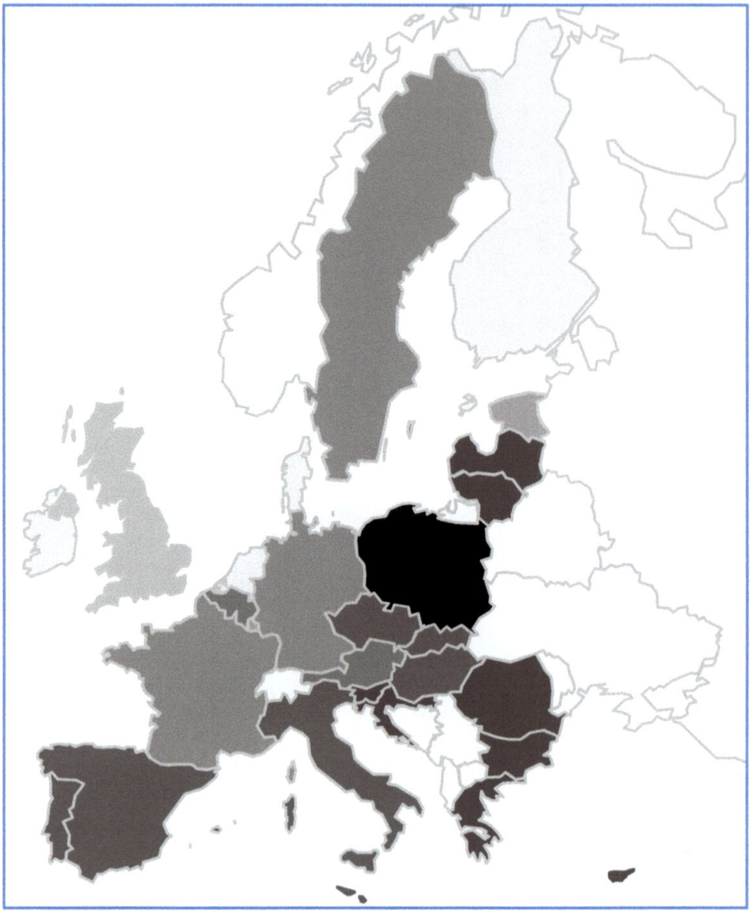

Fig. 5 The eight clusters according to policy and bio-economic opportunities indicators

of municipal waste and for all other resource productivity indicators range between the superior North and the less efficient central/North Sea/Cyprus clusters.

The two single-country clusters of Estonia and Luxembourg both produce very CO_2 intensively. Luxembourg is very resource productive and also uses its energy relatively efficiently—both indicators where Estonia performs very poorly. On the other hand, Estonia produces the least amount of waste per capita and deploys a high share of renewable energy—both indicators where Luxembourg performs very poorly. The single-country cluster of Malta uses its energy and resources highly efficiently and uses little fertilizer, but ranges lowest in the share of renewable energies. Greece, the last individual cluster, produces relatively resource and energy efficiently, but CO_2 intensively. Here, the highest amount of fertilizer is consumed within the EU.

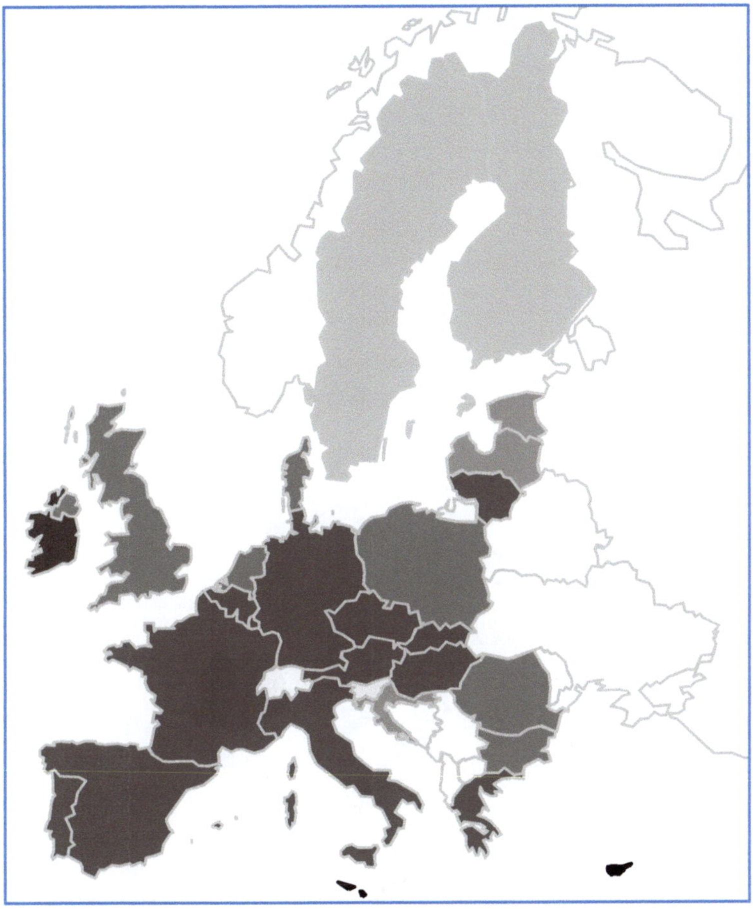

Fig. 6 The seven clusters according to natural asset base indicators

4.2 Knowledge Base

The indicators describing the knowledge base of the countries assessed give a picture of the future capability of the different NIS and show how well nations are prepared to learn and innovate (Balzat and Pyka 2006). We identified six clusters of relatively similar countries (Fig. 4). The North cluster (Sweden and Finland) now merges with the North Sea cluster (Denmark, Ireland, Netherlands and the UK) to the best performing systems in terms of knowledge production and support. Also relatively keen in this respect are the following three clusters, formed by Cyprus, Estonia and Lithuania (high education level of the population and most positive attitudes towards the influence of science and technology on their respective countries), by Austria, Belgium, France, Germany and Slovenia (relatively high proportion of researchers and scientific articles, but only average education level

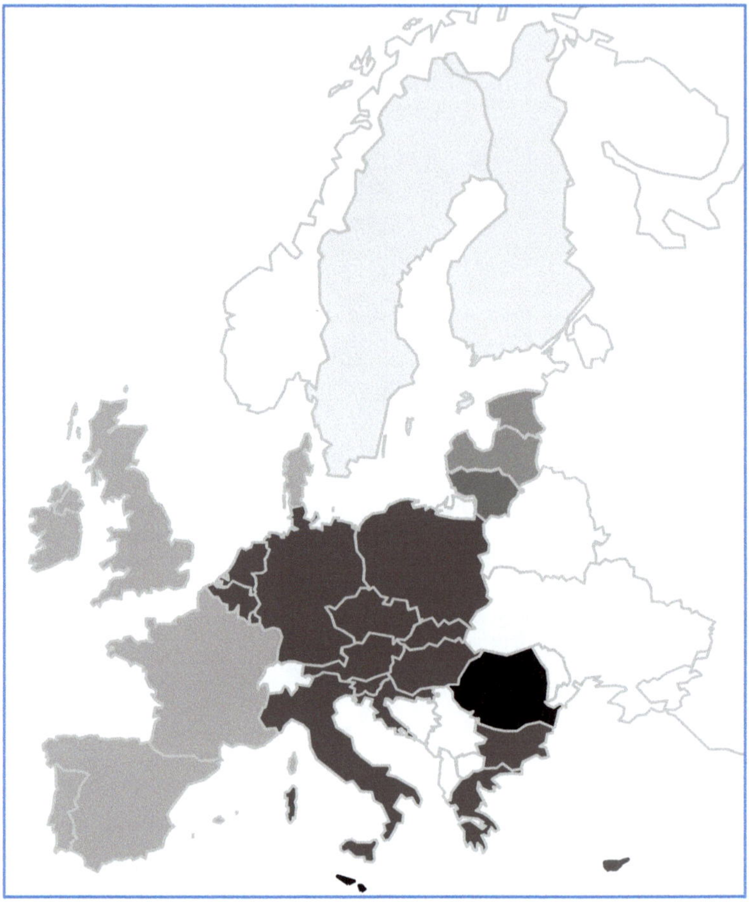

Fig. 7 The seven clusters according to environmental quality of life indicators

of the population and quite negative attitudes towards the influence of science and technology on their respective countries) and by the single-country cluster of Luxemburg with the highest employment share in science and technology but relatively few scientific publications and very little trust of the population in science and technology. The last two clusters, the South-and-East cluster (comprising the rest of Eastern Europe and Greece, Italy and Spain) and the far-South cluster (Portugal and Malta) do not show a strong knowledge base.

4.3 Policy and Bioeconomic Opportunities

This category produces eight clusters (Fig. 5). The Northern cluster—here excluding Sweden and the United Kingdom—proves to be very innovative in general and

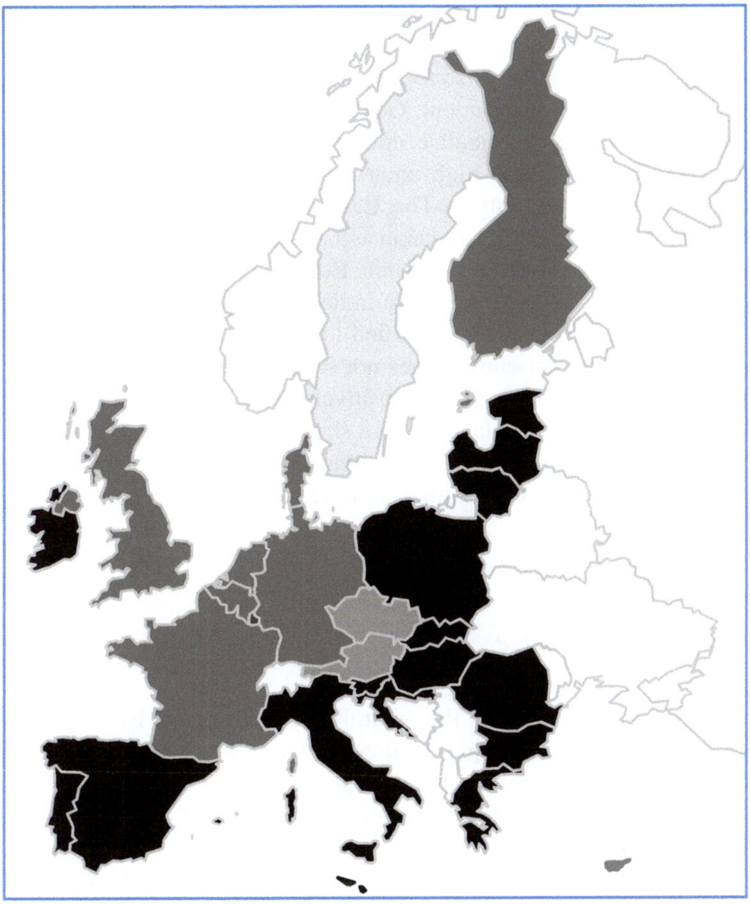

Fig. 8 The five clusters according to socio-economic context indicators

shows a strong political commitment to environmental issues and the bioeconomy at the national level: all countries belonging to this cluster (Denmark, Finland, Ireland and Netherlands) have a bioeconomy strategy in place, raise high environmental taxes and rank high in biotechnology patenting; but they are not very fast in signing international environmental agreements and do not invest much in environmental development assistance. The cluster is closely followed by the single-country cluster United Kingdom. Despite of ranking even higher on the innovation index, the UK produces fewer biotechnology patents and invests less in research and development (R&D). Of all clusters, however, the population of the UK shows the most positive attitudes towards genetic engineering. The two European countries with the highest number of biotechnology patents per million inhabitants form the small-country cluster: Estonia and Luxembourg. The green party has a good share among their EU parliament representatives and their population is very critical when it comes to genetic engineering. Most money is spent for R&D in general and for environmental development projects by the cluster made up of France, Germany and Sweden. These three countries also top all other clusters in the share of green party representatives in the EU parliament. Genetic engineering is not valued positively here. The cluster of the two small economies Austria and Belgium spends quite a lot in R&D, does not raise high environmental and energy taxes and is very slow in notifying EU legislation. The East-and-South cluster (Czech Republic, Hungary, Italy, Malta, Portugal, Slovakia and Spain) and the East cluster (Bulgaria, Croatia, Cyprus, Greece, Latvia, Lithuania, Romania and Slovenia) are comparable in most of their political and innovation indicators, but fundamentally different in their attitudes towards genetically modified food and genetic engineering in crops: The population of the East cluster does not support such technology, whereas the East-and-South cluster regards it very positively. Within the category of policy and bioeconomic opportunities, Poland forms a single-country cluster. Compared to the rest of the EU, it does not focus much on innovation and environmental policy, but Polish people show remarkably positive attitudes towards genetic engineering above the EU average.

4.4 Natural Asset Base

When applying the indicators of the natural asset base of the EU countries, seven clusters emerge (Fig. 6). The single-country cluster of Slovenia has the highest share of protected area and is very rich in freshwater resources as well as in forest cover and standing volume of wood. The North cluster (Finland and Sweden) has even richer forest resources and slightly larger wealth in non-renewable minerals, but the least amount of agricultural land and very few protected areas. With a lot of forest, little agricultural area and medium mineral resource wealth, the East cluster (Croatia, Estonia and Latvia) shows a similar pattern as the North cluster, only a little less wealthy than the same. The cluster formed by the North-Southeast-embracing economies (Bulgaria, Denmark, Netherlands, Poland, Romania and the United Kingdom) comprises the countries with the highest share of agricultural land, the largest mineral wealth and an above average share of protected areas. Water resources and forest are scarcer. Not surprisingly, the islands (the single-country cluster of Ireland and the Southern islands cluster of Cyprus and Malta) are not competitive when it comes to natural wealth with Ireland having the slight geographical advantage of abundant water resources and better conditions for agriculture. The large central cluster comprising the remaining countries takes a medium position between the agriculturally strong the North-Southeast-embracing and the weaker islands.

4.5 Environmental Quality of Life

For the indicators describing the environmental quality of life within the 28 nations, the seven emerging clusters roughly follow a new geographic pattern: in addition to the North-South distribution revealed within the other categories to different degrees, this category shows a West-East distribution (Fig. 7). Next to the North cluster with the by far lowest environmental impairment, the population of the Western cluster enjoys a relatively intact environment. The Baltic economies follow in two clusters (Estonia and Latvia forming one cluster, Lithuania the other one) with some pollution and considerably less access to improved drinking water. The single-country clusters Malta and Romania show the least environmental quality of life, while nevertheless Malta shows the least air pollution by particulate matter and provides its population access to sufficient improved drinking water. The large Central-East cluster with the remaining countries shows average pollution values, provides relatively much improved drinking water, but is strongly polluted by particulate matter in the air.

4.6 Socio-Economic Context

Measuring the socio-economic environments of the national innovation systems leads to the emergence of five clusters (Fig. 8). Sweden and Luxembourg are so different from the other EU countries in this respect that the form their own two clusters: Both countries are very wealthy (Luxembourg's GDP is number one in the EU) and home to a mostly urban and environmentally concerned population with a high employment rate (Sweden's employment rate is the highest in the EU), but little trust in the future. The other three clusters are formed by Austria and Czech Republic, by the central countries Belgium, Cyprus, Denmark, Finland, France, Germany, Malta, Netherland and the United Kingdom and by the rest (Southern and Eastern states plus Ireland), respectively. All clusters except the South-and-East+Ireland cluster produce an above average GDP, have high income equality, environmental awareness and employment rate, relatively little contribution of the agricultural sector to the total GDP and comparably negative attitudes towards the future. In all these aspects, the South-and-East+Ireland cluster shows opposite figures (with important exceptions, such as high GDP in Ireland, very negative attitudes towards the future in Greece and top position of Slovenia in income equality).

4.7 Implications of the Analyses

According to the global analysis and to the six categories defined above, we find some countries to be part of the same clusters over and over again. Such countries usually share geographical, historical, structural, political and/or cultural characteristics and thus have the potential to learn from each other more effectively than countries with differing systems.

Finland and Sweden are one such example. They share geographical and structural as well as cultural characteristics. With their high shares of renewable energies, their strong knowledge base and their wealth in natural resources, they are certainly on a good way towards a knowledge-based bioeconomy. However, their potential and will to innovate and to proceed in technological and institutional terms (as measured in category 3) seems to differ slightly: While Finland was among the first countries to publish a bioeconomy strategy, Sweden has just recently (2012) brought forward a strategy, but nevertheless produces more biotechnology patents and spends more on total R&D. Sweden also emits far less CO_2 than Finland and taxes energy highly compared to general environmental taxes. One last remarkable difference is the population's attitudes towards the future: Of all EU citizens, the Finnish have the strongest trust in the future development, whereas the population of the highly innovative and wealthy Sweden rank at the bottom in their attitudes towards the future and also distrust genetic engineering quite strongly.

Another pair of countries with quite similar patterns is Denmark and the United Kingdom. Both have extensive agricultural land areas that contribute little to the GDP (but consume a lot of artificial fertilizer, especially in the case of the UK). The population of both countries regards environmental protection as very important and attitudes towards genetically modified food are generally positive. In the share of renewable energy in gross final energy consumption, the United Kingdom strongly lags behind, but it produces more resource efficiently and generates less waste per capita than Denmark.

Bulgaria and Poland are two countries found within the same cluster throughout almost all analyses. Those two countries are strongly dependent on their agricultural sectors, produce quite energy and resource intensively, have a relatively low level of education and little focus on research and development. They are wealthy in non-renewable natural resources and have comparably large and productive forests. Poland is very inefficient in the implementation and enforcement of EU legislation and the governments of both countries are failing to implement sufficient environmental policies, e.g. to halt air pollution and guarantee better access to improved drinking water. Unemployment rates are relatively high, but the attitudes towards the future are positive.

This list of comparable bioeconomic innovation systems could easily be extended by discussing similarities between countries as a basis to stimulate reflection and improve the potential to learn from one another (Lundvall and Tomlinson 2000). The mentioned examples shall suffice in the context of this chapter. One last example of particular interest, however, is worth mentioning. Two countries that have great similarities in their historical and geographical background, but have proved to be quite distinctive in the variables measured in the context of our analysis are Estonia and Latvia. The former Soviet states show quite similar systemic patterns in respect to their natural assets, their environmental quality of life and their socio-economic context—three categories most strongly connected to geographical and historical conditions. The dissimilarities become evident when looking at the knowledge base and the policy and bioeconomic opportunities of the two countries. It becomes evident that Estonia has invested much more in those two future-orientated areas than Latvia has: The results are more researchers and human resources in science and technology, a higher level of education among the population and higher expenditures on education (all levels). Estonia has produced the most biotechnology patents per inhabitant, ranks above the EU average on the innovation index and the Estonians have more positive attitudes towards genetic engineering. However, areas of improvement for Estonia remain with regards to its environmental and resource productivity: Across the EU, Estonia's CO_2 emissions are only topped by Luxembourg and it ranks lowest in energy and resource efficiency of production.

For policy planning, the comparisons of innovation systems within the detected clusters can be of use in two respects of different time scales: short-term policy adaptations and long-term policy planning. In the short run, it will surely be beneficial for economies to improve on single areas using benchmark values of individual indicators reached by economies within the same cluster. Sweden, for example, should endeavour to take its population on board of the bioeconomy movement to improve their attitudes towards future technologies and new products by learning how Finland has achieved such positive attitudes. The United Kingdom should be able to create incentives for the industry to put more effort in the development of renewable energies by examining the experiences made by Denmark. In the long run, however, it will not be sufficient for national policies to be geared to structurally similar economies. Long-term policy planning must aim for qualitative and structural change (across current clusters) towards the three focal aims of the knowledge-based bioeconomy: independency from fossil resources, sustainable production and conversion of biological resources and efficient production and dissemination of knowledge. As argued by Kemp et al. (1998), such change will not be achieved by the promotion of certain (new) technologies, but by the inducement of a change towards an integrated system of technologies and social practices. Policy's task is to support such regime-shift by modulating the dynamics of socio-technical change into a desirable direction and thus manage processes instead of defined goals.

5 Conclusions and Outlook

The aim of this study is to analyse the varieties of national innovation systems in their capability to undergo the transition towards a knowledge-based bioeconomy. The underlying empirical variables are chosen to illustrate six areas of national innovation systems: the environmental and resource productivity, the knowledge base, policy and bioeconomic opportunities, the natural asset base, the environmental quality of life and the specific socio-economic context within the 28 countries. With the help of a multivariate cluster analysis we are able to detect similarities and dissimilarities among the countries of the EU. The similarities are of particular interest since similar patterns of bioeconomic innovation systems allow for improved comparability of the outcomes and stimulate mutual learning from experience. In the same vein, the divergence of national innovation systems does not imply that the examined countries are situated on different stages on one defined path towards a functioning knowledge-based bioeconomy. The dissimilarities rather take account of the multitude of approaches towards this goal in their dependence on geographical, historical, structural, political and cultural conditions.

Given the large national differences within the EU, the necessity of a supranational policy planning becomes evident: It might be wise to geographically separate the production of the biomass and research on this production from various fields of refinement or to create centres of specialisation across the EU, thus building a European Innovation System for the knowledge-based bioeconomy. This is not to suggest a political consolidation of the differences between more traditional and agriculturally orientated economies and highly innovative knowledge-based economies, but to take account of the varying natural conditions like climate, space and water availability to name a few that matter when biological production is involved.

Whoever expected a national ranking on bioeconomic performance among the members of the EU from this study has been severely disappointed. The reasons why we cannot provide this are twofold: Firstly, we lack a benchmark. Our definition of the knowledge-based bioeconomy describes an ideal that is not measurable in numbers and figures. Secondly, even if indicators for a well-functioning bioeconomy were defined, they would hardly be assessable empirically due to a lack of data and what is more, they would be subject to continuous change since the nature of innovation is uncertain and path-dependent and benchmarks would have to be adapted based on the innovations introduced during the course of time. This also implies that there are no winners or losers in our comparative analysis, only a variety of innovation systems that are currently more efficient in certain aspects of the bioeconomy and systems that could improve their efficiency by short-term policy learning from other, structurally similar, systems and by a long-term policy-driven adaptation of their innovation systems. Such structural transformations could eventually also serve as models for the transition of less developed economies towards knowledge-based bioeconomies.

To create incentives for the introduction and implementation of political strategies towards knowledge-based bioeconomies and to enable an evaluation of measures and outcomes, it would nevertheless be desirable to develop a theoretical construct of regionally specific indicator values for ideal performance according to which the clusters could be ranked based on their goal realisation level.

Annex

Annex 1

Table 1 Indicators for the analysis of bioeconomic innovation systems

Category	No	Code	Indicator	Year	Sources
1. Environmental and resource productivity	1.1	101	CO_2 emissions (metric tons per capita)	2010	World Bank
	1.2	102	CO_2 intensity (kg per kg oil equivalent energy use)	2010	World Bank
	1.3	103	Resource productivity (GDP in PPS/kg consumed material)	2011	Eurostat
	1.4	104	Energy use (kg of oil equivalent) per $1000 GDP (constant 2005 PPP)	2011	Eurostat
	1.5	105	Share of renewable energy in gross final energy consumption (%)	2012	Eurostat
	1.6	106	Waste generation (kg/capita)	2012	Eurostat
	1.7	107	Recycling rate of municipal waste	2012	Eurostat
	1.8	108	Artificial fertilizer consumption (kilograms per hectare of arable land)	2010	World Bank
2. Knowledge base	2.1	201	Human Resources in Science and Technology (% of active population)	2012	Eurostat
	2.2	202	Researchers FTE (per million inhabitants)	2011	UNESCO
	2.3	203	Scientific and technical journal articles (per thousand capita)	2009	World Bank
	2.4	204	Population with tertiary education attainment (%)	2013	Eurostat
	2.5	205	Population with at least secondary education attainment (%)	2013	Eurostat
	2.6	206	Total public expenditure on education, all levels (% of GDP)	2010	Eurostat
	2.7	207	Attitudes towards the influence of science and technology on the country (% very positive answers)	2013	GESIS/ Eurobarometer

(continued)

Table 1 (continued)

Category	No	Code	Indicator	Year	Sources
3. Policy and bio-economic opportunities	3.1	301	Global Innovation Index	2013	GII 2014
	3.2	302	Number of biotechnology patents (per million inhabitants)	2008	Eurostat
	3.3	303	Total R&D expenditures (PPS per inhabitant at constant 2005 prices)	2011	Eurostat
	3.4	304	Official Development assistance dedicated to environmental issues (% of GNI)	2011	OECD
	3.5	305	Environmentally related taxes (% of GDP)	2012	Eurostat
	3.6	306	Implicit tax rate on energy (Euro per tonne of oil equivalent)	2012	Eurostat
	3.7	307	Years since publication of bioeconomy strategy	2013	BioPro BW
	3.8	308	Years of participation in selected International Environmental Agreements	2010	UNStats
	3.9	309	Transposition deficit of EU legislation (% of directives not yet notified)	2012	Eurostat
	3.10	310	New infringement cases in EU legislation (total number)	2012	Eurostat
	3.11	311	Number of representatives of the green party in EU parliament (share of total national seats)	2013	European Parliament
	3.12	312	Attitudes towards genetically modified food (% agreeing that it should be encouraged)	2010	GESIS/ Eurobarometer
	3.13	313	Attitudes towards genetic engineering on crops (% agreeing that it should be encouraged)	2010	GESIS/ Eurobarometer
4. Natural asset base	4.1	401	Renewable internal freshwater resources (m3 per inhabitant)	2011	World Bank
	4.2	402	Forest total growing stock (m3 per inhabitant)	2010	FAO
	4.3	403	Share of agricultural land cover (% of total land area)	2011	FAO

(continued)

Table 1 (continued)

Category	No	Code	Indicator	Year	Sources
	4.4	404	Share of forest land cover (% of total land area)	2011	FAO
	4.5	405	Terrestrial and marine protected areas (% of total territorial area)	2012	FAO
	4.6	406	Non-renewable natural resources (oil, gas, coal, mineral) rents (% of GDP)	2011	World Bank
5. Environmental quality of life	5.1	501	Population exposed to particulate matter above WHO thresholds 2011 (%)	2011	Environmental Performance Index
	5.2	502	People suffering from pollution, grime or other environmental problems (%)	2011	Eurostat
	5.3	503	People suffering from noise (%)	2011	Eurostat
	5.4	504	Population with access to improved drinking water (%)	2012	Environmental Performance Index
6. Socio-economic context	5.5	505	Forest and other wooded land per capita (ha/inhabitant)	2010	FAO
	6.1	601	GDP per capita in PPS (EU 28=100)	2012	Eurostat
	6.2	602	GINI coefficient of equivalised disposable income (0–100)	2011	Eurostat
	6.3	603	Urban population (%)	2012	World Bank
	6.4	604	Positive attitudes towards future (%)	2012	GESIS/ Eurobarometer
	6.5	605	Attitudes towards importance of environmental protection (%)	2012	GESIS/ Eurobarometer
	6.6	606	Employment rate (% of age 20–64)	2012	Eurostat
	6.7	607	Value added from agricultural sector (% of GDP)	2009	World Bank
	6.8	608	Share of total organic crop area (% of total agricultural area)	2012	Eurostat

Annex 2

Table 2 Cluster overview (Note: the values do not represent weightings and close numbers do not indicate close clusters)

	Global analysis	Environmental andresource productivity	Knowledge base	Policy and bio-economic opportunities	Natural asset base	Environmental quality of life	Socio-economic context
Austria	1	1	1	1	1	1	1
Belgium	1	1	1	1	1	1	2
Bulgaria	2	2	2	2	2	1	3
Croatia	2	2	2	2	3	1	3
Cyprus	2	1	3	2	4	1	2
Czech Republic	2	2	2	3	1	1	1
Denmark	3	1	4	4	2	2	2
Estonia	4	3	3	5	3	3	3
Finland	5	4	4	4	5	4	2
France	1	2	1	6	1	2	2
Germany	1	1	1	6	1	1	2
Greece	2	5	2	2	1	1	3
Hungary	2	2	2	3	1	1	3
Ireland	3	1	4	4	6	2	3
Italy	2	2	2	3	1	1	3
Latvia	2	2	2	2	3	3	3
Lithuania	2	2	3	2	1	5	3
Luxembourg	6	6	5	5	1	1	4
Malta	7	7	6	3	4	6	2
Netherlands	3	1	4	4	2	1	2
Poland	2	2	2	7	2	1	3
Portugal	2	2	6	3	1	2	3
Romania	2	2	2	2	2	7	3
Slovakia	2	2	2	3	1	1	3
Slovenia	1	2	1	2	7	1	3
Spain	2	2	2	3	1	2	3
Sweden	5	4	4	6	5	4	5
United Kingdom	3	1	4	8	2	2	2

References

Altenburg, T., & Pegels, A. (2012). Sustainability-oriented Innovation systems – Managing the green transformation. *Innovation and Development, 2*(1), 5–22.

Balzat, M., & Pyka, A. (2006). Mapping national innovation systems in the OECD area. *International Journal of Technology and Globalisation, 2*(1/2), 158–176.

Benner, M., & Löfgren, H. (2007). The bio-economy and the competition state: Transcending the dichotomy between coordinated and liberal market economies. *New Political Science, 29*(1), 77–95.

BMELV. (2013). *Politikstrategie Bioökonomie. Nachwachsende Ressourcen und biotechnologische Verfahren als Basis für Ernährung, Industrie und Energie.* Retrieved February 8, 2014, from http://www.bmel.de/SharedDocs/Downloads/Broschueren/BioOekonomiestrategie.pdf?__blob=publicationFile

EC. (2009). *Communication from the Commission of the European Union to the European Parliament, the Council, the European Economic and Social Committee of the Regions – Preparing for our future: Developing a common strategy for key enabling technologies in the EU.* Retrieved March 3, 2014, from http://eur-lex.europa.eu/LexUriServ/LexUriServ.do?uri=COM:2009:0512:FIN:EN:PDF

EC. (2012a). *Communication from the Commission of the European Union to the European Parliament, the Council, the European Economic and Social Committee of the Regions – Innovating for sustainable growth: A bioeconomy for Europe.* Retrieved December 2, 2013, from http://ec.europa.eu/research/bioeconomy/pdf/201202_innovating_sustainable_growth.pdf

EC. (2012b). *Commission Staff Working Document, Accompanying the document: Communication on Innovating for sustainable growth: A bioeconomy for Europe.* Retrieved February 14, 2014, from http://ec.europa.eu/research/bioeconomy/pdf/201202_commision_staff_working.pdf

Edquist, C., & Lundvall, B.-A. (1993). Comparing the Danish and Swedish systems of innovation. In R. Nelson (Ed.), *National innovation systems. A comparative analysis.* New York: Oxford University Press.

Eposti, R. (2012). Knowledge, technology and innovations for a bio-based economy: Lessons from the past, challenges for the future. *Bio-based and Applied Economics, 1*(3), 235–268.

Freeman, C. (1987). *Technology policy and economic performance: Lessons from Japan.* London: Pinter.

Freeman, C. (2002). Continental, national and sub-national innovation systems – Complementarity and economic growth. *Research Policy, 31*, 191–211.

Hair, J. F., Black, W. C., Barry, J. B., & Anderson, R. E. (2010). *Multivariate data analysis.* Harlow: Pearson Education.

Hall, P. A., & Gingerich, D. W. (2009). Varieties of capitalism and institutional complementarities in the political economy: An empirical analysis. *British Journal of Political Science, 39*, 449–482.

Hall, P. A., & Soskice, D. (2001). *Varieties of capitalism: The institutional foundations of comparative advantage.* Oxford: Oxford University Press.

Hanusch, H., Pyka, A., & Wackermann, F. (2010). Patterns of future-orientation in Central and Eastern Europe. *Quarterly Journal of the Faculty of Economics at the Corvinus University of Budapest, 15*, 173–190.

Kemp, R., Shot, J., & Hoogma, R. (1998). Regime shifts to sustainability through processes of niche formation: The approach of strategic niche management. *Technology Analysis and Strategic Management, 10*(2), 175–195.

Ketchen, Jr, D. J., & Shook, C. L. (1996). The application of cluster analysis in strategic management research: An analysis and critique. *Strategic Management Journal*, 441–458.

Lundvall, B.-A. (1992). *National systems of innovation: Towards a theory of innovation and interactive learning.* London: Pinter.

Lundvall, B.-A., Johnson, B., Andersen, E. S., & Dalum, B. (2002). National systems of production, innovation and competence building. *Research Policy, 31*, 213–231.

Lundvall B.-A., & Tomlinson M. (2000). *International benchmarking and national innovation systems*. Draft Report for the Portuguese Presidency of the European Union April 2000, communicated by the author 19 May 2014.
Lundvall, B.-A., & Tomlinson, M. (2002). International benchmarking as a policy learning tool. In M. J. Rodrigues (Ed.), *The new knowledge economy in Europe: A strategy for international competitiveness and social cohesion*. Cheltenham: Elgar.
Malerba, F. (2005). Sectoral systems – How and who innovation differs across sectors. In J. Fagerberg, C. Mowery, & R. R. Nelson (Eds.), *The Oxford handbook of innovation*. Oxford: Oxford University Press.
MWK. (2013). *Bioökonomie im System aufstellen. Konzept für eine baden-württembergische Forschungsstrategie "Bioökonomie"*. Retrieved March 1, 2014, from http://mwk.baden-wuerttemberg.de/uploads/tx_ttproducts/datasheet/Konzept_f%C3%BCr_eine_baden-w%C3%BCrttembergische_Forschungsstrategie_Bio%C3%B6konomie.pdf
OECD. (1996). *The knowledge-based economy* (General distribution OCDE/GD(96)102). Paris.
OECD. (2009). *The bioeconomy to 2030: Designing a policy agenda*. OECD International Futures Programme.
OECD. (2011). *Towards green growth: Monitoring progress: OECD indicators*. OECD Publishing. Retrieved March 5, 2014, from http://dx.doi.org/10.1787/9789264111356-en
Patel, P., & Pavitt, K. (1994). National innovation systems: Why they are important, and how they might be measured and compared. *Economics of Innovation and New Technology, 3*(1), 77–95.
Schmid, O., Padel, S., & Levidow, L. (2012). The bio-economy concept and knowledge base in a public goods and farmer perspective. *Bio-based and Applied Economics, 1*(1), 47–63.
Stamm, A., Dantas, E., Fischer, D., Ganguly, S., & Rennkamp, B. (2009). *Sustainability-oriented innovation systems: Towards decoupling economic growth from environmental pressures?* (Discussion Paper 20/2009). Bonn: Deutsches Institut für Entwicklungspolitik.
US Government. (2012). *National bioeconomy blueprint*. Retrieved December 10, 2013, from http://www.whitehouse.gov/sites/default/files/microsites/ostp/national_bioeconomy_blueprint_april_2012.pdf
USDA. (2012). *Biobased economy indicators* (OCE-2010-2, 138 pp.). Office of the Chief Economist, Office of Energy Policy and New Uses (OEPNU), U.S. Department of Agriculture. Prepared jointly by OEPNU and the Center for Industrial Research and Service of Iowa State University.

International Bioeconomy Innovations in Central America

Mercedes Montero Vega and Olman Quirós Madrigal

Abstract This paper presents a first approach to Bioeconomy, its definitions and importance as well as related research in Central America. Since Bioeconomy and its potential is directly linked to agricultural production, research and traditional knowledge on sub-utilized traditional products in these regions were also included. Our main goal is to review Bioeconomy's "state of the art" in this region as well as some perspectives for future development and potential of sub utilized products to be included in the Bioeconomy agenda. Additionally, we have considered that Central America should hold and promote its potential for natural resource supply, such as water and biodiversity. Natural resources enable the region to provide ecosystem services to face food insecurity and develop adapted technologies for climate change. Food and biomass production are clear examples of this potential, nonetheless a re-orientation of the region's economy (and policies) is needed.

1 Introduction

Bioeconomy is not a "sudden hype"; it has spiraled since the late nineties (von Braum 2015, p. 12) when the European Union and the Organization for Economic Cooperation and Development (OECD) first addressed the concept. Nonetheless, this trend has slowly shifted to other regions such as Latin and Central America, in which benefits from Bioeconomy can be linked to agriculture mostly located in rural areas.

Central American countries have a large proportion of rural areas, and therefore rural population. The correlation between rural areas, agricultural production and poverty cannot be neglected. Although rural communities are related to higher poverty rates, these communities often use in their livelihoods, a series of sub-utilized agricultural products which could be included in the Bioeconomy and development agendas.

M.M. Vega • O.Q. Madrigal (✉)
Agricultural Economics and Agribusiness Department, University of Costa Rica, 2060 San José, Costa Rica
e-mail: mercedes.montero@ucr.ac.cr; olman.quiros@ucr.ac.cr

S. Dabbert et al. (eds.), *Knowledge-Driven Developments in the Bioeconomy*, Economic Complexity and Evolution, DOI 10.1007/978-3-319-58374-7_5

In terms of Central American countries, there are some similarities in terms of climate conditions, however poverty rates vary widely and regional generalizations are difficult to make. In terms of poverty rates, the percentage of the population living below the national poverty line are: Guatemala (51%), Honduras (62%), El Salvador (40%), Nicaragua (42.5%), Costa Rica (21.2%) and Panama (23%) (World-Bank 2016).

Linked to poverty, inequality is another particularly important in the region, since Latin America is the most unequal region in the World. Overall, poverty increased during the 2008–2009 economic crisis in all countries; nonetheless it was most notable in the three largest countries, i.e. Guatemala, Honduras and Nicaragua (Hodson de Jaramillo 2011). In this regard, the most affected locations were those inhabited by indigenous people, who also have traditional and unique production techniques.

Inequality, ranked by Gini coefficient range from Panama's 50.7 to El Salvador's 41.8 (World-Bank 2014). Inequality becomes more evident inside each country and between rural and urban areas. *This is evident in the gaps between rural and urban areas, between different social classes and groups, but it mostly affects women, young people, indigenous communities, and Afro-descendant groups, as well as persons with some form of disability* (Hodson de Jaramillo 2011).

Employment in agriculture is a good indicator of people living in rural areas. The largest proportion of agricultural employment is in Honduras 35.8%, while the smallest proportion is in Costa Rica, 12.7% (World-Bank 2016). As indicated by development trends, as economies become more developed or at least with higher economic indexes, the percentage of Gross Domestic Product (GDP) in Agriculture is expected to decrease. On this regard, the most agriculture-dependent economy is Nicaragua's, in which 18.8% of its GDP is from agriculture, on the other hand, the least agriculture-dependent country is Panama in which only 2.89% of its GDP is related to agriculture (World-Bank 2016).

According to MINCyT-CIRAD (2016) *the Latin American and the Caribbean region is particularly well placed to contribute and benefit from the emerging Bioeconomy.* More than 50% of its land has agricultural potential in which for 2050, more than 300 million hectares could be brought into production. This situation, offers a basis for a strong Bioeconomy potential as well (MINCyT, CIRAD 2016), since the region could contribute in terms of food security and clean energy production as well as providing employment opportunities for the rural population.

World population, climate change and the need to protect the environment set important challenges that strive for modifications on production and consumption (von Braum 2015, p. 12). Bioeconomy has gained importance not only on policy debates but in academia as well since it is often mentioned as an important contributor of today's sustainable development in many fields, especially in agriculture-related supply chains, either for food consumption or for energy production.

Although widely used, the term Bioeconomy has had diverse meanings depending on the field of study and the research emphasis; nonetheless Bugge et al. (2016, p. 9) examine the understanding of the concept and provide the following visions to the concept:

1. Bio-technology vision: it refers to the application of bio technology research to commercialization of biotechnology in different fields. Its main goal is economic growth and job creation by new product and their commercialization. The perspective of biotechnology is based on global clusters and/or central regions.
2. Bio resource vision: it emphasizes on the potentials of upgrading and conversion of the biological raw materials. It analyzes biological raw materials in sectors such as agriculture, marine, forestry, and bioenergy, as well as on the establishment of new value chains. Its aim is economic growth and sustainability by conversion and upgrading of bio-resources. Nonetheless in this case the perspective is focused on local and rural areas.
3. Bio-ecology vision: it emphasizes the importance of ecological processes that optimize the use of energy and nutrients, promote biodiversity and avoid monocultures and soil degradation. Its aims are: sustainability and conservation of ecosystems, within a region. Therefore, the perspective is set on local and rural areas.

The negative impacts of a growing population require a greater efficiency in the non-renewable recourses use since it has become a well-known fact that non-renewable resources are getting exhausted. Along with greater efficiency, greater research efforts are also needed for the identification of new energy sources. Biomass derived from food production has created new option for supply chains derived from agro-energy sources that are based on biomass use.

A network approach if often used in the analysis of biomass use within a multidimensional framework so that interrelations and linkages among different supply chains are considered. This network approach ponders diverse products that are and that could be derived from biomass that comes from different sources and different supply chains. This perspective widens the scope of analysis, including agricultural small scale households, as well as markets. A representation of this network approach can be observed in Fig. 1 in which all levels should be integrated vertically and horizontally.

Based on another perspective, Bioeconomy may be achieved by several paths called Bioeconomy Productive Path (CYTED & REBICAMBLI 2014). The main paths are the following:

1. Resource use from biodiversity: economic valuation of biodiversity, therefore, Bioeconomy is perceived as an environmental friendly and profit activity.
2. Eco-intensification: the agricultural practices that improve environmental performance of agriculture without sacrificing current production and productivity. In eco-intensification, microorganisms as well as good agricultural prices can be used to increase production.

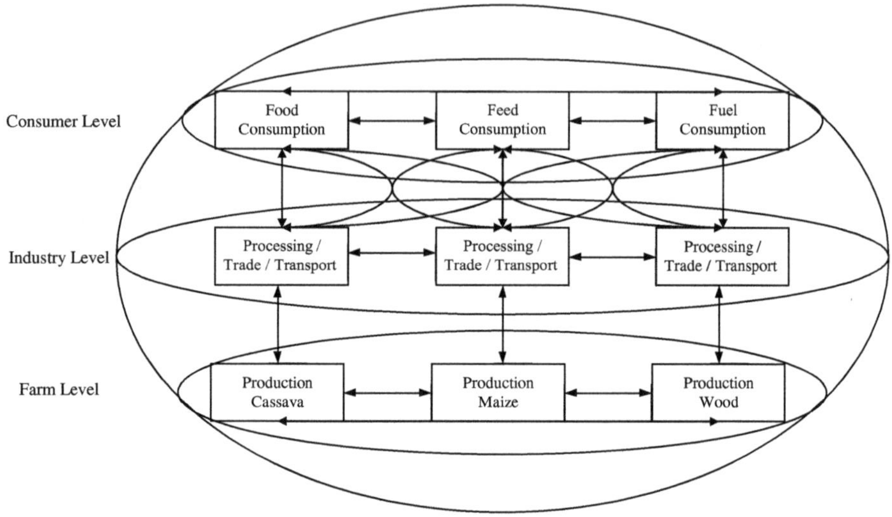

Fig. 1 Network perspective of Bioeconomy. Source. Adapted from Virchow et al. (2014)

3. Bio-technology applications: all sources of technology applications, such as Genetically Modified Organisms (GMOs), molecular techniques, modified enzymes and modified organisms can be considered as biotechnology applications.
4. Bio-refineries and bio products: biomass transformation in a wide range of marketable products as well as energy production.
5. Improving supply chain efficiency: allows the optimization of productive processes in each step of the chain. Efficiency allows for cost reduction, as a result of biomass efficient use. On this regard, what traditional supply chains would waste, Bioeconomy approaches would use as sources of unique products in new markets.
6. Ecosystem services: benefits societies obtain from ecosystems.

On a general perspective of Bioeconomy, El-Chichakli et al. (2016) addresses five cornerstones for a global Bioeconomy, which are mentioned below:

1. International collaboration between governments and public and private researchers is essential for optimizing resource use and sharing knowledge.
2. Measurements of Bioeconomy and its development should be found and implemented. On this same topic, priority targets regarding food security and food loss should be agreed upon internationally so that national measuring systems comply with international global organizations.
3. Bioeconomy initiatives need to be linked more closely with multilateral policy processes and intergovernmental discussions.

4. Educators should collaborate internationally to define the knowledge, skills and competencies required for developing a Bioeconomy that enhances the sustainable use of bio-based materials in manufacturing and in consumer products.
5. Research-and-development support programmes are needed to encourage global collaborations in a few breakthrough projects.

Concerning the perspective of El-Chichakli et al. (2016) there is international collaboration, educators are collaborating internationally and research and development support programs are working towards Bioeconomy goals. One example of this type of initiatives is ALCUE KBBE's collaboration for Bioeconomy in Latin America. However, issues such as Bioeconomy measurements which are more complex are still missing from the conversation, as well as solid international policy.

This paper is structured in the following way: first addressing the discussion of Bioeconomy in Central America, its development, paths and policies. Then, innovations in the region are mentioned and analyzed, in this regard, since there are still low innovations related to the Bioeconomy in the region, potential agro-Bioeconomy products and were also considered for analysis since they contribute to food security and poverty reduction in rural areas.

2 Bioeconomy in Central America

Food supply chains are often studied to achieve a more sustainable production. Research has focused on increasing efficiency, reduction of harvest and post-harvest losses and on the creation of innovative products. According to (Trigo et al. 2013, p. 9) *there is a common mistake in equating the bio economy with sustainability concepts. It must be made clear that bio-based options are not per se more sustainable.* Therefore, sustainability is not a consequence of Bioeconomy, although both concepts could be paired.

The European Union has been leading the Bioeconomy framework in terms of policy applications, including research in other geographical regions. In 2010, the European Union and Latin America launched a EU-LAC Summit in Madrid in which innovation and technology for sustainable development and social inclusion was proposed as a central theme, and a new "Joint Initiative for Research and Innovation (JIRI)" for the Latin America and the Caribbean region was been developed. The collaboration was established in the four following areas:

1. Biodiversity and climate change.
2. Bio-economy, including food security.
3. Energy.
4. Information and communication technologies for meeting societal challenges.

In terms of Latin and Central American development in Bioeconomy issues, the most extended form of assessing Bioeconomy has been through the European Union partnership and the development of knowledge based Bioeconomy This is another term that includes issues related to Food, Agriculture and Fisheries, and Biotechnology. In Latin America, the Knowledge -based Bioeconomy's objectives:

> ... include to exploit new and emerging research opportunities that address social, environmental and economic challenges: the growing demand for safer, healthier, higher quality food and for sustainable use and production of renewable bio-resources, the increasing risk of epizootic and zoonotic diseases and food related disorders; threats to the sustainability and security of agricultural, aquaculture and fisheries production; and the increasing demand for high quality food, taking into account animal welfare and rural and coastal context and response to specific dietary needs of consumers (European-Comission 2008).

Regarding the paths towards Bioeconomy in Latin America, the Knowledge-based Bioeconomy is betting to the following:

1. Resource use from biodiversity: there are many crops in Latin America that have not been exploited because of different reasons that, in the end translate to low (on inexistent) market value. However, nowadays scientific and infrastructure *can contribute to the bio-based economy as new industrial feedstocks or basis for new value chains in the phytotherapeutics, cosmetics, or tropical fruits and other areas* (Trigo et al. 2013).
2. Eco-intensification: these include no-till agricultural practices, precision agriculture strategies, integrated pest and nutrient management, sustainable land management (environmentally friendly production). For example, water treatment in agricultural production is common and often mandatory in some regions in Latin America, nonetheless although regulations are present, these are avoided by farmers in some cases.
3. Bio technology applications: the more likely situation is one of technological "hybridization" and "blending", with a shift from present day energy-intensive technologies to win-win alternatives. Traditional knowledge is site-specific and indigenous populations in the Central America could provide important contributions for biotechnology applications since they have created isolated knowledge in their own communities. This traditional knowledge could be incorporated in the biotechnology research agenda.
4. Bio-refineries and bio products: regional developments in this regards are limited to biofuel production from different potential agricultural sources.
5. The bio-refinery perspective reviews the extent to which bio fuel production can be integrated into agricultural production; According to Mathews (2009, p. 630) Agriculture needs to play its part in this resource-productivity revolution, and biofuel production has the potential to lead the way, both in terms of lowering the need for inputs and improving the yield per input, and in terms of demonstrating and practicing low- or zero-output energy crop production.

6. In 2016, the Costa Rican National Chamber of Industries in collaboration with GIZ, held in Earth University the Second National Forum of Biomass Energy. This forum included the presentation of several success cases in the Costa Rican private sector, including agricultural production such as palm oil and pineapple (EARTH 2016).
7. Improving supply chain efficiency: supply chain efficiency is a challenge specially for developing countries, in which the main food losses are along supply chains. Central America grows fresh exports products whose supply chain strategies could be improved regarding two aspects: a. reduction of food losses, b. alternative uses for lost agricultural products. Both represent market opportunities and therefore income increases for the region, and especially for the agriculture-related economies.
8. Ecosystem services: benefits from ecosystem services include carbon credit systems, eco-tourism and water-management mechanisms. In terms of the regions' advantages, Guatemala and Costa Rica have included payments for environmental services which in both countries have had an important impact on conservation. On the other hand, Eco-tourism is a good opportunity for tourism in all Central American countries because of site-specific natural appeal to world-wide tourist; natural resources located in rural regions. There are some country-specific certifications which intend to certify for good social, environmental and community development goals in host regions.

3 Research in Central America

Research in Central America specifically about Bioeconomy is difficult to encounter mainly because of the lack of regional policies targeted to this subject. Nonetheless, there is plenty of research related the term although it may not be addressed as such. Research encloses topics such as supply chain efficiency, alternative uses of biomass, traditional uses of agricultural products and sub utilized crops amongst others.

The only research article that addresses directly Bioeconomy of Central America, developed a total factor productivity growth that incorporates Bioeconomy components (Zúñiga-González 2013, p. 9). This measurement is called the Bio Economic-Oriented Total Factor Productivity (BTFP) index that can be decomposed into bio economy efficiency change (BEC), and Bio Economic technological change (BTC) components; main results from this research can be observed in Table 1.

Bioeconomy in Central America is an emerging issue and addressing potential benefits for these countries require policy changes in the region. These usually start from a common understanding of the importance of Bioeconomy. The following are some regional (Latin and Central America) efforts to create and promote academic and research dialogue focused on later policy implications.

Table 1 Mean of Bioeconomy technical efficiency for Central American countries 1980–2006

Countries	1980	1990	2006
Belize	1.141	1.106	1.403
Costa Rica	1.304	1.107	3.881
El Salvador	1.053	1.811	0.998
Guatemala	1.03	1.898	0.813
Honduras	1.192	4.052	1.424
Nicaragua	0.526	1.618	1.961
Panamá	1.638	1.423	0.994
Mean	1.126	1.859	1.639

Source: Zúñiga-González (2013)

On 2011 the Latin American and Caribbean Regional Inter-Conference Symposium on the Bio-economy was weld in Cali, Colombia. The main theme of this symposium was focused on Bioeconomy and socioeconomics research agenda targeted to conceptualize the term, its drivers and the mains problems that could be managed by Bioeconomy goals.

Two years later, on 2013, the Symposium on the Bioeconomy in Tropical America was held in San Jose, Costa Rica. This symposium is part of the activities organized by the Latin American, Caribbean and European Union Network on Research and Innovations (ALCUE NET) is association with Costa Rican Ministry of Science, Technology and Communications and the Argentinean Ministry of Science, Technology and Productive Innovation. From the European Union partnership, organization and coordination was provided by the Center of International Cooperation in Agronomical Research for Development (CIRAD).

In this symposium, challenges for Bioeconomy-related policies were addresses, because of the immense challenge it represents for the region. One important issue that what addresses was the large biodiversity along the region; and consequently, the proposal for a starting point for policy creation was a natural inventory as well as the understanding of the potential scope of these products after industry processing. The inventory and the following analysis of the scope of these diverse products would allow each nation to develop country-wide policies in a strategic manner. A general policy framework would therefore be a mistake, since diversity is a central issue that must be considered for policy implications, nonetheless, regional understanding of Bioeconomy and its potential is mandatory.

The main challenges Bioeconomy strives for in terms of policy discussed in this symposium are the following:

1. Develop profitable businesses. This means to translate academic discussions to entrepreneurial businesses that can generate profits and wellbeing from Bioeconomy products and from employment opportunities. Policies should promote these types of entrepreneurs working towards Bioeconomy goals, and therefore, towards development.

2. Define Bioeconomy limits.
3. Develop clear, simple and stable policies.
4. Avoid the creation of generic or standardized policy models for the region (MICITT 2013).

On 2015 another symposium, the Latin America and the Caribbean Bioeconomy was held and organized by The Economic Commission for Latin America in Santiago, Chile. This Symposium's objectives were similar to the ones presented before. It started by presenting challenges and definitions but emerged towards a collaboration discussion in Latin American's Bioeconomy development. It included experiences of Bioeconomy initiatives in Europe and some regions of Latin America, the promotion of sustainable development through Bioeconomy and the de-carbonization of economies.

4 Innovations in Central America

As it was abovementioned, it is particularly important to leverage biomass demand to reduce poverty in rural regions. According to Schneider (2015) in the world there are more than 400 million small farmers then can be integrated into biomass production.

On the other hand, global demand for food is rapidly increasing (Tillman et al. 2011) and meeting this demand in a sustainable way is a major challenge. Analysis of food supply chains worldwide leaded by the Food and Agriculture Organization (FAO) has estimated that approximately one third of global food production is wasted. FAO has developed two types of research agendas related to Bioeconomy: the analysis and reduction of food loss and food waste and the analysis of Bioenergy. Both related to food security.

Efficient supply chain management is particularly important in the agri-food supply chain since food loss and waste account for 1/3 of all produced food (24% of all calories produced) (Lipinski et al. 2013). By efficient supply chains that tackle losses, resources use and food losses along the supply chain can be reduced. These affects are expected to rise prices for small farmers, to improve sustainability goals and to holistically create sustainable food systems.

Food loss is proportionally more relevant in developing countries while food waste in developed countries. According to (FAO 2011) in Latin America, losses are distributed in the following way: 28% in production, 22% in handling and storage, 6% in processing, 17% in distribution and marketing and 28% in consumption.

On the other hand, according to (FAO 2010): Bioenergy potentially offers poor countries many advantages.

> Firstly, bioenergy developments offer the opportunity for enhanced energy security and access by reducing the dependence on fossil fuels and providing a localized solution. Increased energy security in turn can have positive effects on food security. Secondly, a bioenergy sector can create a new market for producers as well as offer new forms of employment that will positively affect agricultural and rural incomes, poverty reduction and economic growth. Thirdly, bioenergy has the potential to contribute to environmental objectives including the reduction of greenhouse gas emissions. Not surprisingly, bioenergy has been placed high on the policy agenda of developing countries.

Household agriculture plays an important role in food security, especially in rural areas and play an important role in agriculture in Latin America. In most Latin American countries, household agricultural production surpasses 30% of total value of agricultural production (FAO 2011).

Recent changes in adverse agro-climate conditions could increase economic and nutritional vulnerability in agricultural households in Central America, especially in those families that depend directly from agriculture either for their income or for auto-consumption. However, diversification from rural and agricultural families is an important pillar for increasing resiliency of these households, it reduces risk from climate and market conditions and therefore collaborates with food security.

Identification of sub-utilized products is necessary as a first step to create policies to promote this type of agricultural production since traditional, as well as marketable products can both collaborate to achieve Bioeconomy goals: traditional products by biomass marketable options and supply chain modification and traditional products by Resource use from biodiversity as well as for further biotechnology applications from incipient research. Some of these products with potential for Bioeconomy goals in Central America can be observed in Table 2.

Marketable products have developed biomass uses because of the recent need of agricultural enterprises to manage wastes. On the other hand, traditional products are useful for food security goals as well as for biotechnology innovations; nonetheless, these require research efforts to bring traditional knowledge into academia.

5 Two Examples: Pineapple and Jicaro (*Crescentia cujete*)

Pineapple in one of the most important exports products in Honduras and especially in Costa Rica. Production and therefore exports have busted in the last 20 years and consequently there have also been plenty of controversy because of environmental issues related to intensive pineapple production. Figures 2 and 3 show the traditional pineapple production process and the new bio-value chain approaches.

The pineapple bio-value chain would be a case of a combination of Bioeconomy paths that include not only better supply chain efficiency but also bio-energy modifications and opportunities as well as potential for bio technology. Bio ethanol and wood pallets are no longer of food supply chain but part of a holistic chain from a Bioeconomic perspective.

Table 2 Market and traditional agricultural product in Central America with potential for Bioeconomy goals

Type of product	Product	Regions/Country (ies)	Traditional uses	Bioeconomy uses
Market	Sugar cane	All countries (except El Salvador)	Sugar, dulce/rapadura	Biomass: Energy production, biofuel (ethanol)
	Coffee	All countries	Traditional: beverage. Today: different uses	Organic fertilizer, energy, biogas, alcohol
	Pineapple	Costa Rica and Honduras	Fresh fruit, juice, dry fruit	Biomass: Alcohol, organic fertilizer, fiber
	Banana	All countries (except El Salvador)	Fresh fruit	Biomass: organic fertilizer
	Palm oil	All countries (except El Salvador)	Oil	Biomass, biodiesel
	Rice	All countries (except El Salvador)	Rice	Biomass: electricity (energy)
Traditional	[a]Jícama (Pachyrhizus spp.)	Central America (and Mexico)	Fresh consumption	Food security
	[a]Árbol del pan (Artocarpus altilis)	Tropical regions all over the World	Fresh consumption	Food security
	Chan (Hyptis suaveolens)	Costa Rica	Fresh consumption	Food security
	Jícaro (Crescentia cujete)	Nicaragua and Costa Rica (northern region)	Fresh consumption	Food security

[a]FAO's traditional crops. Available at: http://www.fao.org/traditional-crops/en/

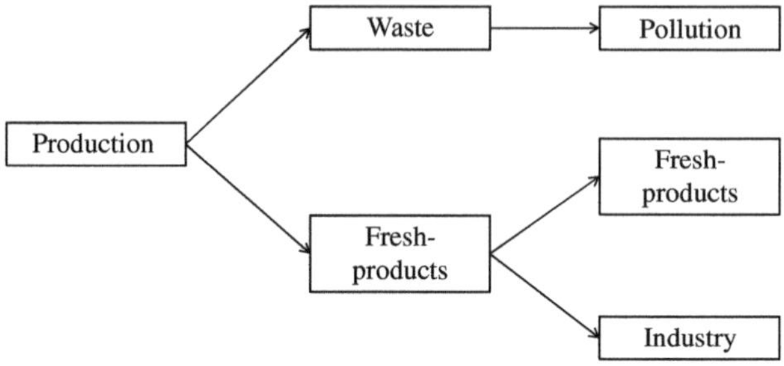

Fig. 2 Traditional value chain approach for pineapple production. Source. Adapted from Arauz (2013)

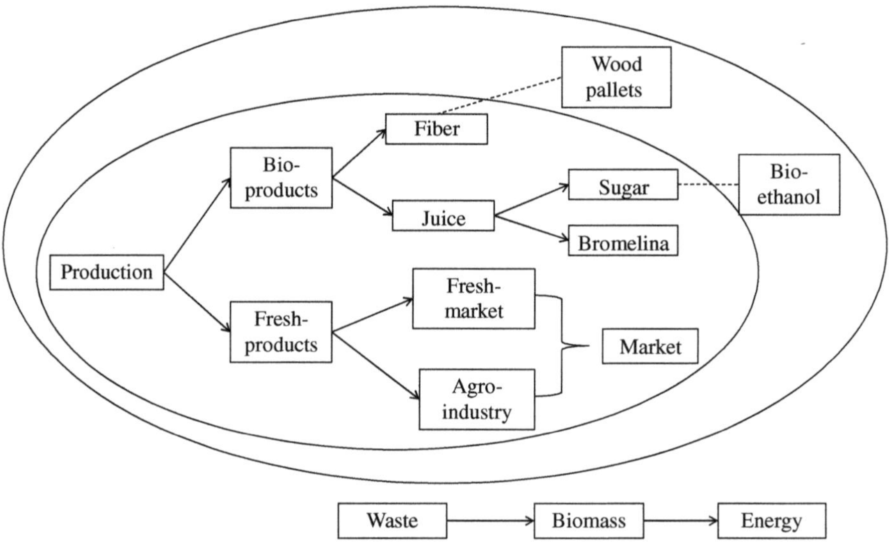

Fig. 3 New bio-value chain approach for pineapple production. Source. Adapted from Arauz (2013)

On the side of traditional crops of Central America, *jícaro* is the best adapted silvo-pastoral tree with high potential of minimizing the climate change effects in the region, especially in the pacific coast of Central America, where it is endogenous. These characteristics make it a potential contributor to food security especially in rural areas in the region: fruits are used for consumption and carbon can be obtained from the fruit peel. It is mostly used to prepare a traditional drink called *Horchata* (Quirós 2014).

Nowadays, reproduction in Costa Rica is done naturally while in Nicaragua there is some preparations for production but it grows naturally most of the time. Nicaragua has some rudimentary industry in which the product can be transformed into a powder for later preparations or into a powder-mix drink. In Nicaragua, the products extend to local markets however in Costa Rica it is only sold in Restaurants and local markets, in both countries, consumption does not extend further, geographically.

In terms of nutrition's content, it has higher protein content and fat contents than soy seed and it can produce the equivalent of 4 metric tons per hectare per year, which is about the same protein content and non-saturated fatty acids of 100,000 liters of milk. Further research is needed for addressing how to increase market potential for this product, as well as its economic and social impacts in rural societies.

6 Conclusions

Paths towards Bioeconomy are diverse and agricultural production can be adapted to produce in a bio-efficient way. Supply chain innovations are one of the most common types of innovations in food production systems, cases such as pineapple and *jícaro* can certainly increase the market value of the production in two ways. First of all, due to an increase in the number of marketable products (what was traditionally considered waste in now available for sale) and second as a sustainable market strategy in which small companies (and family-owned enterprises) can promote their approach on how they have challenged the traditional food system to a more sustainable one.

The re-activation of traditional food systems can decrease the effects of price volatility in World markets and agricultural households can depend not only in their marketable product's sales, but also in their local markets. This system would create better opportunities in terms of income and in terms of food security, especially in crises in rural Central America.

Further academic and private research is needed to create new options to decrease food loss and food waste. However, policy and economic changes in the sector are needed to promote more sustainable agriculture systems Central America.

References

Arauz, L. (2013). *Ecosistemas agroindustriales: oportunidad para nuevos emprendimientos en agricultura tropicalment ante las crisis globales*. First Agricultural Economics Congress. San José, Costa Rica: Universidad de Costa Rica.

Bugge, M. M., Hansen, T., & Klitkou, A. (2016). What is the Bioeconomy? A review of the Literature. *Sustainability, 8*(7), 691. doi:10.3390/su8070691.

CYTED, & REBICAMBLI. (2014). *Estado del arte de la bioeconomia y el cambio climático*. León, Nicaragua: UNAN.

EARTH. (2016). *II. Foro Nacional de Energía Biomásica*. San José, Costa Rica: EARTH.

El-Chichakli, B., von Braum, J., Long, C., Barben, D., & Philip, J. (2016). Policy: Five cornerstones of a global bioeconomy. *Nature, 535*(7611), 221–223.

European-Comission. (2008). *Cooperation theme 2: Food, agriculture and fisheries, and biotechnology* (Work Programme 2009). Retrieved from http://ec.europa.eu/research/participants/data/ref/fp7/88711/b_wp_200901_en.pdf

FAO. (2010). *Bioenergy and food security. The BEFS analytical framework.* Retrieved from http://www.fao.org/3/a-i1968e.pdf

FAO. (2011). *Global foos losses ans doos waste.* Dusseldorf, Germany. Retrieved from http://www.fao.org/docrep/014/mb060e/mb060e.pdf

Hodson de Jaramillo, E. (2011). *Towards a Caribbean & Central American knowledge based bioeconomy: visions, pathways and priorities.* Retrieved from https://issuu.com/mercadeoepuj/docs/hacia_una_bioeconom__a_en_am__rica__49dbf1713c43c8

Lipinski, B., Hanson, C., Lomax, J., Kitinoja, L., Waite, R., & Searchinger, T. (2013). Reducing food loss and food waste. *Working Paper. Installment 2 of Creating a Sustainable Food Future. Washington.*

Mathews, J. A. (2009). From the petroeconomy to the bioeconomy: Integrating bioenergy production with agricultural demands. *Biofuels Bioproducts and Biorefining, 3*, 613–632. Wiley InterSciencce. doi:10.1002/bbb/181.

MICITT. (2013). *Memoria simposio de Bioeconomia en América tropical proyecto.* San José, Costa Rica

MINCyT-CIRAD. (2016). *Senior Officials Working Group on Bioeconomy. V EU-CELAC Senior Officials* Retrieved from http://alcuenet.eu/assets/SOM%20Thematic%20Report-%20Bioeconomy%202016.pdf

Quirós, O. (2014). *Aprovechamiento del jícaro (Crescentiaalata) en sistemas silvopastoriles tradicionales de América Central para la producción competitiva de alimentos de alto valor nutricional y funcional.* Escuela de Economía Agrícola y Agronegocios. Universidad de Costa Rica.

Schneider, R. (2015). Diseño de la bioeconomía para la seguridad alimentaria. In CEPAL (Ed.), *Bioeconomía: nuevas oportunidades para la agricultura* (pp. 31–33). CEPAL: Santiago, Chile.

Tillman, D., Balzer, C., Hill, J., & Befort, B. L. (2011). Global food demand and the sustainable intensification of agriculture. *PNAS, 108*(50), 20260–20264.

Trigo, E. J., Henry, G., Sanders, J., Schurr, U., Ingelbrecht, I., Revel, C., ..., & Rocha, P. (2013). *Towards bioeconomy development in Latin America and the Caribbean.* Retrieved from https://agritrop.cirad.fr/567934/1/document_567934.pdf

Virchow, D., Denich, M., Kuhn, A., & Beuchelt, T. (2014). *A biomass-based value web as analytic approach to optimize the agricultural sector's efficiency and resilience.* Paper presented at the IFPRI 2020, Addis Ababa, Ethiopia.

von Braum, J. (2015). Bioeconomía y desarrollo sostenible - dimensiones. In CEPAL (Ed.), *Bioeconomía: nuevas oportunidades para la agricultura* (pp. 11–13). CEPAL: Santiago, Chile.

World-Bank. (2014). *GINI Index (World Bank estimate).* Retrieved from http://data.worldbank.org/indicator/SI.POV.GINI?locations=SV-CR-GT-PA-HN-BZ-NI

World-Bank. (2016). *Data for Latin America.* Retrieved from http://data.worldbank.org/?locations=CF-ZJ-CR-BZ-GT-SV-HN-NI-PA

Zúñiga-González, C. A. (2013). Total factor productivity and the bio economy effects. *Journal od Agricultural Studies, 1*(1), 1–29.

Innovation Under the Bioeconomy Context in Brazil

Alcides L. Leao, Ivana Cesarino, Suresh Narine, and Mohini Sain

Abstract This chapter discusses the possibilities of using biomass for several industrial applications, including biopolymers, bio-based materials and biofuel. It also covers the potential conflict between materials/energy and food, cascading concept of biomass utilization and the use of residues. Finally, it discusses the state of art of biomass and agriwastes utilization in Brazil, going from chemical feedstock, biofuels to bio-based materials, at macro, micro and nanoscales.

1 Introduction

Innovation in Brazil is established on very low levels, even compared to other emerging economies. Nowadays, Brazil invests about 0.9–1.2% of the GNP (depending on the source of the information), compared to 2.4% in OECD—The Organization for Economic Cooperation and Development—countries. This index represents the effort of a country, including private and governmental funds for science and technology development. Amongst Asian countries, Korea is the leader with 4.4%. And it is also important to mention the concept of transnational versus multinational companies or corporations. Specifically, the research and development, related to innovation, which are carried out at the company's headquarter, meaning in most cases, developed countries. One important factor, that reduces the investment in Brazil, is the low profitability of the companies, due to the high level of taxes (federal, state and municipal) and too many social benefit. Another issue in Brazil is related to intellectual properties. The Brazilian institute

A.L. Leao (✉) • I. Cesarino
Department of Biotechnology and Bioprocess, College of Agricultural Sciences, UNESP – São Paulo State University, 18610-307 Botucatu, Brazil
e-mail: alcidesleao@fca.unesp.br; ivana@fca.unesp.br

S. Narine
Trent University, Peterborough, ON, Canada
e-mail: sureshnarine@trentu.ca

M. Sain
University of Toronto, Toronto, ON, Canada
e-mail: m.sain@utoronto.ca

S. Dabbert et al. (eds.), *Knowledge-Driven Developments in the Bioeconomy*, Economic Complexity and Evolution, DOI 10.1007/978-3-319-58374-7_6

in charge of patents is very slow in releasing the final grant decision. It takes an average of 10.8 years. Finally, one of the main aspects is the fund for research, which, even in many developed countries, either it does not draw investments, or the budget is low, but in some others the cost for such funds is very low and even without interest. The cost for funding in Brazil is high and the funds in form of grants are negligible in terms of the amount available for the whole country.

Brazil is amongst the world's eighth largest economy and is pursuing to be competitive as an industrial exporter. Innovation is an important tool to ensure that it does not fall behind on research and development, since it is clear that the country cannot depend on being a simple manufacturer or producer of low value commodities. The country's industrialization, or recently, the deindustrialization phenomenon that has been happening in the last few years, depends on the increase of its productivity and the reduction of its costs, by becoming more productive through innovation and the use of technology. Brazil has emerged as a significant financial and industrial power in recent years, and it is poised to become a significant player on the field of biotechnology internationally, by taking advantage of circumstances not available in other countries, in particular its native biodiversity (Zuniga et al. 2016).

The country is facing the "Dutch disease"—meaning that most of the country's revenue is from commodity exports, which leads to a deindustrialization phenomenon The World Economic Forum classifies Brazil as in the final stage of transition towards a developed economy—the phase of evolving from an efficiency-driven growth model to an innovation-inspired one. Although about 3/4 of the GDP growth in the recent years was due to the increase of the working force and only 1/4 was due to productivity gains, which is the opposite to other fast developing economies. Nevertheless, there are several attempts in innovation, research, and development, with several successful cases, an example is Embraer, which is the third-largest commercial jet producer in the world by units sold. Embraer is innovative in both business model and aircraft designs. Also the Brazil's automotive sector is responsible for important innovations like the "Flex" engine—capable of running on ethanol or petrol and local designed models including Chevrolet's Meriva minivan and Volkswagen's Fox, models that are commercialized globally. In the last case, three interior parts were developed made of natural fibers in partnership with UNESP.

Brazil's spendings on research and technology were rising at nearly 7% a year, a figure second only to China's 11%. Brazil ranks the 75th position in the World Economic Forum's global competitiveness index, compared with Korea at 26th, China at 28th, Russia at 45th, South Africa 49th and India 55th. For the pillar innovation, the rank is Korea 19th, China 31th, South Africa 38th, India 42th, Russia 68th and Brazil 84th. Still, the Forum announced that "Brazil continues its downward trend, dropping 18 places from last year to 75th amid low prospects of growth and deteriorating terms of trade. The country's performance is uneven across the Index. Brazil's most important competitiveness strength is its extremely large market size (7th). It benefits from a relatively high level of technological readiness (54th), especially ICT use, along with sophisticated businesses (56th), and it registered a significant improvement in the quality of its air transport and infrastructure (95th, up 18 places). However, it deteriorated in nine out of the twelve pillars. With a large fiscal deficit and rising inflationary pressure, Brazil's weak macroeconomic performance (down 32 places to rank 117th) is negatively impacting the country's competitiveness" (World Economic Forum 2016).

Nassif et al. (2013), discussed some of the Brazilian economical data. The total Brazilian exports, basic products surpassed those of manufactured goods in 2010 with an increasing trade deficits knowledge-based manufacturing sector. Also the technological gap dramatically increased, showing Brazil farther down from the edge science and technology showing that Brazil is not only in a process of early de-industrialization but also falling behind comparatively with developed and even emerging economies.

Primary products, including minerals, agricultural and other low values commodities constitute more than 50% of Brazil's total exports. Exports of manufactured goods were once the most important for the Brazilian economy in Brazil, with a higher value than primary and semi-manufactured goods combined. Lately the agribusiness has been growing and Brazil became one of the world's leading exporters of soy, sugar, meat, chicken, coffee, tobacco, and orange juice. The percentage of the industry in the GDP was reduced from over 25 to 23% in the last 2 years, confirming the deindustrialization process happening in Brazil. Over the past 7 years, the value of commodity exports has quadrupled. Brazilian exports have been falling systematically since July of 2014 due to a slump in commodity prices. Exports in Brazil are reported by the *Ministério do Desenvolvimento, Indústria e Comércio Exterior* (Ministry of Development, Industry and Exterior Trade) (Fig. 1) Ministério da Ciência, Tecnologia e Inovação (MCTI) (2012).

The following export products groups represent the highest dollar value in Brazilian global shipments during 2015. Also shown, is the percentage share each export category represents in terms of overall exports from Brazil, valued in US dollars.

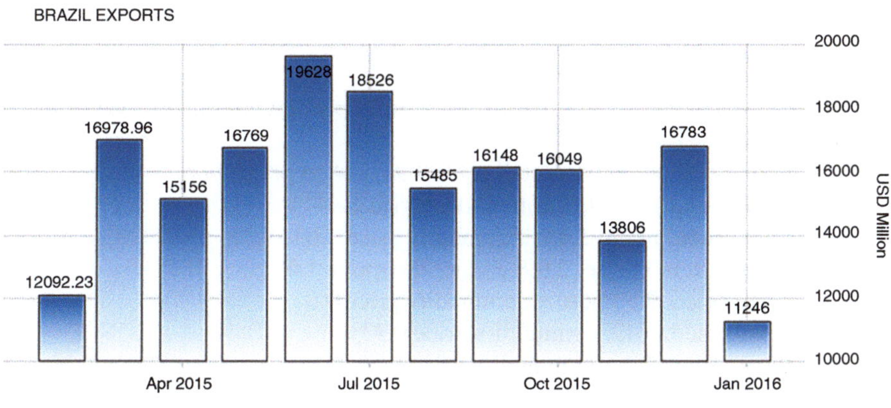

Fig. 1 Brazilian export total in 2015/2016. Source: http://www.tradingeconomics.com MINISTÉRIO DO DESENVOLVIMENTO, INDÚSTRIA E COMÉRCIO EXTERIOR

1. Oil seed: $21.2 billion (11.1% of total exports)
2. Ores, slag, ash: $16.7 billion (8.7%)
3. Oil: $13.7 billion (7.2%)
4. Meat: $13.1 billion (6.8%)
5. Machines, engines, pumps: $11.4 billion (5.9%)
6. Vehicles: $9.6 billion (5%)
7. Iron and steel: $8.9 billion (4.7%)
8. Sugar: $7.8 billion (4.1%)
9. Food waste, animal fodder: $6.2 billion (3.2%)
10. Coffee, tea and spices: $6 billion (3.2%)

Oil seed was the fastest-growing among the top 10 export categories, up in value 28.3% for the 5-year period starting in 2011. In second place for improving export sales was Brazilian food industry, waste and animal, posting a 3.1% gain.

Commodity prices tend to be quite volatile. Economies tied to them become so, too: when prices are high, the economy booms. The risks of commodity dependence goes beyond the possibility of lost export revenues. Brazil's manufacturing sector is facing a sharp decrease particularly in high-tech exports, losing ground to other developing countries, most of all to China. Brazilian manufacturing has lost ground in both global and domestic markets. Commodity success over the past decade has also reduced the pressure for much-needed structural and institutional investments and reforms, including increased investments in infrastructure, research and development (R&D), and education.

Brazil's lack of progress in developing its own knowledge economy is manifested in a number of indicators. Unfortunately, most of the R&D in Brazil is funded by government agencies. In 2013 about 60% of the total investment in research came from public funding, which represented roughly USD21 billion. In comparison, in the USA private companies invests 2% and the government 0.7% of the GNP. The Brazilian federal government had set a goal of 0.9% of the GNP to be invested by the private sector in 2014, which was not met and would take more than 60 years to reach that objective.

The solution is adding value into the biomass chain. Several examples can be listed (prices in USD):

- Coffee sachet for the *expresso* machine: multiply the commodity price by 25 times
- T-shirts for the Super bowl: Retail price 9.94; cotton price 0.24
- Gasoline: Retail price: 2.05; commodity price 1.44
- Coffee cup: Retail price: 2.45; commodity price 0.03
- Chicken legs: Retail price: 2.99; commodity price 1.69
- Beer: Retail price: 0.83 bottle; commodity price 0.01
- Beef: Retail price: 4.49; commodity price 0.33
- Corn: Retail price: 2.50; commodity price 0.06

Source: The Wall Street Journal, January 31/February 1, 2015 B1.

Bio-based Economy or Bio Economy is an economy in which crops, residues and waste from agriculture, food industry, other industries and households are used for food and non-food applications, an economy in which biomass is used for the production of materials, chemicals, fuels, plastic, energy (electricity and heat) and even for food. In the Bio-based Economy organic materials replaces fossil fuels. As an example, the OECD reccomends that both the public and the private sectors must take active roles in designing such approach in order to maximize the full potential of the bioeconomy (CEPI 2012).

Another concept that is important was defined by Anastas and Warner (1998), is the green chemistry. Green chemistry comes close to biobased economy in many ways and its full meaning should be only this, excluding green process or green products that come from man-made resources or fossil based. Among the 12 principles listed, the number seven, which is listed as "a raw material or feedstock should be renewable rather than depleting the non-renewable natural resources, whenever technically and economically practicable", should be the "sine qua non" for the green chemistry. Therefore the biobased economy, green chemistry and the other one, *ecomenia*, should work together.

It is important to develop the concept of *ecomenia* (from the Greek, oikos-menes–meaning environmental pathway). By definition, an exploratory process of low environmental impacts, or the use of natural resources based on an ecological intuition of non-linear biological processes, resulting in profitability and sustainability. This concept can pave the road for the bio-based products or the bioeconomy.

The following definitions are important for a better understanding of full biomass utilization:

- Bio-based: derived from biomass
- Biomass: material of biological origin excluding material embedded in geological formations and/or fossilized.
- Bio-based product: product wholly or partly derived from biomass.

The bio-based product is normally characterized by the bio-based carbon content or the bio-based content.

A survey was made at the INPI (National Institute of Industrial Properties), using the several keywords found in the title only. The access was made in February 2016, using the Portuguese language, as listed in Table 1. In the right column are the numbers of patents granted. As can be observed, the number of patents granted in the field of bioeconomy is very small compared to other countries with much less natural resources, such as Germany, Sweden, Korea, Canada, USA, etc. This shows the need of the Brazilian government to stimulate the private sector to invest more in R&D, mainly for bio-based products.

The opportunity value of biomass is determined by the revenue from the various products or by-products on the market and the production costs including capital and operation costs. Therefore a biomass producing country should implement some measure to add value to the biomass chain. In most of the cases products with a relative high market value are associated with high production costs, and vice versa (Langeveld et al. 2010).

Table 1 Patents granted by Brazilian INPI related to bioeconomy products or processes

Keywords	# Pat.	Keywords	# Pat.	Keywords	# Pat.	Keywords	# Pat.
Biomaterials	35	Green PP	35	Vegetable fibers composites	305	Extractives	4
Biopolymers	49	Green PE	978	Bioprocesses	1	Nanocellulose	11
Bioplastics	3	WPC	1	Green chemistry	53	Nanocrystal	1
Biorefinery	7	Wood plastic	142	Rubber	2111	Natural Fibers	57
Polyol	18	Green composite	110	Cellulose	1188	Renewable resources	10
PHB	5	Renewable composite	1935	Lignin	10	Biofuel	69
PHA	14	Lignocellulosics composites	104	Hemicellulose	31	Bioenergy	6

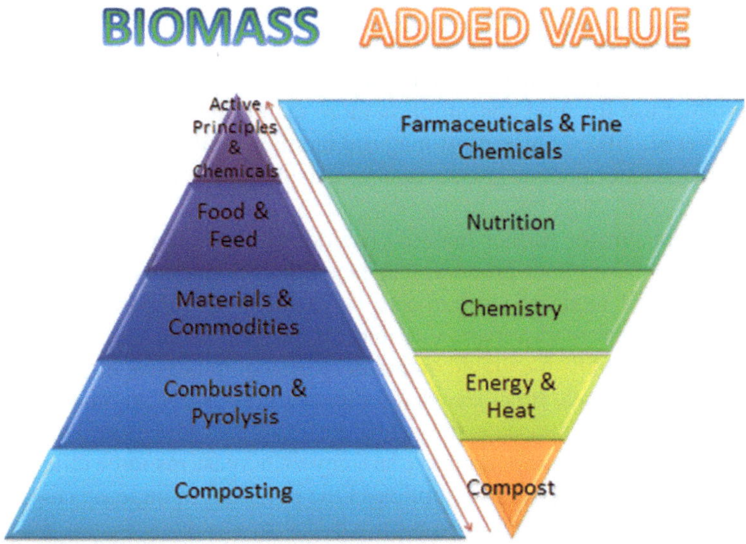

Fig. 2 The pyramid value concept for the biomass utilization. Adapted from Van Haen (2013)

The cascading use of biomass can be an alternative to bring development to the country, transforming it from a simple exporter of low value commodities, such as iron ore, soybeans, etc., to a technological global player. Cascading use of biomass takes place when biomass is processed into a bio-based final product and this final product is used at least once more either for materials or energy. If one starts a production of any feedstock from a low cost raw-material, there is more room for either adding value or profit margin. This can be represented by the figure below. Cascading use of biomass is described as single-stage, when the bio-based final product is directly used for energy. Cascading use of biomass is described as multi-stage when biomass is processed into a bio-based final product and this final product is used at least once more as a material. It is only after at least two uses as a material that energy use is permitted. In 'material use' biomass serves as a raw material for the production of all kinds of goods, as well as their direct use in products. This distinguishes it from energy use, where biomass serves purely as an energy source, and the use for food and feed purposes (Raschka and Carus 2012). For example, cascade utilization of wood: wood, as a raw material, is used sequentially for materials and then for energy use or for composting (e.g. solid wood furniture, chipboard, recycled chipboard, burning, and compost). Figure 2 clearly distinguishes between single-stage and multi-stage cascading use.

The above-mentioned definition of cascading use (Raschka and Carus 2012), leads to the following:

- All sources of biomass being the starting point of a cascade would be treated equally, whether they are be mainly biogenic, by- or intermediate products from primary agricultural, marine or forestry production, material or energy use, or from the food and feed industries.

- Requiring a "final product" for material use, the above-stated definition stresses the product use for final consumption more than other definitions (e.g. those of Kosmol et al. 2012). Since most material biomass chains include numerous and often traded intermediate products, the consideration of the "final product" is important.
- The requirement for further material use deliberately avoids a requirement for a second bio-based final product. It means that it is not only material recycling that counts, but material use can also be satisfied by intermediates, fillers and excipients that are incorporated into material process chains.
- Energy use should be the viable option at the end of the cascading use at least for the majority of cases. It is clear that a direct energy use of biomass (energy use without prior material use) is not considered as a cascading use. Understanding cascading principle is important as a strategy to increase resource efficiency. New ways of biomass material use implicates the potential to increase cascading use (Essel et al. 2014).

Value Pyramid

Another view on the bio-based economy is the so called 'value pyramid', or bio-cascading principle. This bio-based economy's value pyramid indicates that biomass value is determined by its application. In a proper functioning market, the value and price of biomass will be reflected by its application value. For example, pharmaceutical products add a lot of value per unit of product but in a small volume, whereas energy carriers adds little value per unit of product but in large quantities. Therefore, it justifies one important argument in favor for not directly burning biomass.

Below the definitions can be listed for the use of biomass as a raw material for industrial applications:

- Product—A good, idea, method, information, object or service created as a result of a process and serves a need or satisfies a demand. Main activity or product released from an enterprise: e.g. sugar cane from a sugar cane mill.
- Co-product—Product manufactured along with a different product, in a process in which both are required in the production of another product. Something with value or importance compatible to the main product: e.g. sugar and ethanol.
- By-product—Something produced usually in an industrial or biological process in addition to the principal product: e.g. sugar cane trash.
- Residue or waste—Usually a small amount of something that remains after a process has been completed or a thing has been removed; the amount of something valuable that is left, after or during, a production process. A situation in which something valuable is not being used or is being used in a way that is not appropriate or effective. Everything that serves for an own productive process or for a third party: e.g. vinasse, rich in potassium.
- Reject—By own definition, something presently useless and with no value, which only application is final disposal: e.g. landfill or incineration.

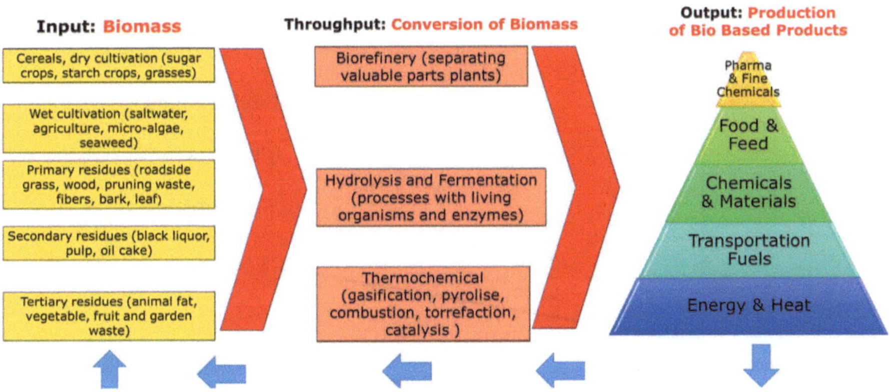

Fig. 3 The cascade cycle of biomass adding value strategy for the biomass. Adapted from: http://nbso-brazil.com.br/living-lab-biobased-economy/value-pyramid

In Fig. 3 is possible to observe the increasing value of a raw material versus an energy recovery process. Just one piece of biomaterial made of lignocelullosic material can represent tons of biomass for energy production. For a country with large biomass capacity production the strategy should be to produce more valuable goods from the biomass instead of using the lowest value alternative that is conversion into energy or heat.

One of the main task for a R&D team should be transform rejects into, at least residue or even a by-product, since the generation of reject in an industrial process implied in use of man power, energy, equipment time, logistics, etc., meaning cost that will be transferred for the final product. This concept can be seen at Fig. 4. Residues can be used for application in products with a (presumably) lower value than, such as feedstock or production of second-generation biofuels for road-transport, marine or aviation. Remaining residues, then, can be converted into energy, in particular electricity. The sum of the highest possible economic values of all the various components of biomass makes the biomass a product that can have a higher value for the producers than if the entire product is used only for production of electricity or heat, which should be the last option for the biomass utilization, instead of the nowadays approach of direct burning. That is the case for the sugar cane bagasse and sugar cane trash, although only the bagasse burning in Brazil with co-generation process corresponds to one Itaipu hydropower or 14.000 MW of installed power. It is equivalent to 3,000,000 oil barrels/day or 6.4×10^{15} kJ/day (UNICA 2011).

In Fig. 5, the red line, below the X-axis show raw material with negative cost, meaning something that the owner has to pay for an adequate final disposal in landfills or an expensive thermal solution. Therefore any R&D that use these raw material has more chance to succeed, since there are more room for investment in terms of technology or additives.

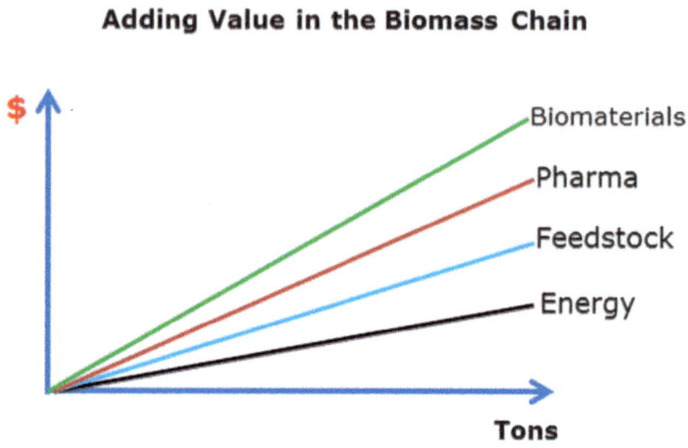

Fig. 4 The relation between raw material and final product cost

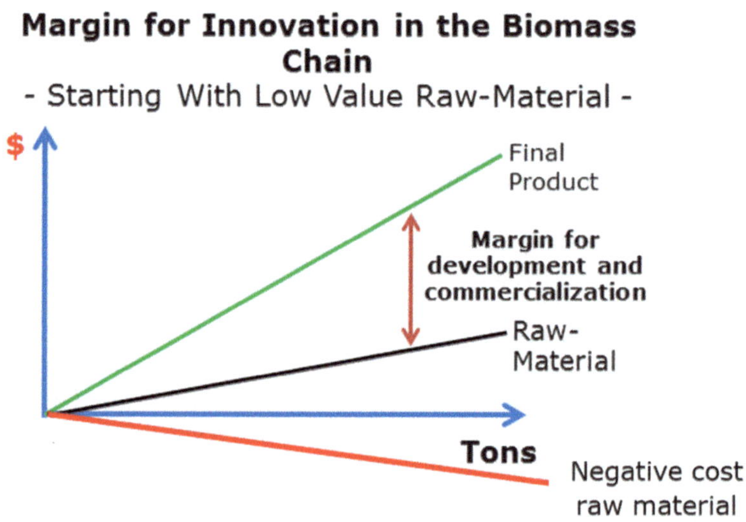

Fig. 5 The use of low cost raw-material can be an advantage for a bio-based product

2 Biofuels in Brazil

The presence of large amounts of oxygen in plant carbohydrate polymers means the pyrolytic chemistry differs sharply from these other fossil resources. Wood and other plant biomass is essentially a composite material constructed from oxygen-containing organic polymers. The direct burning is limited by the low fixed carbon content, which in the case of sugar cane bagasse is less than 9%, depending on the sources of data and the sugar cane variety, local conditions, etc. Proximate Analysis determined the average values for the trash components (dry and green leaves and tops) that presented practically the same composition in volatile material (80%), ashes (4%) and fixed carbon (15%) expressed on dry basis. These figures are similar to those obtained for bagasse, except for ash that is lower in bagasse (2%). The water content was approximately 13% for dry leaves, 65% for green leaves, 80% for tops and 50% for bagasse (Hassuani et al. 2005).

3 Biobased Materials in Brazil

3.1 Natural Fibers Industrial Applications

Since ancient Egypt natural fibers are part of our civilization. The traditional applications are baskets, clothing, sacks, bags, ropes, rugs, paper, medicines, etc. Several articles are listed in the literature describing the advantages of natural fibers compared to its counterpart the man-made fibers, such as glass fiber and polyester in composites applications. Lightweight, strength with stiffness and mainly low-cost are specially poised to replace glass fibers and mineral fillers in the plastic industry.

Combining natural fibers with fossil plastics is a great economic opportunity, since the price of virgin polypropylene in Brazil is around USD2.0/kg while its recycled counterpart costs about USD0.8/kg, which can result in economic gain for the transformer.

There are several reasons for an extensive use of natural fibers use in composites, mainly for the automotive industry, which represents the largest market niche in Brazil for these materials. The technique of injection molding is well known in the plastic industry and represents billions of dollars worldwide. However, the injection molding of natural fibers reinforced materials represents a new segment in industry. Only 5% of the worldwide market of reinforced plastic with natural polymers (fibers/flour) is dedicated to injection of molding applications, with the majority for extrusion of profiles, mainly for construction (decking, door and window frames, etc.). Several reasons are listed to motivate the use of the natural fibers in the composites industry, such as:

- Environmentalist pressure for more utilization of Natural Renewable Resources;
- Better efficiency in converting raw-materials in products compared to other man-made fibers;
- Products based on Life Cycle Analysis (ISO 14000);
- National strategy to create rural jobs in economically deprived areas;
- Good mechanical properties relations: Weight versus Resistance;
- Recyclability;
- Composites/Ecomenes;
- Reduction of the Greenhouse Effect;
- Renewable source of raw material;
- Excellent specific strength and high modulus. High flexural and tensile modulus—up to 5′ base resin, high notched impact strength—up to 2′ base resin;

- Reduced density of products;
- Lower price of polymer composites reinforced with natural fibers than those reinforced with glass fiber;
- Reduced tool wear; and
- Safe manufacturing processes, no airborne glass particles, relief from occupational hazards. Reduced dermal and respiratory irritation. No emission of toxic fumes when subjected to heat and incineration.

3.2 Biopolymers

The World Global Economics Forum in its 2016 report says that plastics has become the ubiquitous workhorse material of the modern economy. Yet, while delivering many benefits, the current plastics economy has drawbacks that are becoming more apparent by each day. Significant economic value is lost after each use, and given the projected growth in consumption, by 2050 oceans are expected to contain more plastic than fish (by weight), and the entire plastic industry will consume 20% of total oil production and 15% of the annual carbon budget.

Plastics due to its possibility of molding, processing and storage is appropriate for packaging applications around the world. Although, the traditional plastic is fossil based, not biodegradable and remain in nature for centuries, being one important item in the landfills, littering streets and rivers and potentially harmful to the environment when incinerated. Today, more than 80% of synthetic polymer production involves six major commercial materials: polyethylene terephthalate (PET), high-density polyethylene (HDPE), poly (vinyl chloride) (PVC), low-density polyethylene (LDPE), poly (propylene) (PP), and polystyrene (PS). If these plastics were not available to the $400 billion/year packaging industry, the production and transportation of replacement packaging materials would contribute an additional carbon footprint equal to the output of over 12 million automobiles per year (Pilz et al. 2005). Therefore its replacement by the bioplastics for food packaging is an ecological way to reduce the environmental impacts of these materials, with gains for the society. Among these bioplastics, it can be listed the starch based, oxo-degradables and the PHB, that degraded when exposed to the biological or physical active environment. In Brazil the PHB is produced from the sugar cane, a renewable resource, of which Brazil has the highest technology in the world. The polyhidroxybutirate (PHB) is the simplest and most common polymer from the hidroxyalcanoatos (PHA) family. It is a polyester homopolymer largely available in nature due the fact that it is synthesized by a large number of bacteria produced from sacarose, glycose, fructose and lignin (*Ralstonia*, *Burkholderia* and some *Pseudomonas*).

Technically the PHB is very similar to polypropylene and it is water resistant, has a good stability for UV radiation, it is not permeable to gases, it is biocompatible, has a high regularity in the carbon chain and high molar mass. Regarding its biodegradation, it is a polymer that can easily be degraded by the microorganism's action in the environment, releasing just carbon dioxide and water. It is a hard, fragile polymer and suffers degradation to each thermal cycle that is submitted to.

The PHB is accumulated by several different bacteria. Also PHB can be obtained from the lignin contained in the sugar cane. Polyalkylenehydroxybenzoates (PAHBs) are a new class of thermoplastic, biorenewable aromatic/aliphatic polyester having the general formula H-[O-(CH_2)n-O-4-(3-X, 5-Y-C_6H_2)-CO]z-OH (Mialon et al. 2011).

With the production of biopolymers from sugar cane molasse, the sugar cane industry is entering a new field, supplying the biomedical sector with a renewable material used for patch, surgical threads and even some cardiac prosthesys (Cherian et al. 2012). The product took 15 years to get to the market and was developed from a fusion of a bacterium with the sugar cane molasse. The advantages of this material is that speeds the cicatrization processes for burned skins and can be naturally absorbed by the human body in a not harmful way.

4 Bioenergy and Biofuels

Hydrocarbons (oil, natural gas and derivatives) represents about 55% of global energy consumption and due to its short coming shortage with the consequent price increase, Brazil has chosen an energy route that includes a long term development of a majority of the green energy matrix. By the last BEN (National Energy Balance 2016), the Brazilian emission per capita of CO_2 is about 2.35 tons, which is much less than China (5.1 tons per capita), EU (7.1 per capita) and USA, (16.9 per capita). In this context, the sugar cane represents, in the biomass group, the most important component (Empresa de Pesquisa Energética (Brasil) 2016). In 2011 the sugar cane bagasse was equivalent to 42.8 Mtep versus 47.1 Mtep from 2010, a sharp reduction due to weather conditions. In average, about 50% of the sugar cane harvested is dedicated to ethanol production and the rest 50% for sugar production. About 2/3 of the sugar produced is for export and the rest is for internal consumption. For the ethanol the situation is different, where 85% goes for internal market (from data, 90% are for fuel) and only 15% for export.

Brazil is in a very unique position to lead the biomass revolution in the world, based on the facts that has extensive crop fields, including the sugar cane; the biggest biodiversity in the planet; intense solar radiation; water in abundance; clime diversity; pioneering in the production of ethanol biofuel. The country has all the conditions to receive most of the world's investments in this area, if some constraints are solved. The quantity of lignocellulosic residues generated annually is about 350 million tons, and based on sugar cane bagasse composition, one can estimates 170 for cellulose; 100 million for hemicellulose; and 80 million for lignin.

Figure 6 shows the changes in the Brazilian energy matrix that occurred sustainable energy.

For electrical energy production it has been considered a value of 12.5 kWh/t of sugarcane and 258 kcal/kWh, considering an efficiency of 30% for the Rankine

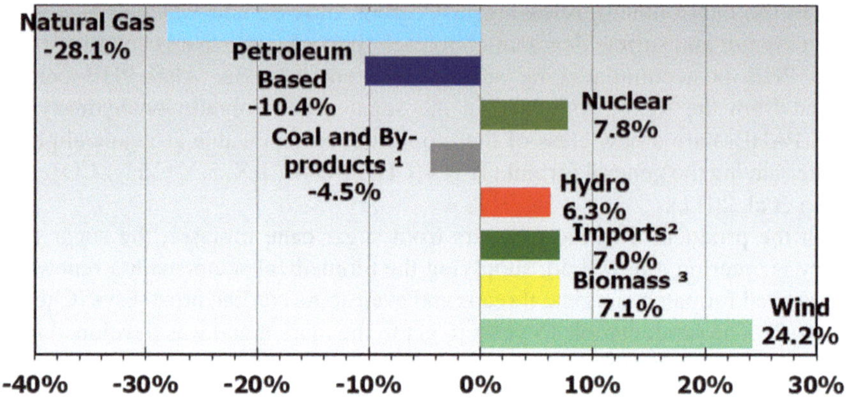

[1] Includes biodiesel

[2] Includes only gasoline A (automotive)

[3] Includes black liquor, fuel oil, gas flare, mineral coal, coke and charcoal, among others.

Fig. 6 Changes in the Brazilian energy matrix from 2010/2011. Source: BEN (2012)

cycle (corresponding to 860 kcal/kWh), since this industry figures represents the majority of Brazilian sugar-ethanol industries. In Table 2 the energetic value can be observed for some of the sugar cane components.

Instead of direct burning, the country should move towards pyrolysis. Pyrolysis dates back to at least the ancient Egyptian times, when tar for caulking boats and certain embalming agents were made by pyrolysis. Pyrolysis processes have been improved and are now widely used with coke and charcoal production. Biomass pyrolysis products are a complex combination of the products from the individual pyrolysis of cellulose, hemicellulose, and extractives, each of which has its own kinetic characteristics. In the 1980s, researchers found that the pyrolysis liquid yield could be increased using fast pyrolysis where a biomass feedstock is heated at a rapid rate and the vapors produced are also condensed rapidly (Farag et al. 2002). The accelerated climate change raises the incentive for all countries to produce energy and chemicals from biomass aiming the replacement of non-renewable sources. In addition, the burning of fossil fuels, which produces carbon dioxide, has serious environmental consequences. In contrast to fossil fuels, the use of biomass for energy provides significant environmental advantages.

Most processes that convert biomass to liquid fuels begin with pyrolysis, followed by catalytic upgrading of the resulting biocrude liquid. Another approach is focused on the use of catalysts for biomass cracking (in situ upgrading) to generate chemicals, mainly phenolic resin. Furthermore, the wood preservative industry is interested on finding a low-cost and an environmentally friendly mean to dispose treated wood products. This represents another opportunity to convert waste to fuels

Table 2 Gross calorific value of sugar-cane harvest materials and residues (UNICA 2011)

Materials and residues	Energetic values (Mcal/t)
Remaining material	3800.0
Sugar cane (losses)	1070.0
Industrializable material	4054.5
Impurities	3400.0
Bagasse (50% humidity)	1800.0

or chemicals with environmental benefits. The maximum utilization of our forest productivities is becoming increasingly essential. Logging generates considerable residues, which, to date, have been discarded as waste or simply not used.

The pyrolytic breakdown of wood produces a large number of chemical substances. Some of these chemicals can be used as substitutes for conventional fuels. Thermal degradation processes include liquefaction, gasification, and pyrolysis. Liquefaction in a reducing medium also generates solids and gases. Gasification produces hydrogen, carbon monoxide, carbon dioxide, and water by partial combustion. Gasification also produces hydrocarbons, particularly in the lower temperature ranges in fluidized-bed reactors. Pyrolysis converts organics to solid, liquid and gas by heating in the absence of oxygen. The amount of solid, liquid, and gaseous fractions formed is dependent mainly dependable on the process variables, as are the distribution of products within each solid, liquid, and gas phase produced. Processing carbonaceous feedstocks to produce heat, chemicals, or fuels offers an alternative to landfills and provides a supplement to fossil fuel use.

Although, if residues from wood processing are increasingly used as raw materials of bio-based products, the wood industry sector may become more dependent on external energy resources (Sokka et al. 2015), which can be positive on a long term period, since the use of lignocellulosics as a source of energy is not efficient, by the reasons discussed previously.

4.1 Biofuels

Rhodia, now controlled by the Belgian Solvay will build its first factory in Brazil overall bio n-butanol, in partnership with the American Cobalt. The product, which will be extracted from the bagasse of sugar cane in a process similar to the production of second generation ethanol, is widely used by the company in the production of solvents, aimed at the segment of paints for the automotive industry. Companies do not report the contributions in this business. The estimated value found, however, that investments of this size are at US$200 million. A memorandum of understanding between the two companies was signed at the end of 2012. SGBio, a joint venture between the Solvay Group and GranBio, has acquired assets from Cobalt, a US biotechnology company and leading developer of technology for n-butanol, acetone, ethanol and butane made from biomass. The purchased assets include the bank of microorganisms and intellectual property related to patents, trademarks, processes, operating procedures and know-how. With the acquisition, SGBio boosts its proprietary knowledge, now having access to a full suite of higher performance technologies utilized in the production of biochemicals. The robustness and flexibility of the microorganisms have been proven at scale, demonstrating high productivity levels, which considerably reduces risks. Brazil is not self-sufficient in the production of butanol, whose demand per year is estimated at 50,000 tons, moving nearly $100 million. The country imports two thirds of its annual needs. Butanol is produced from fossil raw materials. The aim of Rhodia and Cobalt is to get the same product from renewable sources. Butanol is used for the manufacture of automotive paint solvents and production of acrylic resin. In practice, the production of bio n-butanol follows a process similar to the second generation ethanol. Companies through their technology can produce renewable butanol from the sugar extracted from sugar cane bagasse. That route allows the production of ethanol used as fuel.

Another example of new developments in Brazil is Butamax™ Advanced Biofuels, a joint venture between DuPont and BP oil company, announced the production of bio i-butanol, using different technology from that used by Rhodia and Cobalt, for biofuel production, an innovative biobutanol production technology offering a low-cost, high-value drop-in biofuel for global transportation fuels supply. Butamax technology is designed to convert the sugars from various biomass feedstocks, including corn and sugarcane, into biobutanol using existing biofuel production facilities.

4.2 Bio-oil

Bio-oils are dark brown, free-flowing organic liquids that are comprised of highly oxygenated compounds (Czernik and Bridgwater 2004). The synonyms for bio-oil include pyrolysis oils, pyrolysis liquids, bio-crude oil (BCO), wood liquids, wood oil, liquid smoke, wood distillates, pyroligneous acid, and liquid wood. Throughout this report, the terms pyrolysis oil, tar, pyrolytic tar, and bio-oil will be used. Pyrolysis liquids are formed by rapidly and simultaneously depolymerizing and fragmenting cellulose, hemicellulose, and lignin with a rapid increase in temperature. In contrast to petroleum fuels, bio-oil contains large oxygen content, usually 45–50%. The presence of oxygen is the primary reason for the difference in the properties and behavior between hydrocarbon fuels and biomass pyrolysis oils (Oasmaa and Czernik 1999).

The lower heating value (LHV) of bio-oils is only 40–45 wt% of that exhibited by hydrocarbon fuels. The LHV of bio-oils on a volume basis is ~60% of the heating value of hydrocarbon oils, because of the high oxygen content, the presence of water, and the higher bio-oil density. A typical heating value of bio-oil is ~17 MJ/kg. Bio-oil is miscible with ethanol and methanol but immiscible with hydrocarbons. Bio-oil can be stored, pumped, and transported in a manner similar to that of petroleum-based products and can be combusted directly in boilers, gas turbines, and slow- and medium-speed diesel engines for heat and power applications (Czernik and Bridgwater 2004).

5 Biochar

Brazil is the largest producer of charcoal, accounting for about 40% of the world's production. In Brazil, the legal production of charcoal, in most of the cases, comes from eucalyptus (*Eucalyptus* sp.) plantations, and is important for the metallurgical industry, and this process is categorized as a Clean Development Mechanism (CDM) under the Kyoto Protocol. In the traditional process of charring, approximately 35% of the carbon is fixed in the charcoal, and the remainder is emitted into the environment as smoke and non-condensable gases, such as CO_2, carbon monoxide (CO) and methane (CH_4). The carbon contained in biochar is composed mostly of aromatic structures, which are characterized by linkages in the form of benzene rings of C atoms with oxygen (O) or hydrogen (H). These links between CO and CH aromatic structures govern the stability of biochar and are used to measure the degree of aromaticity of the compounds (Petter and Madari 2012). One of the most important use of biochar is as adsorbents in the separation and purification of gases, liquids and solid compounds and thus referred as activated carbon. Activated carbon is largely used in Brazil for water treatment by the water utilities companies, and is produced from coconut shells, fast growing trees and hardwood solid residues (Leao et al. 2015).

6 Biobased Materials: Nanocellulose

The advantages of the presence of nanoparticles in composites includes: reduced weight, improved mechanical properties and better stress transfer, reducing the amount of dead load in many applications, mainly automotive and aeronautical.

The chemical, physical and biological properties of materials at the nanoscale have fundamental differences about their properties at the conventional level, because of quantum mechanical interactions at the atomic scale. Nanotechnology brings multidisciplinary innovations in all areas of knowledge (chemistry, physics, agriculture, modern biology, among others) (Cherian et al. 2012).

In the production of pulp and paper, an average of 35% of the input material becomes waste in form of general waste (sludge, lime and ash from the boiler) and currently promoting some of these 56% of energy needed in the industry. The sludge waste is generated in two stages in the process of wastewater treatment. The primary sludge is obtained after the clarification process which is carried by primary sedimentation, and dissolved air flotation in which waste solids are removed. The secondary sludge comes from secondary treatment that is usually a biological process, in which microorganisms converts the waste into carbon dioxide and water while consuming oxygen. Disposal of waste from pulp and paper is a difficult environmental issue since its majority is directed to landfills (Cherian et al. 2012).

The use of fibers from the primary sludge for the production of composite materials can generate a lightweight, durable, non-abrasive, renewable, biodegradable and recyclable material allowing competition with other artificial materials with less availability or environmental disadvantages and still helps with the disposal of industrial waste (Cherian et al. 2012).

The cellulose is composed mainly of crystalline regions which will result in important mechanical properties when dispersed in polymer for the production of nanocomposites. Recent advances in producing bio-fibers, microfibrillated or nano-sized fibers with high-strength and surface area offer manufacturing high-performance composites (Cherian et al. 2012).

Cellulose fibers attract considerable interest as reinforcing fillers for thermoplastic polymers especially those with a relatively lower melting point like polypropylene, high and low density polyethylene. Sludge from paper mills consists mainly of two components, fine cellulose fiber and inorganic materials, and can offer a number of benefits, as a substitute for the typical inorganic reinforcing fillers in manufacturing thermoplastic polymer composite (Cherian et al. 2012).

6.1 *Biomaterials*

Different reinforcement agent, scaffold and composites processes can be used based on the several strategic points, such as availability, seasonability, cost, environment and social impact, future trends, etc. The reinforcement agents can be used at macro, micro and nano scales and the final properties of such compounds depending heavily on the processing conditions. Considering all of these aspects we could list some of the potential raw materials to be used as a source of nanocellulose:

- *Eucalyptus* residue (sawdust): large availability, but requires some preparation such as screening, milling and drying.
- *Eucalyptus* bark: it is a disposable material nowadays with almost no application, but available in large quantities almost everywhere in Brazil.
- Black liquor: today it has been used for co-generation and its availability in the market depends on further negotiations.
- Rice husk: it has a high lignin and silica content, which can be a problem in the composites processing. It has a local market today and there are plans to use this resource for energy generation or biorefinery.
- Sugar cane trash (straw): it is largely available nearby the sugar cane mills and its price nowadays is higher than the sugar cane tonnage itself. We consider an availability of around 6–7 tons/ha of trash that MUST be harvested from the field. However, the lignin content, inorganics and seasonability are negative aspects.
- Coir fibers: it is a by-product from the coconut industry, with large plantations in the Northeast of Brazil

Table 3 Brazilian agro production worldwide

Main products	Production ranking	Exports ranking
Sugar	1st	1st
Coffee	1st	1st
Orange juice	1st	1st
Soybean	2nd	2nd
Beef	2nd	1st
Tobacco	2nd	1st
Ethanol	2nd	1st
Broiler	3rd	1st
Corn	4th	3rd
Cellulose	4th	1st
Sisal (natural fibers)	1st	1st

- Sugar cane bagasse: it is a commodity nowadays, but its potential destination will be hydrolysis for later fermentation and may be biorefinery. As a reinforcement agent for composites applications.

7 Conclusions and Recommendations

For several agricultural commodities in the world, Brazil takes either the first or second place in terms of quantity, production cost or efficiency. It has the lowest cellulose pulp production cost in the world as can be seen in Table 3.

Nevertheless, the country needs to increase its spending on innovation to stop the deindustrialization process currently under way today. Also, it is worth to mention that the Municipal Solid Waste represents the most important source of biomass in Brazil and it has not been tapped for either energy production through direct burning, or pyrolysis, which would be the most efficient alternative.

References

Anastas, P. T., & Warner, J. C. (1998). *Green chemistry: Theory and practice* (p. 30). New York: Oxford University Press.

CEPI – Confederation of Paper Industries. (2012). *Biobased for growth a public-private partnership on biobased industries* (p. 43).

Cherian, B. M., De Oliveira, G. M., Costa, L. M. M., Souza, S. F., & Leao, A. L. (2012). Protein-based polymer nanocomposite for regenerative medicine. In J. J. Maya & S. Thomas (Org.), *Natural polymers composites* (1st ed., Vol. 1, pp. 255–293). Cambridge: The Royal Society of Chemistry.

Czernik, S., & Bridgwater, A. V. (2004). Overview of applications of biomass fast pyrolysis oil. *Energy Fuels, 18*, 590–598.

Empresa de Pesquisa Energética (Brasil). (2016). *Balanço Energético Nacional 2015: Ano Base 2014* (p. 292). Rio de Janeiro: Empresa de Pesquisa Energética. Brazilian Energy Balance 2016 Year 2015, 2016.

Essel, R., Breitmayer, E., Carus, M., Fehrenbach, H., Von Geibler, J., Bienge, K., & Baur, F. (2014). *Discussion paper defining cascading use of biomass*. R&D-project "Increasing resource efficiency by cascading use of biomass – from theory to practice". Huerth, 14 March 2014.

Farag, I. H., Laclair, C. E., & Barrett, C. J. (2002). *Technical, environmental and economic feasibility of bio-oil in New-Hampshire's north country*. UNH Bio-oil Team Report (Final Report submitted to NH IRC), August 2002.

Hassuani, S., Leal, J. M. R. L. V., & Macedo, I. C. (2005). *Biomass power generation sugar cane bagasse and trash* (1st ed., 217 p.). Brazil: PNUD – Programa das Nações Unidas para o Desenvolvimento CTC – Centro de Tecnologia Canavieira Piracicaba.

Kosmol, J., Kanthak, J., Herrmann, F., Golde, M., Alsleben, C., Penn-Bressel, G., Schnitz, S., & Gromke, U. (2012). *Glossar zum Ressourcenschutz*. Dessau-Roßlau: Umweltbundesamt.

Langeveld, H., Sanders, J., & Meeusen, M. (2010). *The biobased economy: Biofuels, materials, and chemicals in the post-oil era*. London: Earthscan.

Leao, A. L., Cherian, B. M., Narine, S., Sain, M., Souza, S. F., & Thomas, S. (2015). Applications for nanocellulose in polyolefins-based composites. In S. Mohanty, S. K. Nayak, B. S. Kaith, & S. Kalia (Org.), *Polymer nanocomposites based on inorganic and organic nanomaterials* (1st ed., Vol. 1, pp. 215–228). Hoboken, NJ: Wiley.

Mialon, L., Vanderhenst, R., Pemba, A. G., & Miller, S. A. (2011). Polyalkylenehydroxybenzoates (PAHBs): Biorenewable aromatic/aliphatic polyesters from lignin. *Macromolecular Rapid Communications, 32*(17), 1386–1392.

MINISTÉRIO DA CIÊNCIA, TECNOLOGIA E INOVAÇÃO (MCTI). (2012). *Estratégia Nacional de Ciência, Tecnologia e Inovação 2012–2015* (p. 220). Balanço das Atividades Estruturantes. Brasília, DF.

Nassif, A., Feijo, C., & Araujo, E. (2013, October). *Structural change and economic development: Is Brazil catching up or falling behind?* (UNCTAD Discussion Paper 211, p. 28).

Oasmaa, A., & Czernik, S. (1999). Fuel oil quality of biomass pyrolysis oils state of the art for the end users. *Energy Fuels, 13*, 914–921.

Petter, F. A., & Madari, B. E. (2012). Biochar: Agronomic and environmental potential in Brazilian savannah soils. *Revista Brasileira de Engenharia Agrícola e Ambiental, 16*, 761–768.

Pilz, H., Schweighofer, J., & Kletzer, E. (2005). *The contribution of plastic products to resource efficiency* (229 p.). Final Report Updated version 1.1. Plastics Europe (formerly APME) – Association of Plastics Manufacturers.

Raschka, A. & Carus, M. (2012). *Industrial material use of biomass basic data for Germany, Europe and the world* (28 p.). R&D project "Environmental innovation policy: More efficient resource use and climate protection through sustainable material use of biomass".

Sokka, L., Koponen, K., & Keränen, J. T. (2015). *Cascading use of wood in Finland – with comparison to selected EU countries* (26 p.). Research Report VTT-R-03979-15.

UNICA. (2011). *União da Indústria de Cana-de-Açúcar*. Dados e Cotações–Estatísticas. Accessed August 2015, from http://www.unica.com.br/dadosCotacao/estatistica/

Van Haen, R. (2013). *Innovations across the entire value pyramid of the biobased economy*. Amsterdam: Bio Based Press.

World Economic Forum. (2016). *Global competitiveness index (GCI) the global competitiveness report 2015–2016*. Accessed February 2016, from http://reports.weforum.org/global-competitiveness-report-2015-2016/

Zuniga, P., Negri, F., Dutz, M. A., & Pilat, D. (2016). *Conditions for innovation in Brazil: A review of key issues and policy challenges*. Workshop on innovation for productivity growth in Brazil IPEA.

Tasmania's Bioeconomy: Employing the Seven Capitals to Sustain Innovative and Entrepreneurial Agrifood Value Chains

Holger Meinke, Laurie Bonney, Katherine Evans, and Morgan Miles

Abstract Tasmania, Australia's southernmost and smallest island state, depends strongly on its bioeconomy. Currently the farm gate production of Tasmania's bioeconomy contributes around 7.4% to the overall Gross State Product (GSP). This figure is considerably higher than for Australia, where the bioeconomy contributes 2.5% to the overall Gross Domestic Product (GDP). Based on this measure, Tasmania's economy is more in line with the economies of Brazil (5.7%) or New Zealand (7.2%). It is estimated that Tasmania's bioeconomy currently contributes 16–20% of overall economic output, when taking into account the economic impact of related value chains that reach from agricultural suppliers to retailers. Government policy for economic growth in Tasmania aims to build up this sector over the following decades. To achieve the stated growth targets, technologies must be combined with business capabilities in order to effectively and efficiently commercialize innovation while maintaining sound environmental practices. A technology-driven, irrigation-led transformation is currently underway in the state, turning Tasmania's bioeconomy into a highly knowledge-intensive sector of the economy. To fully realize the economic, environmental and social potential of investment in irrigation infrastructure, there must be similar investments in research, knowledge creation, marketing, value chain innovations and capability development.

H. Meinke (✉)
School of Land and Food, Tasmanian Institute of Agriculture, University of Tasmania, Hobart, TAS, Australia

Centre for Crop Systems Analysis, Wageningen University, Wageningen, The Netherlands
e-mail: holger.meinke@utas.edu.au

L. Bonney • K. Evans
School of Land and Food, Tasmanian Institute of Agriculture, University of Tasmania, Hobart, TAS, Australia
e-mail: lawrence.bonney@utas.edu.au; katherine.evans@utas.edu.au

M. Miles
School of Management and Marketing, Charles Sturt University, Bathurst, NSW, Australia
e-mail: mmiles@csu.edu.au

S. Dabbert et al. (eds.), *Knowledge-Driven Developments in the Bioeconomy*, Economic Complexity and Evolution, DOI 10.1007/978-3-319-58374-7_7

1 Introduction

The bioeconomy[1] underpins all economic growth and development. Without the development of agriculture over 10,000 years ago, the astounding transformational shift in human behaviour that resulted in the creation of our civilisations would not have been possible. Agriculture provided the foundation on which other sectors of our economies could develop and grow. The efficiencies created by agriculture—the ability to reliably feed growing populations with fewer and fewer farmers—meant that no society has ever turned away from agricultural practices (Leith and Meinke 2013). This situation is true despite a range of undesirable impacts brought about by the agricultural revolution [see Harari's (2011) comments on 'history's biggest fraud']. As a consequence of these efficiencies, the current contribution of the bioeconomy to large and highly developed economies is only about 1 to 3% of their GDP (Table 1).

The relatively low contribution of bioeconomy to the GDPs of developed nations is a direct result of the efficiencies created by agriculture. Efficient food and fibre production permitted labour resources to be deployed elsewhere; this, in turn, created new industry sectors that now dwarf agriculture's economic value. In other words: while the *relative* economic importance of agriculture has diminished over time, the strategic importance of the bioeconomy to sustain nearly eight billion people remains. Moreover, a renewed interest in the agrifood sector has important, underlying drivers: a secure food supply; sustainable production of healthy and safe foods, together with other attributes demanded by consumers; and, increasingly, a place-based need for a social licence to operate (Prno and Slocombe 2012).

The economic, social and environmental impacts of agriculture shape our societies. Important debates about the role of agricultural systems are part of political discourse everywhere. Controversies about environmental degradation caused by agricultural production, versus agriculturists as stewards for our managed landscapes, are everyday occurrences. It is the strategic importance of agriculture as a pillar of our societies that requires particular attention in terms of policy support for research, development and education.

Here we focus on the island of Tasmania, where proportionally the bioeconomy plays the most important role of any Australian state. We argue that Tasmania is a microcosm offering insights for other societies and economies striving for higher innovation and entrepreneurial potential that leads to more profitable and sustainable production of bio-based products. Tasmania's bioeconomy is presented and interpreted in relation to its history, geography and current socio-economic

[1]Here we use the definition by the European Commission that defines the bioeconomy as *the sustainable production of renewable resources from land, fisheries and aquaculture environments and their conversion into food, feed, fibre bio-based products and bio-energy as well as the related public goods. The bioeconomy includes primary production, such as agriculture, forestry, fisheries and aquaculture, and industries using/processing biological resources, such as the food, pulp and paper industries, and parts of the chemical, biotechnological and energy industries.*

Table 1 Percent of agriculture as a contributor to gross domestic product (GDP) based on farm-gate value as well as current population numbers for a range of countries (Australian Bureau of Statistics 2013; World Bank 2015)

Country	% of GDP	Population (million)	As of
UK	0.7	64	2013
Germany	0.9	81	2013
Japan	1.2	127	2012
USA	1.3	316	2012
France	1.7	66	2013
Netherlands	2.0	17	2013
Australia	2.5	23	2013
Brazil	5.7	200	2013
New Zealand	7.2	4	2010
China	10.0	1357	2013
Fiji	12.2	>1	2013
Indonesia	14.4	250	2013
India	18.2	1252	2013
Vietnam	18.4	90	2013
Papua New Guinea	36.3	7	2012

status. Opportunities and challenges associated with agricultural intensification and new irrigation infrastructure are then explored in terms of the need to address the resource stress nexus. Contemporary innovation systems and value-chain theory are used to frame the way forward, with theories-of-change, capitals accounting and entrepreneurial thinking introduced as tools to design and implement effective interventions and innovation platforms.

2 Wheels Within Wheels: Tasmania's Bioeconomy Within an Australian Context

Australia's agricultural farm gate production contributes about 2.5% to the annual GDP (Table 1; Australian Bureau of Statistics 2013; World Bank 2015). This contribution increases to around 12% or AUD$155 billion when accounting for the value-adding processes of food, fibre and other bio-based products. The value of all economic activities that support farm production includes farm inputs; food manufacturing; transport and logistics; wholesaling and retailing; and the food service sector (National Farmers Federation 2015). The farm gate contribution of 2.5% to Australia's GDP is at the higher end for a fully developed economy, indicating the importance of renewable, primary production for Australia's bioeconomy.

Australian farmers are amongst the most efficient agricultural producers in the world. Agriculture in Australia is a knowledge-intensive sector, characterised by a high degree of mechanisation, and increasingly, automation as a result of high labour costs and often extensive landholdings. These are consequences of Australia's

Table 2 Economic snapshot of the bioeconomy's contribution to each Australian state and territory, and Australia as a whole; 2013 data: Gross State Product (GSP), GSP per person, farm gate value of the bioeconomy (Australian Bureau of Statistics 2013)

	GSP ($ million)	GSP per person	Farm gate value of bioeconomy ($ million)	Bioeconomy's contribution to GSP (%)	Population (million)
TAS	24,191	47,222	1790	7.4	0.51
SA	94,210	56,674	4805	5.1	1.66
QLD	294,548	63,840	7953	2.7	4.61
VIC	333,393	58,682	8001	2.4	5.68
NT	19,860	83,828	338	1.7	0.24
NSW	471,354	64,098	6599	1.4	7.35
WA	252,999	102,232	2530	1.0	2.47
ACT	34,414	90,631	0	0.0	0.38
Australia (GDP)	*1,524,969*	*66,549*	*38,124*	*2.5*	*22.91*

biophysical conditions (a large and dry continent with high urbanisation and a low population density) and a market economy with little or no subsidisation.

Agriculture's contribution of 2.5% to the nation's GDP masks considerable variance across the eight Australian states and territories. Economic activities are unevenly distributed and each state's contribution to Australia's GDP varies considerably, as indicated by their Gross State Products (GSP; Table 2). The GSP of each state for the 2012/2013 financial year varied from 1% for Western Australia to 7.4% for Tasmania (excluding the Australian Capital Territory (ACT) of Canberra, where no primary production takes place). There are many reasons for this diversity that go beyond the scope of this chapter. The post-farm gate contribution of Tasmania's bioeconomy to the overall economic performance of the state is estimated to range between 16% and 20% (Bennett 2015), which is proportionally much higher than for any other state in Australia.

Agrifood products from Tasmania vary from traditional commodities based on dairy, beef, sheep, vegetables, wine, fruit (such as cherries, berries and nuts), oysters, abalone and salmon to rather unique produce such as leatherwood (*Eucryphia lucida)* honey, medicinal opium poppies (Tasmania produces about 50% of the world's medicinal opiates such as morphine, codeine and thebaine), pyrethrum (75% of current world demand for pyrethrum is serviced from Tasmania) and various essential oils.

Tasmania faces constraints in transporting commodities. It is separated from mainland Australia by Bass Strait, a 350 km wide and 500 km long, relatively shallow, but often rough stretch of sea (max. depth 83 m; average 60 m) with the central bathymetric Bass Basin 120 km wide and 400 km long (Jennings 1959). Bass Strait presents an barrier to the movement of perishable primary produce that requires the coordination of multiple modes of logistics to reach national or international markets. Two state-run ferries transport tourists and freight supplemented by a private freight service with an additional two ships. The more profitable tourists compete with freight during the summer and the service lacks economies of scale to reduce costs, making it as expensive to ship a container across Bass Strait as it is to move it from Melbourne to Scotland. In January 2016, the fragility of the service was illustrated when storm damage to port facilities temporarily reduced capacity of the state-run ferries by 60% at the peak of the harvest season.

3 Geography, Climate and Soils

Tasmania is comprised of about 68,000 km^2, roughly the size of Sri Lanka or Ireland and about twice the size of Taiwan (Fig. 1a, b). Located in the 'roaring 40s', between 40°S and 44°S, 144°E and 149°E in the Southern Ocean, Tasmania has a temperate maritime climate ideally suited for a wide variety of crops, pastures, livestock production and aquaculture. Mean maximum temperatures are 18–23 °C in summer and 9–14 °C in winter (ACE CRC 2010). Average annual rainfall ranges from 2700 mm in some highland locations (as a consequence of predominately westerly winds and orographic lift) to 450 mm in parts of the central midlands,

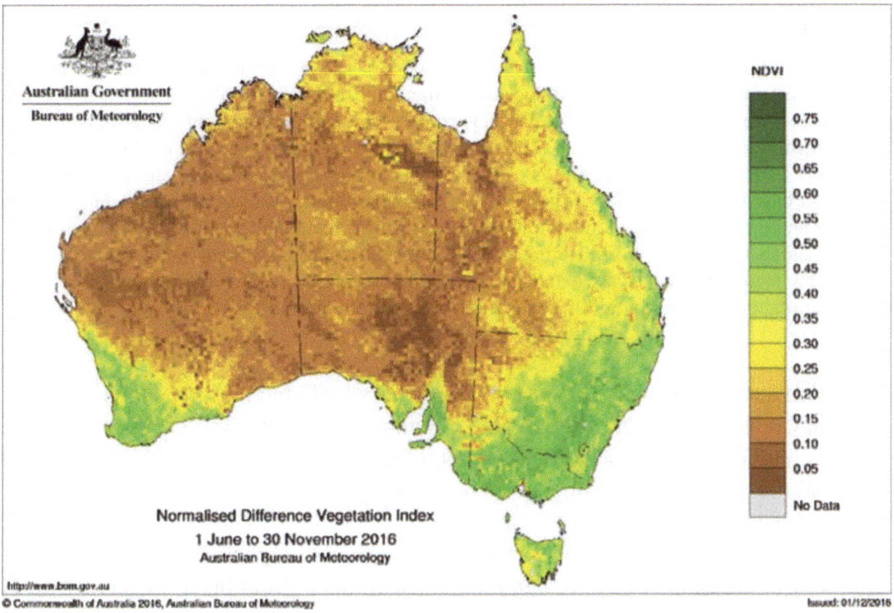

Fig. 1 Satellite image originally processed by the Bureau of Meteorology from the polar orbiting satellite NOAA—14 operated by the National Oceanographic and Atmospheric Administration (NOAA). (**a**) Normalised Difference Vegetation Index (NVDI) for the whole of Australia, six monthly average (1st June 2016 to 30th November 2016). The NVDI is a measure of vegetation

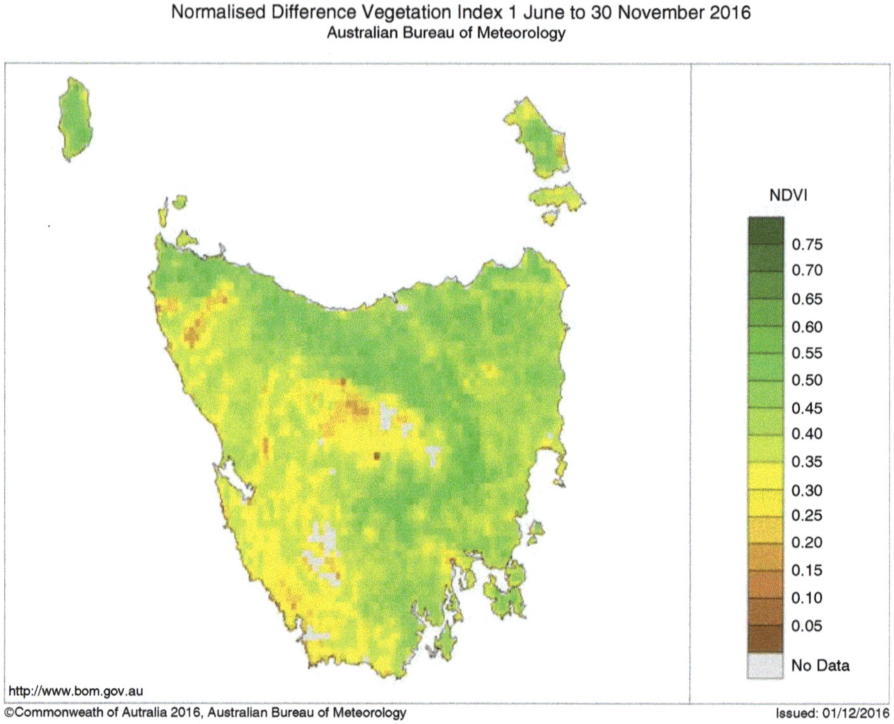

Fig. 1 (continued) cover and photosynthetic greenness based on satellite data, with cover ranging from highest to lowest. Tasmania is located to the south of the Australian mainland. Note the darker area with <25% plant cover—Australia is the driest inhabited continent on Earth, being 75% arid or semi-arid (Ummenhofer et al. 2009). (**b**) NDVI six monthly average (1st June 2016–30th November 2016) for Tasmania. The NVDI is a measure of vegetation cover and photosynthetic greenness based on satellite data, with cover ranging from highest to lowest. The *lighter* areas in the west and east are largely mountainous, forested or the uninhabited south west. National Parks, the Tasmanian Wilderness World Heritage Area, Reserved and Management Areas comprise about 45% of the state. The *darker* areas indicate the highest vegetation cover and photosynthetic rate, reflecting plant conditions in traditionally arable agricultural land in Tasmania

located in a rain shadow to the east of Tasmania's highland regions. However, these average values mask a very high degree of annual variability. Scott (1956) remarked that Tasmania's rainfall variability is greater than experienced in some other regions of the globe with similar climates such as the UK, British Columbia or the South Island of New Zealand.

Like the rest of Australia, Tasmania is strongly impacted by the El Niño–Southern Oscillation (ENSO) phenomenon, with most droughts associated with El Niño (e.g. 1914, 1965–1967, 1972, 1982–1983, 1997, 2002–2003, 2006 and 2015–2016) and floods associated with La Niña seasons (e.g. Bureau of Meteorology, http://www.bom.gov.au/climate/enso/lnlist/). In early 2016, Tasmania was in the grip of another El Niño drought following the hottest and driest spring on record (Australian Broadcasting Corporation 2015). By 2050, climate modelling suggests climate-induced increases in rainfall over Tasmania's coastal regions and reduced rainfall over central Tasmania and in the north-west (ACE CRC 2010).

Tasmania's soils are diverse due to its geological history and variations in climate, landscape and vegetation. The Tasmanian landscape is dominated by an old, erosion resistant geology that emerged about 650–1000 million years ago. Several periods of submersion under Gondwana's seas and metamorphic folding has produced rocks rich in mica and quartzite and other minerals that have been the basis for the mining industry of the west and north-west coasts. During the Tertiary Period, basins formed in the Tasmanian landscape as a result of the separation of Antarctica and the New Zealand sub-continent. This produced shallow soils on the hard dolerite hills of the Midlands and Derwent Valley in the south, and the unstable sandstone and mudstone-based soils susceptible to tunnel and gully erosion in the south-east and northern parts of the state.

Agriculture generally occurs on slopes of less than 20% below 300 m altitude in all areas of the state except the south-west wilderness region. The north-west and north-east of the state are characterised by igneous basalt and high annual rainfall (>750 mm), which has produced the characteristic brown or red ferrosol soils as well as deep, well-weathered, well-drained and friable soils that originally supported dense forests. Despite their rich colouration and apparent depth, the ferrosols are not highly fertile but provide an ideal medium for intensive cropping of vegetables, some berry and pome fruits, and the grazing of dairy and meat livestock (Scanlon et al. 1990; Doyle and Farquhar 2000; Sustainable Development Advisory Council 2002).

The rugged, incised topography and Tasmania's history of development (described below) have resulted in landholdings being relatively small (100–250 ha) in the fertile north and of marginal size for livestock commodity production in the central midlands, south and east despite gradual aggregation over the latter half of the twentieth century.

4 Some Historical Context

Tasmania was colonised by Europeans in 1803 as a penal settlement for Britain's overflowing prisons, leading to protracted confrontation and conflict with the indigenous population. Approximately 65% of the current population are descended from those convicts (Rubio et al. 2002). The average convict transported for life was freed within four years to make their own way, often thriving by their own efforts in an amazingly rugged landscape. When the transport of convicts ended in 1853, the eastern half of the state was developed on the public purse with government land grants, food for 3 years and free convict labour. Land grants were large as were the sandstone manors built by convict labour.

Until the mid-1850s, the rugged west and north-west were regarded as 'wastelands' and used as a security buffer against the very worst criminals that had been sent to an isolated west coast penal settlement. By the late 1850s, the area was starting to develop through unofficial mining, forestry and agriculture to supply the Victorian goldfields. Finally, with government sanction, the north-west developed without the support of free convict labour and was relatively neglected by the newly established bi-cameral government, enduring cycles of boom and bust through the mid to late nineteenth and early twentieth centuries (Stokes 1969; Morgan 2003; Boyce 2008; Alexander 2010).

Tasmania's recent history of European settlement exerts strong influences on the modern culture and recent development of the state. Tasmania is considered to be ethnically and socially homogeneous but suffers from highly parochial attitudes between the south, which hosts the seat of government, the north which has the wealth derived from agriculture, and the north-west which feels neglected and isolated from power. The origins of these attitudes can be traced to Tasmania's history of development over a mere six to seven generations of European settlement. Parochialism drives contemporary expectations of government support and affects policy priorities for public development (Bonney et al. 2013b).

Historically, farmers have learned to cope on their small farms and the associated risk of boom and bust cycles by producing a range of commodities. Pre- and post-World War II there were long-term investments in the state's fertile agricultural north by large multi-national food processors and marketers. Despite this investment, 'mixed farming' prevailed and opportunistic behaviour became entrenched as farmers used the leverage created from the threat of switching commodities and outlets to improve the prices they received. This led to Tasmania's agriculture being largely focused on small-scale commodity production for processing, transactional spot markets and a deeply held commitment to opportunistic behaviour.

New agribusiness companies exacerbated this situation by developing a paternalistic protection of farmers as a means of maintaining their share of the raw material market, which has shielded farmers from the change imperatives brought by globalisation. This behaviour has reinforced the impression that current global pressures for change are ephemeral because throughout history the 'busts' have always been followed by 'booms'. Farmers in Tasmania endeavour to 'wait-out' the downturns in anticipation of another upturn in economic conditions. However, in recent years, agribusiness companies have sporadically attempted to facilitate some change initiatives that have only been supported by a progressive minority of farmers.

Today, two thirds of farms employ the owner-operator only and are reliant on labour-hire contractors using international back-packers and itinerant labourers. Around 63% of Tasmanian farms have an estimated value of agricultural output (EVAO) of less than $150,000 and farm businesses rely on one or more family members working off-farm to survive. There are probably less than 1500 economically viable farms in the state and education levels are low relative to other industries, with around 5% of farmers having under-graduate qualifications and 15–20% having vocational certificate three or four qualifications (Australian Bureau of Statistics 2007).

A large proportion of the current generation of farmers are approaching retiring age. Consequently most are unwilling to change their current business models or consider alternative business structures or practices to achieve the economies of scale that would enhance their cost-competitiveness and enable access to new markets, develop innovative value-adding or new products (Bonney 2006, 2011). It appears that for some agricultural commodities the combination of farmer attitudes, demographic aging and processor paternalism has resulted in a transactional or resistant form of supply chain 'followership' in response to an increasingly transformational leadership by agribusiness and paradigmatic global change (Defee 2007; Bonney 2011). Transformational followership might sometimes be more important than leadership in developing adaptive, high performing value chains during times of rapid change (Defee 2007).

Given this history, the state's population has developed a highly resilient, independent but parochial culture, focused on gaining government support for their endeavours. To this day, despite in-migration, belief systems in Tasmania often manifest as an 'entitlement culture', risk averse and isolated, but with economically important pockets of entrepreneurship and innovation. Notwithstanding these constraints, the state's farmers are highly efficient in commodity production and technically advanced due to a history of high quality publicly funded research and extension. A welcome addition has been the more recent development of a large and innovative agricultural consulting sector (Bonney et al. 2013b, 2016).

Tasmania's bioeconomy is characterised by diversity as a consequence of its geography, history, climate and other geo-political factors. Its relatively small size, surrounded by the pristine waters of the Southern Ocean, and the distance from mainland Australia means that extensive agriculture based on low-value bulk commodities are generally not economically viable.

5 Tasmania's Awakening: Opportunities and Challenges

Tasmania is undergoing a phase of unprecedented intensification and transformation of its primary production sector by rapidly developing a reputation for high quality, often niche products, value adding, agri-tourism, fine food and beverages founded on a reputation for having "clean and green" food safety. A further increase in the profitable and sustainable production of these and other bio-based products requires entrepreneurism; functional, co-innovative and transparent value chains; innovative business models; proactive risk management; and knowledge creation and collaboration to achieve market access. Future prosperity in Tasmania requires transformational change that encompasses both the technical and social domains and focuses on delivering superior value to consumer segments.

Hence, the challenge for the future of Tasmanian agrifood producers encompasses overcoming the socio-cultural as well as geographic and economic constraints. Meadows (1999) identified that the most successful system interventions are those that jump the paradigm, change the mindset (values, attitudes, goals, structure), the rules of the system (incentives and dis-incentives), and/or the structure of information flows. The generational change that is currently underway in the industry offers the opportunity for new, non-traditional farmers, many not from Tasmania and often without any agricultural background. These new farmers bring new ideas, new attitudes, new ways of working, and new means of funding to address Tasmania's challenges. Tasmanian-born professionals returning to the state after careers elsewhere, and often for lifestyle reasons, are also shaping the state's future.

It is also incumbent on those who research and support the agricultural system to develop a new paradigm of engaging with industry through openness and responsiveness to develop new, practical systems-focused research outcomes that meet both business and educational needs. Modern approaches to agricultural innovation involve the development of an 'agricultural innovation system', a set of principles, analyses and actions that facilitate the identification, design and implementation of investments, approaches and interventions that promote innovation. Hence, consistent with systems theory, all relevant actors are affected by changes in the system (Ashby 1957; Von Bertalannfy 1968).

6 Adding Water Is Not Enough

Here we briefly outline some of these challenges using, by way of example, the current rollout of new irrigation schemes across Tasmania and in the context of current government policy. More than 150 GL of new irrigation water will be available when all the schemes become fully operational. Already irrigation contributes to approximately 60% of the gross value of agricultural production.

The Tasmanian Government's AgriVision 2050 policy (Tasmanian Liberals, https://www.tas.liberal.org.au/sites/default/files/policy/Cultivating%20prosperity%20in%20agriculture.pdf) sets a substantial stretch target for Tasmania's bioeconomy, namely increasing the farm gate to an annual value of AUD$10 billion by 2050, up from AUD$1.8 billion in 2012/2013 (Table 2). Although this vision is underpinned by significant investment in irrigation infrastructure—about AUD$500 million of private and public funds have already been invested in new irrigation schemes—realising and sustaining the benefits will require substantial investment in knowledge infrastructure, innovation platforms, value chain approaches, benchmarking and monitoring. These efforts have already resulted in an increase of nearly 5% of agrifood exports from Tasmania in 2015 to a total value of $2.74 billion (Parliament of Australia 2016). In part, this is the result of the once-off increase due to the increased irrigation capacity (Fig. 2).

Continuous value adding is now required to keep up the momentum and to maintain the growth rates required for achieving the governments vision (Fig. 2).

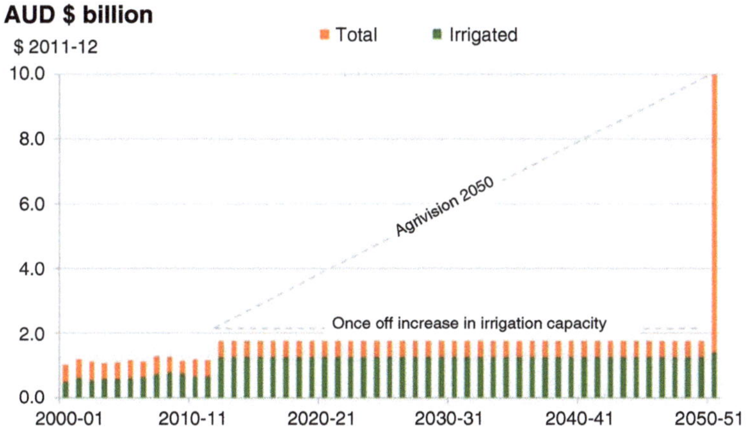

Fig. 2 Tasmania's research and innovation challenge: the AUD$10 billion target for 2050, commencing from the increase in irrigation capacity for 2010–2011. Innovation via research, development and expansion, in addition to increased irrigation capacity, is essential to facilitate investment and expansion in Tasmania's bioeconomy. Source: Tasmanian Government's AgriVision 2050 policy (http://bit.ly/1MxuovX) and R. Nelson, pers. com. 2015

This necessitates a dramatic increase in the value derived from each litre of irrigation water. For example, if 80% of the $10 billion target is to be achieved via irrigated agriculture, the value generated from irrigation water has to increase from currently AUD$3500 to $16,000 per mL of water. This would require an extensive step-change in productivity. But is that the only solution? Or, does the answer lie to a large extent in the agrifood system moving into a new paradigm of creating consumer value through advances in linked production, logistics and marketing into new, targeted niches around the world?

The obvious pathway for achieving such an ambitious vision is through applied and highly industry relevant research, development and extension. Researchers from multiple disciplines must work with industry, community and policy makers to achieve such a transformation. A key question for Tasmania is how such intensification can be economically, socially and environmentally sustainable. Using an Agricultural Systems Research (ASR) approach, industry experts, academics, farmers, policy makers and representatives of Tasmania's broader civil society are jointly investigating improvements in four key areas:

1. On farm systems (productivity, management systems, precision technologies, new crops and processes);
2. Business models and value chains (innovations, entrepreneurship and exporting);
3. Natural resource management (landscape health, ecosystem production, maintaining soil productivity, drainage, waterlogging, salinity, interactions between on-farm and landscape scale, biosecurity); and
4. Research, development, extension and education (arrangements and institutions, effective innovation, education and adoption).

The conceptual model 'just add water' is unlikely to result in the desired growth in economic development and value chain creation from irrigation. A value chain is comprised of linked businesses where the chain partners decide to co-innovate in order to create and deliver value for which their customers and consumers will pay a premium price (Bonney 2011). Entrepreneurial value chains must be created by aligning the strategic interests of knowledgeable and technically skilled farmers, input suppliers, value adding processors and marketers through public-private partnerships that are based on trust, shared values and co-innovation.

7 Systems Within Systems: Integrating Knowledge, Innovation and Entrepreneurship

Tasmania's situation exemplifies how modern agriculture and aquaculture are now high knowledge intensive systems that can no longer rely on single transformational innovations such as the ones that powered the green revolution of the 1960s and 1970s. Norman Borlaug's contribution to agricultural science and plant breeding at the time resulted in high-yielding, disease resistant crops that saved about a billion people from starvation. Borlaug and colleagues managed to find a very effective technological fix to overcome resource limitation. Much research, thinking, knowledge and insight went into the creation of these green revolution technologies resulting in an unprecedented increase in food production. Yet their application was relatively simple and little additional knowledge was required to deploy these technologies at farm level.

Now, during the first quarter of the twenty-first century, the challenge to our agricultural and food systems is different. This time it is not only about increasing yields per area. Instead the challenge is to increase productivity, rather than just production, but without additional resources and without negative environmental or social impacts. Today we are confronted with what the Shell Oil's scenario planning group (Shell Scenarios Team 2013; Bentham 2014) terms 'resource stress nexus (RSN)'. The RSN refers to increasing pressures on water, food, and energy resources to meet the demands of an expanding global population. These demands include rapid changes in consumer preferences requiring more of all three of these interlinked resources. The RSN impacts on food and energy production as well as the viability of urbanisation. All these sectors increasingly compete for the same resources.

Tasmania is uniquely positioned in the Australian bioeconomy to be able to exploit the RSN with abundant water, arable soils, and an economy based on a renewable hydro-electric power grid. Further, modern Tasmania is characterised by strong, cooperative partnerships across the research, education, policy and private sectors via a dynamic joint venture between the Tasmanian Government and the University of Tasmania. This vibrant partnership has led to the establishment of innovation networks that span the public and private sectors and serve as an example of institutional innovation (OECD 2016; Tasmanian Institute of Agriculture 2016). Tasmania's small but mobile population and accessible government foster this relationship and the science-government-industry interactions that are needed for effective, economy-wide research and development. The island's agrifood system provides an ideal platform that demonstrates the effectiveness of boundary organisations, i.e. small groups of committed scientists, policy-makers and industry leaders who are all concerned with translating science into action (Cash et al. 2003).

Exploiting the RSN presents issues that go beyond production. There is a growing need to harness both bio-physical and social dimensions in system solutions that focus on increasing total economic yield for whole value chains. The locus of competition in modern value chains is increasingly shifting from single businesses to whole chains. Products are only as competitive as the whole chain that delivers them to a consumer. As a result, value chains are now regarded as recursive, interconnected networks that reflect the hierarchy of emergent properties in the overall food system (Collins 1999; Li and Wang 2007). Whilst they are frequently depicted as linear sequences of processes, it is now recognised that value chains are actually complex networks of relationships, both internal and external to the chain (Moore 1993; Lazzarini et al. 2001). These relationships assist the chain participants in acquiring the tangible and intangible resources needed to innovate, collaborate and compete, often simultaneously (Allee 2008; Fig. 3). A permutation of the value-chain network is the use of digital technologies to bring the consumer closer to the producer through novel marketing strategies, social media and/or direct investment by the consumer in business development. What could be more important for Tasmania's bioeconomy, given the island's sparse population and logistical challenges, than effectively managing value chain relationships and interdependencies?

Continuous innovation requires resources beyond the capabilities of a single farm. Hence, producers within chains are collaborating vertically to solve their shared problems and to exploit their opportunities. They are also collaborating horizontally with governments, research institutions and even their own competitors where they do not directly compete, and there is sufficient common interest. For example solving transport problems such as getting goods from Tasmania across Bass Strait to the port of Melbourne (Mason et al. 2007).

While food security is a fundamental driver, vertical and horizontal integration also serves to incorporate the unprecedented diversity in consumer demand and interest in food quality. The production, preparation and consumption of food is inextricably related to social identity. Food knowledge is entering a new era when diffuse, ill-defined and often misleading concepts such as 'organic', 'local', 'wild', 'sustainable', 'healthy' and 'national' are being influenced and are influencing global food consumption and hence the nature of production, transport and marketing (Rhea and Bettles 2012). The Asian food boom now has tangible, local and global impacts, with demand for some products (e.g. baby formula) outstripping supply (SBS News 2015). Geopolitics has put Australia at the forefront of these developments. Impacts such as increased demand are particularly noticeable in Tasmania due to its low population and high reliance on the agrifood sector.

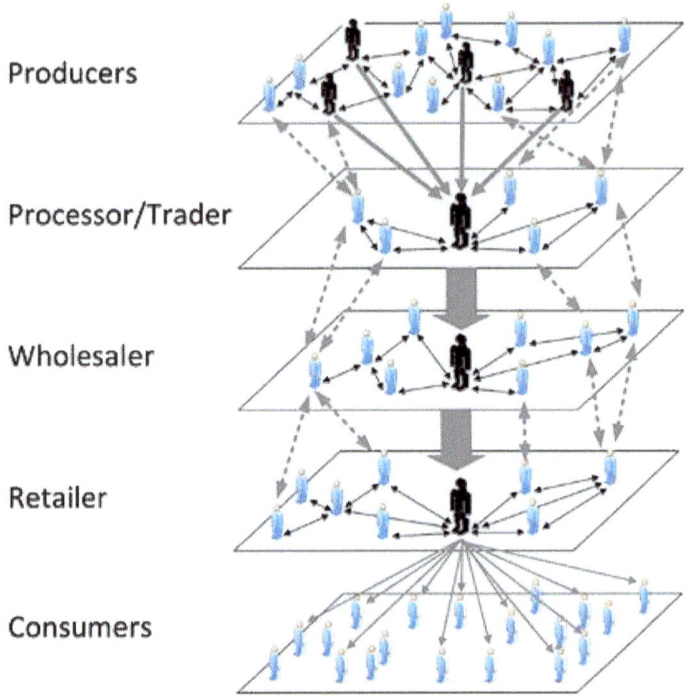

Fig. 3 A value chain network depicting the vertical and horizontal interactions between producers and consumers. Source: derived from netchain (Lazzarini et al. 2001); value network (Allee 2008); and business eco-systems concepts (Moore 1993)

Agricultural value chains will need more skilled intermediaries to foster knowledge flows and to build trust and productive relationships. Those seeking knowledge-based services will need to make active choices about why, how and what they access. In Tasmania the passive receipt of information and historical sense of entitlement to publically-funded services for private benefit will become increasingly uncommon. Attitudinal changes are inevitable. Institutional arrangements need to: (1) foster a more effective linkage between the research capabilities at the University of Tasmania and the needs of farmers, advisors, agribusinesses and communities; and (2) recognise that innovation in agriculture is facilitated through value chains that co-innovate for mutual benefit.

Innovation relating to Tasmania's bioeconomy requires an environment conducive to the interplay between society, producers and industry. The current needs of society are embodied in market choices, the regulatory environment and a social licence for agriculture. The practical needs and concerns of Tasmanian producers in securing markets and creating profit will ultimately influence receptivity to change and disruption. Agribusiness systems will be strongly shaped by information and innovation that influences the options available.

Producers themselves have aspirations and capacity that will affect the uptake of innovation. Such entrepreneurism requires the right attitude and skills from everybody involved in Tasmania's value chain. While some attitudes will and need to change, skills must also be developed. This relies on a supportive, accessible and inclusive education system that caters for all educational needs—from primary school to vocational training to associate, graduate and post-graduate qualification. It also requires a supportive and responsive policy environment that clearly articulates the Government's role in this process and helps to overcome a deep seated entitlement culture that has far too often stifled entrepreneurial spirit in Tasmania (West 2013).

8 Solutions to Complex Problems: Innovation Platforms and a Theory of Change

Rural Australia and Australia's agrifood sector faces unique, complex problems that require development approaches emphasising endogenous or local rather than external interventions (Khisty 2006; Tomaney 2010) to address the 'wicked problems' (Rittel and Webber 1973) i.e. problems that can only ever be partially resolved. One approach is to use 'place-based' strategies that identify how the unique attributes of individual places determine the constraints, and the tangible and intangible assets that influence the development of comparative advantages and fosters innovation (Ryser and Halseth 2010; Woods 2012).

Traditional approaches have focused on economic analyses with little consideration of the non-traded, knowledge-based intangibles that may be used to construct regional advantage. A proportion of this knowledge is tacit, meaning it cannot be fully codified or documented. Taking such tacit knowledge into account could create meaningful government interventions that go beyond fragmented policies and simple 'place-branding', instead focusing on coordinated, holistic strategies that facilitate supportive, top-down, regional, whole-of-government policy support and community cooperation (Bachtler 2010). In particular, rural-urban networking and capacity have been found to be important to regional innovation (Dabson 2011; Pritchard et al. 2012) as long as the influence of individual values, beliefs and norms on adaptive behaviour in regional change are well understood and accounted for in the planning stage (Raymond et al. 2011). Long-term approaches that focus on innovation, facilitate the active involvement of stakeholders and develop human capital are essential (Tomaney 2010).

Innovation platforms are a critical success factor for constructive dialogues and capability development (Ekboir and Rajalahti 2012). They draw on networks of diverse public/private actors who voluntarily contribute the necessary resources and facilitate innovation. Such innovation platforms are often a consequence of effective boundary organisations forming action-oriented communities of practice (here we expanded the boundary organisation model discussed by Guston (1999) by including private sector actors as a third partner to the science—government model). A very practical outcome of innovation platforms is the construction of competitive advantages (McCall 2009; Eversole and McCall 2014) by creating business ecosystems. Bonney et al. (2016) advocated this approach based on Emery and Flora's (2006) seven capitals (natural, cultural, human, built, financial, political and social capital) as the basis for analysis and the development of a more entrepreneurial, innovative agrifood industry in Tasmania. The capitals framework is a typology to understand and analyse a community's fundamental building blocks of development from a systems perspective, and provides a lens to consider both the inhibitors and enablers of innovation in the agricultural sector.

Continuing research into regional agricultural entrepreneurship and innovation, that has compared seven Australian regions, suggests that there are similarities and differences in regional factors that drive innovation. Even the most common factors have a varying emphasis from region to region (Bonney et al. 2013b, 2015, 2016). The research also suggests that there is a sequential influence of capitals in 'place-based development' that moves from 'foundational' natural, cultural and social, to the 'enabling' political and financial capitals. When these influences align, they can create the human capital necessary to produce and deliver place-based products and services.

These broader systems approaches are particularly important as Tasmania experiences an entrepreneurial renewal with an emerging cultural tourism industry driven by the Museum of Old and New Art or MONA (Lehman et al. 2016; MONA 2016). Against the background of these cultural changes, a portfolio of new, agricultural-based ventures is also emerging. These ventures market high value, premium agricultural products such as organic farmed Atlantic salmon, ultra-fine wool, Wagyu and organic beef, artisan cheese, leatherwood honey, flowers, pyrethrum, pharmaceuticals, cherries, high quality whisky, gin and vodka, fine wines, craft beers and ciders.

We know from experience that more knowledge doesn't necessarily lead to better action; the 'know-do gap' within agriculture and food systems has been widely documented. For example, knowledge about the causes of diet-related diseases does not necessarily change peoples' eating habits. Behaviour change requires empowerment through shared knowledge and individual attitudes, and a supportive culture to convert intention to action (Fishbein and Ajzen 1975, 2010). Individuals must be given the capacity to actively contribute to their aspired outcomes (such capacity includes the availability, affordability and acceptability of, for instance, healthier alternatives). Empowerment leads to informed decision-making and in doing so creates value. Hence, empowerment can lead to well-reasoned action or in-action.

Cooke (2007), an architect of regional innovation systems in Europe, has suggested a framework for policy platforms to assist the development of 'constructed advantage'. It is a process of further developing existing social capital that produces not only product innovation but also local governance that enables innovation to occur. According to Cooke (2007) this involves a strategic focus on:

- **Economy**: proactively 'constructing' future sources of economic competitiveness in the region with: inter-firm interactions; integration of knowledge generation; and, both local and global business networks.
- **Multi-level governance**: seeking out governance mechanisms that support a proactive approach to the region's future by: alignment of stakeholder interests and management of expectations; strong policy-support for innovators; enhanced budgets for outcomes-focussed research; and vision-led, principled and ethical policy leadership.
- **Knowledge Infrastructure**: the active involvement of knowledge-based organisations in constructing advantage in the region through horizontal co-innovation solving shared problems.
- **Community and culture**: community and public, cultural orientation toward proactivity, entrepreneurship and innovation.

Coordination of these policy instruments in order to achieve desired outcomes is often one of the biggest challenges for an institution. This is where well-planned and effective interventions based on a theory of change (ToC) and effective foresighting can help.

The ToC approach is '... *a comprehensive description and illustration of how and why a desired change is expected to happen in a particular context ... it does this by first identifying the desired long-term goals and then works back from these to identify all the conditions (outcomes) that must be in place (and how these relate to one another causally) for the goals to occur ...*' (Centre for Theory of Change 2016).

In practice this can be achieved by strategic foresight which anticipates future events by articulating possible, plausible, probable and preferable futures as demonstrated for the rollout of irrigation infrastructure in Tasmania (OECD 2016). Foresight illuminates the implications of present actions thus helping to avoid problems and develops plans to achieve the preferred future (Voros 2003; Slaughter 2004). More importantly though, the process facilitates the development of anticipatory individual mental models and group cultures that enable an agility to cope with the environmental drivers, critical uncertainties and wildcards (low probability, high impact events) which drive their emergent future trajectory. In doing so, foresighting facilitates unified, coordinated action in a general direction with an ability to cope with the unexpected (Ingvar 1985; MacKay and McKiernan 2004). In this context, foresighting effectively becomes a 'construction of the future' (McCall 2009) rather than allowing serendipity to prevail or, more to the point, its antonym, zemblanity to dominate (i.e. situation where humanity constructs their own misfortune in the systems designed to avoid it; Giustiniano et al. 2016).

Hence, stakeholder engagement must happen at the onset and, ideally, throughout the development process to generate additional means and ends. A useful entrepreneurial method is effectuation logic (Sarasvathy and Venkataraman 2011) in which actors use the status quo ("what I know, who I know, and who I am") as a starting point for the creation of a preferred, new future. Adaption of this method to policy making in Tasmania suggests that policies become most effective when they are co-created with stakeholders allowing a preferred future to emerge as contingencies are leveraged and new partnerships are created.

Effectuation consists of five principles (Society for Effectual Action 2016):

1. "Bird-in-Hand" or always start with the means you can control;
2. "Affordable Loss" to control the downside risk by using partners and pre-committed stakeholders to co-create a policy draft;
3. Using wildcard surprises in policy to create a new insight into systems dynamics and then use this to create new policy opportunities;
4. Co-create a policy draft via strategic conversations with stakeholders willing to put "skin in the game" and risk financial, political, and reputation loss; and,
5. A "Pilot-in-the-plane" approach based around a philosophy that the future is created rather than predicted and as such can be shaped to create a better future for all.

Effectuation leads to a divergent and expanding effectual cycle that is recursive, dynamic and flexible, and that results in both new outcomes and new means. It takes into account the dynamic of messy, human interactions and relationships that exist in real life (Bonney et al. 2013a).

In summary, methods such as strategic foresighting and effectuation logic allow people to think ahead and consider, model, explore, create and respond to future eventualities. The process includes questioning ingrained assumptions and (often limiting) beliefs that underpin current strategy. Usually behaviour change only occurs once a series of pre-conditions are met. Particularly in group settings, practice change is contingent on positive experiences for any change in participants' knowledge, attitudes, skills and/or aspirations (KASA). KASA-level change is regarded as a pre-requisite to practice change.

The effectiveness of these methodologies was acknowledged by the OECD's Observatory for Public Sector Innovation as part of their 'Stakeholder Engagement for Inclusive Water Governance' series (OECD 2016). In this series the OECD has recognised the leadership of the Tasmanian Government and the Tasmanian Institute of Agriculture, a boundary organisation that conducts extensive stakeholder consultations across a broad range of local actors (business, service providers, farmers, civil society, etc.). In response to three divergent foresighting scenarios, the team established well-defined and agreed irrigation research priorities supported by all stakeholders. This research and development (R&D) coalition is now delivering the knowledge infrastructure needed to compliment hard investments in water resource infrastructure. The coalition also needs to establish how Tasmania can avoid the problems that have historically plagued irrigation: salinization, water logging, erosion and the over exploitation of water resources.

In pursuing food security as a pillar of Tasmania's bioeconomy, we need to recognise that there are top-down constructs that will be shaped and influenced by policy and institutional settings. In the end, however, embedded food-systems will be implemented by farmers, agri-business leaders and processors pursuing economic ends. Non-food products could play an increasingly important role in Tasmania's bioeconomy as the current examples of medicinal alkaloids and pyrethrum production already demonstrates. Technological disruptors such as synthetic biology and the use of microbes to produce plant-derived chemicals will further test the adaptive capacity of the agrifood sector. This will be an area of rich social narrative, and in the process, norms, values and world views are and will be challenged and nearly every proposed 'solution' is likely to be contested at some level (Leith and Meinke 2013). Trade-offs will be inevitable, and compromises will have to be reached, particularly in instances where a farmers' economic viability is often driven by short term gains that can compromise their long term sustainability. There will be a need to resist short term solutions that more readily attract resourcing than longer, but more sustainable approaches.

Tasmania's bioeconomy is a microcosm that offers insights for other societies and economies in transition. The ability for Tasmanian agricultural value chains to be innovative and entrepreneurial is derived from its natural capitals such as its land and water resources. Effective innovation (OECD 2005) that creates and commercialises "new or significantly improved product (good or service), process, new marketing method or a new organisational method in business practices" strongly depends on such natural capital and needs to be paired with adequate human, financial, and social community capital. While Tasmania's natural capital requires appropriate governance and protection (e.g. sound and enforced environmental protection laws and effective biosecurity measures), the cultural and human capitals require development and nurturing. For Tasmania this means a particular emphasis on improved levels of education in order to overcome some of the deeply ingrained cultural impediments to sustainable development. A focus on education and its governance is critical for an island that has a worryingly high rate of functional illiteracy and welfare dependency (Rigney 2013).

9 Conclusions

Over the last decades Tasmania's bioeconomy has moved from a situation where knowledge came embedded in the inputs delivered to the farm (e.g. hybrid seeds, mineral fertilisers, etc.) to a situation where farmers now need to be highly skilled, knowledgeable, technologically savvy and digitally connected if they want to partake in the bio-based revolution that is taking place. Opportunities abound but engagement and investment decisions are not simple, markets and value chains are globalised and production methods are more scrutinised which determines market access. The challenges ahead will increasingly be characterised by technical complexity, uncertainty, a mix of social, economic and biophysical drivers, abundant data and information of variable quality. Contested issues among diverse stakeholders will create additional challenges.

Tasmania can have a very bright and vibrant future. With good governance of all the components that make up Tasmania's agriculturally-based value chains, the island may get close to the vision articulated by Government in 2015: a more than fivefold increase in the value of Tasmania's bioeconomy by 2050. To make this future a reality, the relevant actors need to first imagine it. The community need to agree on what they want and then jointly figure out and commit to pathways that will get them there. Arguably, the age of business-as-usual with occasional change management is over, necessitating structures that enable ongoing adaptation, knowledge and risk management. Tasmania has all the ingredients and tools for this task, especially the natural capital and human potential. Here we have outlined some of the well-tested principles and approaches to research, industry development, knowledge creation and policy development that can achieve the desired vision, create acceptable compromises, build socio-economic resilience and, ultimately, create a better future. All we need to do is make it happen.

References

ACE CRC. (2010). *Climate futures for Tasmania general climate impacts: The summary*. Antarctic Climate and Ecosystems Cooperative Research Centre, Hobart, Tasmania. Accessed February 21, 2016, from http://acecrc.org.au/climate-futures-for-tasmania/

Alexander, A. (2010). *Tasmania's convicts: How felons built a free society*. Crows Nest: Allen and Unwin.

Allee, V. (2008). Value network analysis and value conversion of tangible and intangible assets. *Journal of Intellectual Capital, 9*(1), 5–24.

Ashby, W. R. (1957). *An introduction to cybernetics* (2nd ed.). London: Chapman and Hall.

Australian Broadcasting Corporation. (2015). *Weather Bureau confirms the driest spring on record for Tasmania*. http://www.abc.net.au/news/rural/2015-12-01/weather-bureau-confirms-the-dryest-spring-season-on-record/6990794

Australian Bureau of Statistics. (2007). *Census of population and housing 2006*. Australian Bureau of Statistics, customised dataset commissioned in 2008. Accessed February 21, 2016, from http://abs.gov.au/websitedbs/censushome.nsf/home/historicaldata2006

Australian Bureau of Statistics. (2013). Accessed February 27, 2016, from http://www.abs.gov.au/AusStats/ABS@.nsf/MF/5220.0

Bachtler, J. (2010). Place-based policy and regional development in Europe. *Horizons, 10*(44), 54–58.

Bennett, M. (2015). *Tasmanian food and beverage industry scorecard 2012–13*. Hobart, Tasmania: Department of Primary Industries, Parks, Water and Environment.

Bentham, J. (2014). The scenario approach to possible futures for oil and natural gas. *Energy Policy, 64*, 87–92.

Bonney, L. B. (2006). *Industry training demand profile: Agriculture*. Hobart, Tasmania: Government of Tasmania. Accessed February 27, 2016, from http://www.education.tas.gov.au/vet/employers/industradvice/training_demand_profiles

Bonney, L. B. (2011). *Insights into "mysterious processes": Incentivising co-innovation in agrifood value chains*. Hobart, Tasmania: University of Tasmania.

Bonney, L., Collins, R., Verreynne, M. L., et al. (2013a). A short note on entrepreneurship as an alternative logic to address food security in the developing world. *Journal of Developmental Entrepreneurship, 18*(3). doi:10.1142/S1084946713500167.

Bonney, L., Castles, A., Eversole, R., et al. (2013b). *A regional agricultural development strategy for North West Tasmania: Milestone 3 Report*. RIRDC PRJ-008839 Rural Industry Research and Development Corporation, Canberra, ACT.

Bonney, L., Castles, A., Eversole, R., et al. (2015). *Accounting for agriculture in place-based frameworks for regional development: A value assessment and development framework and toolbox for building constructed advantage in agriculture based regions 15/002*. Canberra: Rural Industries Research and Development Corporation. ISBN: 978-1-74254-792-3.

Bonney, L., Adhikari, R., Woods, M., et al. (2016). *Rapid assessments of regional primary industry competitiveness: Entrepreneurial value assessment and development (EVAD) framework for regional development*. Canberra: Rural Industries Research and Development Corporation.

Boyce, J. (2008). *Van Diemen's Land: A history*. Melbourne: Black Inc.

Bureau of Meteorology. *La Niña – Detailed Australian analysis*. http://www.bom.gov.au/climate/enso/lnlist/

Cash, D. W., Clark, W., Alcock, F., et al. (2003). Knowledge systems for sustainable development. *Proceedings of the National Academy of Sciences, 100*, 8086–8091.

Collins, R. (1999). *Developing supply chains in new agricultural industries: A social systems approach*. Paper presented at Country Matters national conference on the role of the social sciences in developing policy and programs for Australia's rural industries, Canberra, ACT, May 20–21, 1999.

Cooke, P. (2007). To construct regional advantage from innovation systems first build policy platforms. *European Planning Studies, 15*(2), 179–194.

Dabson, B. (2011). Rural regional innovation: A response to metropolitan-framed place-based thinking in the United States. *Australasian Journal of Regional Studies, 17*(1), 7–21.

Defee, C. C. (2007). *Supply chain leadership*. Knoxville: University of Tennessee.

Doyle, R., & Farquhar, D. (2000). Tasmanian viticultural soils and geology. *New Zealand Society of Soil Science, 3*, 43–44.

Ekboir, J., & Rajalahti, R. (2012). *Agricultural innovation systems: An investment sourcebook, International Bank for Reconstruction and*. Washington, DC: Development/International Development Association or The World Bank.

Emery, M., & Flora, C. (2006). Spiralling-up: Mapping community transformation with community capitals framework. *Community Development Journal, 37*(1), 19–35.

Eversole, R., & McCall, T. (2014). Constructing advantage in the Cradle Coast region, Tasmania: Knowledge partnering as a regional development platform approach. *Regional Science Policy and Practice, 6*(3), 251–263.

Fillis, I., Lehman, K., & Miles, M. P. (2016). The museum of old and new art: Leveraging entrepreneurial marketing to create a unique arts and vacation venture. *Journal of Vacation Marketing*, 1–12. doi:10.1177/1356766716634153.

Fishbein, M., & Ajzen, I. (1975). *Belief, attitude, intention and behavior: An introduction to theory and research*. Reading, MA: Addison-Wesley.

Fishbein, M., & Ajzen, I. (2010). *Predicting and changing behavior: The reasoned action approach*. New York: Psychology Press/Routledge Taylor Francis Group.

Giustiniano, L., e Cunha, M. P., & Clegg, S. (2016). Organizational zemblanity. *European Management Journal, 34*(1), 7–21.

Guston, D. (1999). Stabilizing the boundary between US politics and science: The role of the Office of Technology Transfer as boundary organization. *Social Studies of Science, 29*, 1–25.

Harari, Y. N. (2011). *Sapiens – A brief history of humankind* (p. 498). New York: Penguin Random House.

Ingvar, D. (1985). Memory of the future: An essay on the temporal organization of conscious awareness. *Human Neurobiology, 4*, 127–136.

Jennings, J. (1959). The submarine topography of Bass Strait. *Proceedings of the Royal Society of Victoria, 71*(1), 49–72.

Khisty, C. (2006). A fresh look at the systems approach and an agenda for action: Peeking through the lens of Churchman's aphorisms. *Systemic Practice and Action Research, 19*(1), 3–25.

Lazzarini, S. G., Chaddad, F. R., & Cook, M. L. (2001). Integrating supply chain and network analyses: The study of netchains. *Journal on Chain and Network Science, 1*(1), 7–22.

Lehman, K., Fillis, I., & Miles, M. P. (2016). Innovative new art venture creation: A case study analysis. *Journal of Vacation Marketing*.

Leith, P., & Meinke, H. (2013). Overcoming adolescence: Tasmania's agricultural history and future. *Griffith Review, 39*; Tasmania: The tipping point? (Online ed.). Accessed February 27, 2016, form http://griffithreview.com/edition-39-tasmania-the-tipping-point/overcoming-adolescence

Li, X., & Wang, Q. (2007). Coordination mechanisms of supply chain systems. *European Journal of Operational Research, 179*(1), 1–16.

MacKay, R. B., & McKiernan, P. (2004). The role of hindsight in foresight: Refining strategic reasoning. *Futures, 36*, 161–179.

Mason, R., Lalwani, C., & Boughton, R. (2007). Combining vertical and horizontal collaboration for transport optimisation. *Supply Chain Management, 12*(3), 187–199.

McCall, T. (2009). *Spatial innovation in Tasmania: Constructing advantage through regional development platform methods (RDPM) Australian Innovation Research Centre*. Hobart, Tasmania: University of Tasmania.

Meadows, D. (1999). *Leverage points: Places to intervene in a system*. Hartland, VT: The Sustainability Institute.

MONA. (2016). Accessed February 21, 2016, from https://themonaeffect.wordpress.com/

Moore, J. F. (1993). Predators and prey: A new ecology of competition. *Harvard Business Review, 71*(3), 75–83.

Morgan, S. (2003). *Land settlement in early Tasmania: Creating an antipodean England*. Cambridge: Cambridge University Press.

National Farmers Federation. (2015). Accessed February 21, 2016, from http://www.nff.org.au/farm-facts.html

OECD. (2005). *Oslo manual: Guidelines for collecting and interpreting innovation data* (3rd ed.). OECD and European Commission.

OECD. (2016). Accessed February 27, 2016, from https://www.oecd.org/governance/observatory-public-sector-innovation/innovations/page/participatoryfore-sightingforirrigationrdplanning.htm#tab_description

Parliament of Australia. (2016). *Questions without notice: Electorate of Bass: Trade* (Nikolic, Andrew, MP) by The Hon Malcolm Turnbull, PM, Government of Australia. Accessed February 27, 2016, from http://parlinfo.aph.gov.au/parlInfo/search/display/display.w3p;db=CHAMBER;id=chamber/hansardr/aaec4b82-2411-445e-9df4-a787c30c60c2/0140;query=Id%3A%22chamber/hansardr/aaec4b82-2411-445e-9df4-a787c30c60c2/0000%22

Pritchard, B., Argent, N., Baum, S., et al. (2012). Local–if possible: How the spatial networking of economic relations amongst farm enterprises aids small town survival in rural Australia. *Regional Studies, 46*(4), 539–557.

Prno, J., & Slocombe, D. S. (2012). Exploring the origins of 'social license to operate' in the mining sector: Perspectives from governance and sustainability theories. *Resources Policy, 37*(3), 346–357.

Raymond, C. M., Brown, G., & Robinson, G. M. (2011). The influence of place attachment, and moral and normative concerns on the conservation of native vegetation: A test of two behavioural models. *Journal of Environmental Psychology, 31*(4), 323–335.

Rhea, M., & Bettles, C. (2012). Future changes driving dietetics workforce supply and demand: Future scan 2012-2022. *Journal of the Academy of Nutrition and Dietetics, 112*(3), S10–S24.

Rigney, S. (2013). Big thought and a small island. *Griffith Review 39*; Tasmania: The tipping point? (On-line ed.). Accessed February 21, 2016, from https://griffithreview.com/articles/big-thought-and-a-small-island/

Rittel, H. W. J., & Webber, M. M. (1973). Dilemmas in a general theory of planning. *Policy Sciences, 4*(2), 155–169.

Rubio, J. P., Bahlo, M., Butzkueven, H., et al. (2002). Genetic dissection of the human leukocyte antigen region by use of haplotypes of Tasmanians with multiple sclerosis. *The American Journal of Human Genetics, 70*(5), 1125–1137.

Ryser, L., & Halseth, G. (2010). Rural economic development: A review of the literature from industrialized economies. *Geography Compass, 4*(6), 510–531.

Sarasvathy, S., & Venkataraman, S. (2011). Entrepreneurship as method: Open questions for an entrepreneurial future. *Entrepreneurship Theory and Practice, 35*(1), 113–135.

SBS News. (2015). *Baby formula shortage: Major retailers further restrict sales*. Special Broadcasting Service Television Nov 17, 2015 – 8:18pm.

Scanlon, A., Fish, G., & Yaxley, M. (1990). *Behind the scenery–Tasmania's landforms and geology*. Hobart, Tasmania: Department of Education and the Arts.

Scott, P. (1956). Variability of annual rainfall in Tasmania. *Proceedings of the Royal Society of Tasmania, 90*, 49–57.

Shell Scenarios Team. (2013). *New lens scenarios*. Accessed February 27, 2016, from http://www.shell.com/energy-and-innovation/the-energy-future/shell-scenarios.html#vanity-aHR0cDovL3d3dy5zaGVsbC5jb20vc2NlbmFyaW9z

Slaughter, R. A. (2004). Futures beyond dystopia: Creating social foresight (1st edn.). London: RoutledgeFalmer

Society for Effectual Action. (2016). Accessed February 21, 2016, from www.effectuation.org

Stokes, H. J. W. (1969). *North-West Tasmania 1858–1910: The establishment of an agricultural community*. Canberra: Australian National University.

Sustainable Development Advisory Council. (2002). *Agriculture and soils in Tasmania*. Canberra, ACT: Australian Bureau of Statistics.

Tasmanian Institute of Agriculture. (2016). Accessed February 27, 2016, form http://www.utas.edu.au/tia/about-us

Tasmanian Liberals. *Cultivating prosperity in agriculture*. https://www.tas.liberal.org.au/sites/default/files/policy/Cultivating%20prosperity%20in%20agriculture.pdf

Tomaney, J. (2010). *Place-based approaches to regional development: Global trends and Australian implications*. Sydney: Australian Business Foundation.

Ummenhofer, C. C., England, M. H., McIntosh, P. C., et al. (2009). What causes southeast Australia's worst droughts? *Geophysical Research Letters, 36*(4), 1–5.

Von Bertalannfy, L. (1968). *General systems theory*. New York: Brazillier.

Voros, J. (2003). A generic foresight process framework. *Foresight, 5*(3), 10–21.

West, J. (2013). Obstacles to progress. *Griffith Review 39*; Tasmania: The tipping point? (Online ed.). Accessed February 21, 2016, from https://griffithreview.com/articles/big-thought-and-a-small-island/

Woods, M. (2012). Rural geography III. *Progress in Human Geography, 36*(1), 125–134.

World Bank. (2015). Accessed May 30, 2015, from http://data.worldbank.org/indicator/NV.AGR.TOTL.ZS

Agricultural Biomass Utilisation as a Key Driver for Malaysian Bioeconomy

Ismail Norli, Ariffin Fazilah, and Ismail Mohamad Pazli

Abstract This chapter address some key opportunities and challenges for the bioeconomy development in Malaysia with an emphasis on biomass utilization, energy and industry applications. A few related areas have been reported and the discussion includes various resources including issues in supply chain pertaining to biomass sustainability and availability. Some examples of the food and food related ingredients project outcomes (rubber industry, banana, palm oil industry) were exhibited and their significance in Malaysian bioeconomy is proposed. Case studies and examples are provided to illustrate both driving forces and constraints (past, present and future) of bioeconomy development in Malaysia. A few bioeconomy projects in Malaysia which directly or indirectly contribute to international policy related to climate change, food technology and technology transfer are reported. An overview of the selected research project that was conducted by Universiti Sains Malaysia's researchers in relation to biomass resource development and minimization of the environmental impacts was depicted through an integrated research flow diagram.

I. Norli (✉)
Environmental Technology Division, School of Industrial Technology, Universiti Sains Malaysia, Penang, Malaysia
e-mail: norlii@usm.my

A. Fazilah
Food Technology Division, School of Industrial Technology, Universiti Sains Malaysia, Penang, Malaysia
e-mail: fazilah@usm.my

I. Mohamad Pazli
Northern Skills Development Centre Sdn. Bhd., Penang, Malaysia
e-mail: mfazlie@ymail.com

S. Dabbert et al. (eds.), *Knowledge-Driven Developments in the Bioeconomy*, Economic Complexity and Evolution, DOI 10.1007/978-3-319-58374-7_8

1 Introduction

We are living in the era which is facing a pressing concern towards moving from heavy dependence on fossil fuels to opening up new doors for bio-based material. Therefore, bioeconomy is under the spotlight of current research trend. In 2005, National Biotechnology Policy was drafted in Malaysia with the aim of transforming Malaysian economy into a high-income, inclusive and sustainable economy. This then led to the launching of its very own bioeconomy program known as Bioeconomy Transformation Program (BTP) in the year 2012. The program provided a platform by the government for the private sector to maximize commercial opportunities. The program is also designed specifically to benefit Malaysian society through cultivating commercialization of bio-based products from the industry (Albrecht and Ettling 2014).

The Malaysian economy is showing its potential, according to the Ease of Doing Business Report. Malaysia has risen from twelfth to the sixth position among 189 economies that were appraised. Moreover, Bioeconomy Community Development Programme (BCPD) was also introduced to boost the BTP. The BCDP was designed to transform rural communities via adoption of bio-based technologies. Malaysian Biotechnology Corporation (BiotechCorp) was selected in this regard to be the prime mover in the efforts towards formulation and implementation of strategies to improve those areas with vast potential. While the Ministry of Science and Technology and Innovation (MOSTI) was chosen to facilitate interdepartmental coordination between the relevant Bioeconomy ministries (Annual Report Bioeconomy Malaysia 2014). Figure 1 illustrates an overview of the road map to bioeconomy in Malaysia.

In BTP, up to 20 entry point projects (EPPs) were introduced in 2012 and in October 2015, a total of 48 EPPs were introduced which encompassed projects for three major areas including agriculture, healthcare and industry (Annual Report Bioeconomy Malaysia 2014). These EPPs are focusing in the area of bioremediation, biomaterials, bio-based chemicals, industrial bio-inputs, bio-based farm inputs, high-value food varieties, high-value bio-ingredients, stem cells and regenerative medicine, molecular screening and diagnostics, drug discovery and preclinical services and biopharmaceuticals.

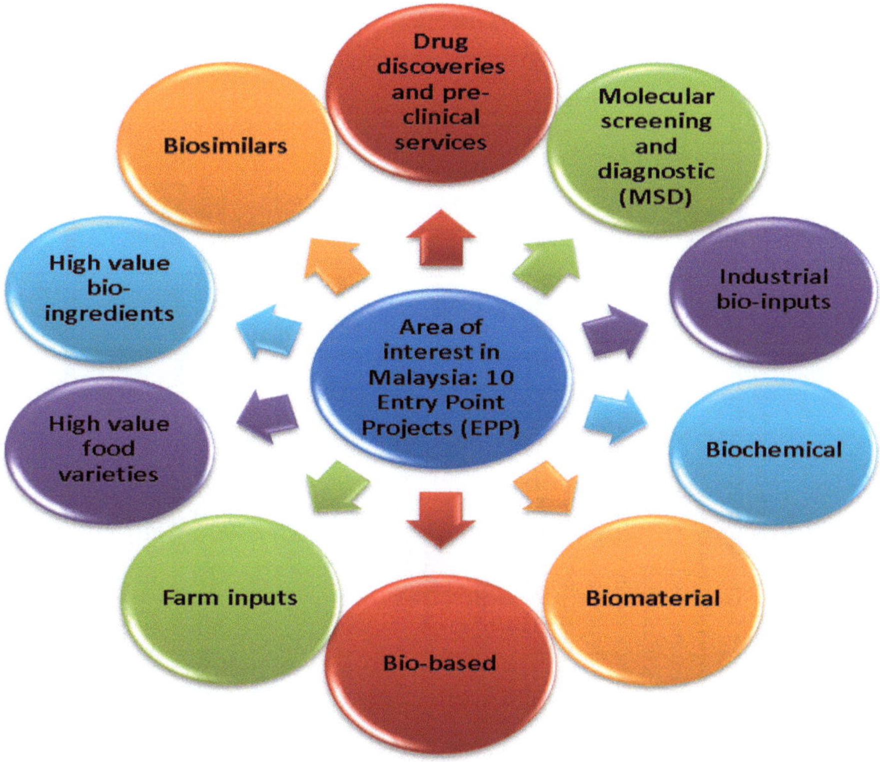

Fig. 1 Bioeconomy in Malaysia: insight, initiative and potential (Annual Report Bioeconomy Malaysia 2014)

According to Bioeconomy Malaysia, the focus of agricultural entry point projects is divided into three categories which are (1) secure local-based farm inputs (2) identify, produce and commercialize high value local bio-ingredients and (3) develop and commercialize high value of food varieties. List of projects in lieu of these categories is recorded in Table 1.

Table 1 Major areas covered by EPPs (Agriculture, Industry, Healthcare)

Agriculture	*List of projects*
Secure local-based farm inputs	• Production of bio-fertilisers through integrated waste treatment plants • Scale up production and commercialisation of bio-feed for the livestock industry • Conversion of palm oil waste into bio-fertilisers through the implementation of the integrated waste treatment plant
Identify, produce and commercialize high value local bio-ingredients	• Develop sustainable collection, extraction and commercialisation of mangosteen • Scale up the plantation, extraction and commercialisation of stevia as an alternative sweetener • Research, development and manufacture of emulsifiers and stabilizers • Commercialization of bee farming, processing of honey and by-products and training on bee handling • Cultivation and processing of *Haematococcus pluvialis* biomass for the production of astaxanthin • Mass cultivation of *Haematococcus pluvialis* for the production of natural astaxanthin • Production of Advanced Bio-Therapeutic Extract (ATE) from *Cocos nutrifera* milk for health and wellness products
Development and commercialize high value of food varieties	• Premium edible bird's nest and downstream products • Develop, produce and commercialize indigenous hybrid seeds • Scale up production of high value mushroom varieties • Development of fully integrated multi-platform aquaculture facilities • Liquid Immersion Bioreactor (LIB) method to commercialise MD2 pineapple plantlets • Development of fully integrated aquaculture facilities • Production and commercialization of MD2 pineapple • Expansion of shiitake mushroom production • Development and commercialisation of high-value tropical abalone aquaculture • Setting up high-value fresh mushroom facility
Industry	*List of projects*
Industrial bioeconomy upstream inputs	• Production and utilisation of Compressed Bio-methane Gas (CBG) for transportation and industrial sectors • Setting up a biogas power generation plant at Felda Kahang based on the Feed-in-Tariff model • Establishment of an energy crop plantation as input for the industrial bioeconomy • Production of syngas for steam generation from solid biomass • Bio-compressed natural gas (BioCNG) from palm oil mill effluent • Setting up 14 biogas power generation plants at Felda palm oil mills based on the Feed-in-Tariff model • Setting up a biogas power generation plant at Kapilit palm oil mill

Biochemicals from renewable resources	• Bio-based chemical production using renewable palm oil derivatives • Renewable production of L-methionine and thiochemicals • Production of cellulosic sugars from woody biomass produced by energy crop plantations • Production of isobutanol from cellulosic feedstocks • Integrated bio-refinery complex at Palm Oil Industrial Cluster in Lahad Datu, Sabah • Bio-ethanol and bio-methanol production from wood chips • Expansion of cGMP stearic acid and Medium Chain Triglycerides (MCT) powder plants
Biomaterials from renewable resources	• Agro-based bio-resin production for bioplastics use • Production of bio-polyols for bio-polyurethanes from palm oil derived oleic acid • Establishment of commercial production of polyhydroxybutyratehexanoate (PHBH) from palm oil • Scale up production and promote usage of biodegradable and compostable packaging products from sustainable agro-waste • Development and manufacturing of biodegradable packaging products from agricultural biomass waste materials
Bioremediation	• Setting up bio-based waste management facilities in Iskandar, Malaysia
Healthcare	*List of projects*
Develop and commercialize biopharmaceuticals	• Develop and commercialize bio-similars/biologics drugs • Expansion of sterile infusion solution plant • Integrated bio-based and synthetic pharmaceutical plants for intermediates and finished products
Develop drugs discovery and pre-clinical services ecosystem	• Develop drugs discovery and pre-clinical services ecosystem
Scale up innovative molecular screening and diagnostics	• Scale up innovative molecular screening and diagnostics (MSD) products and services • Research, development and manufacturing of rapid test kits for in vitro diagnostic application
Scale up stem cells and regenerative medicine	• Scale up stem cells and regenerative medicine

BiotechCorp (2015)

2 Bioeconomy Research Scenarios in Malaysia

Various studies are being conducted in universities and institutes across Malaysia which relate to the objectives of Bioeconomy Transformation Program (BTP). Some of the BTP related areas shall be highlighted in the following section.

2.1 Energy and Environmental Related Research

National Biofuel Policy (NBP) was introduced in 2006 and was aimed at introducing alternative energy in place of fossil fuels (Abdul-Manan et al. 2014). In Eighth Malaysia Plan, renewable resources (biomass, biogas, municipal waste, solar and mini-hydro) were given priority as an effort in promoting the so-called fifth fuels.

Hosseini and Wahid employed a newly designed lab-scale flames reactor named "self-preheated flameless combustion (SPFC)" system. The flameless chamber is used as a heater for preheating an oxidizer over the self-ignition temperature of the fuel. The results obtained indicated power generation of 10.8 MW utilizing Palm Oil Mill Effluent (POME). However, the downside observed was the high rate of pollutant emissions (CO_2 and NO_3). In Malaysia, oil palm biomass is converted into various value added products. Mesocarp fibre and palm kernel seeds were used as fuel for boilers to generate electricity and steam for palm oil extraction (Subramaniam et al. 2008). However, high moisture content in empty fruit bunches (EFB) makes it impossible to be used as a fuel since high amount of energy is required for drying it before it can be used as fuel (Kelly-Yong et al. 2007). Palm kernel seeds, on the other hand, have lower moisture content (17%) and is a decent biomass fuel (Prasertsan and Prasertsan 1996). The biogas generated in the mill can be employed in boilers as co-combustion fuel which can reduce the usage of other biomass as fuel.

Ying et al. (2014) reported that when EFB and kenaf core fibres were treated with water, acid and alkaline mediums, they enhanced the hydrolysability. Hence, the conversion of sugar from cellulose was improved and was crucial for increased production of cellulosic ethanol. Oil palm ash from the mill boiler was reported to have low levels of toxicity and contained potassium, and so it is being used as fertilizer (Yin et al. 2008; Ying et al. 2014). Nevertheless, tonnes of EFB were also reported to be used as raw material in brown paper production (Saleh et al. 2009). Empty fruit bunches (EFB) are a promising source of sugar for production of bioethanol, a second generation biofuel which is generated from non-edible cellulosic biomass (Kong et al. 2014). Malaysian Palm Oil Board (MPOB) had carried out a study to convert EFB into papermaking pulp. Chlorine-free methods were also used for bleaching the pulp for paper production (Gurmit et al. 1999; Kamarudin et al. 2009). They were used for making cigarette paper and bond paper (Agency 2001). Various studies have been carried out on other lignocellulosic materials in manufacturing binderless board. However, Sulaiman et al. considered oil palm as a high potential, non-wood lignocellulosic materials in producing wood based panels. Binderless particle board produced from oil palm trunk was reported to have high mechanical strength. In recent studies, both young and old palm trunk were used in producing binderless particle board. However, the binderless panel made from young trunk showed superiority in terms of mechanical and physical properties (Sulaiman et al. 2011; Lamaming et al. 2014; Embrandiri et al. 2015).

In oil palm mill that processes 60 t/h of EFB, up to 2.4 t of methane gas can be derived from POME per year and if written in terms of calorific value, the amount of POME is equal to about 3.4 million litres of diesel (Chin et al. 2013). Hence, the biogas recovered in the mills has a high potential to work as replacement for diesel. On the other hand, as an effort in conjunction with the implementation of Entry Point Project No 5 (EPP 5) under National Key Economic Areas (NKEA), by 2020, the Malaysian government has aimed to accomplish the installation of biogas facilities in all palm oil mills in the country (MPOB 2014).

Biochar can be created when biomass is heated to high temperature (300–1000°C) under limited or zero oxygen condition. Biochar is also defined as charred organic matter and has the tendency to improve soil properties when used. A collaboration was made between Universiti Putra Malaysia (UPM) and Nasmech Technology Sdn Bhd in building a biochar production plant with the aim of converting EFB and other residue materials into biochar in large quantities (20 t daily) (Bernama 2010). However, one of the challenges faced by the biochar production is the absence of long term partnership contract between individual producers and oil palm biomass suppliers in the market. Setting up collection centres to improve better collaboration may be needed for increasing the acceptance and usage of value added products derived from palm oil mill waste (Sulaiman et al. 2011). Palm shell has been used as fuel in the boilers of Malaysian cement companies, and the results indicated a reduction in CO_2 emissions by 366.26 thousand metric tonnes in the year 2006 alone (Dit 2007).

Production of hydrogen through gasification of oil palm waste (EFB, fibre, shell, trunks and fronds) has also been reported as promising green technology (Azali et al. 2005; Kelly-Yong et al. 2007). Hydrogen as a transportation fuel provides higher engine efficiency as well as zero emissions (Neef 2009). Hence, utilizing palm oil waste as an alternative to fossil fuel in Malaysian industries may still be at an early stage, but, there is no doubt that it is exhibiting massive potential. However, cost benefits analysis should also be conducted before introducing any new technology into the market as many of the products and technology are merely at the research and development stage. A model of future hydrogen demand in the transportation of peninsular Malaysia was created by Kamarudin et al. (2009), and the production of hydrogen from biomass resources has received attention from government and researchers in Malaysia. However, further investigation and research are still needed.

2.2 *Food and Food Related Ingredients*

The production of the agricultural produces in Malaysia is a lot, however, the abundant amount of agricultural biomass is not efficiently utilised. Thus, it is crucial to mitigate this matter in order to enhance the Malaysian bioeconomy that indirectly contributes to strengthen the economy of the country and vary the utilisation and consumption of the agricultural biomass. Few agricultural biomasses as food products will be discussed further in this section.

2.2.1 Rubber Industries

Rubber (*Hevea brasiliensis*) industries are known as one of the important agricultural industry in Malaysia, however, the use of rubber biomass is limited compared to other agricultural crops (Ratnasingam et al. 2015). Rubber can only be grown in tropical climates of Malaysia and few other countries. Total rubber consumed by the industry increased from 187,592 t to 579,248 t for the previous 17 years (1990–2007) (Abu-Jarad et al. 2011). For food application, the biomass utilisation is focussing on the rubber seed. Figure 2 illustrates the different maturity stages of the rubber seed.

Fig. 2 Rubber seed that potential as food product (**a**) The young rubber seed (**b**) The old rubber seed (**c**) Fruit of rubber seed (**d**) The kernel of rubber seed (Ishak et al. 2015)

Table 2 The utilisation of rubber tree biomass

Biomass	Summary of research	References
Rubber seed	Protein source Rubber seeds were dehulled, and the kernel was milled using a household dry blender (50 Hz)	Eka et al. (2010)
	Production of rubber seed oil	Abdullah and Salimon (2009)

The rubber wood biomass commonly contributes to biomass energy and production of furniture wood for Asian, European or American markets (Zafar 2015). In Malaysia about 62% of the rubber wood is used as fuel, 5% is used for other purposes such as furniture, and there is no information about the remaining 30–35% which could mean that such amount would be left unused (Lim 1986). To date, there is scarce information on the development of agricultural biomass waste as food products in Malaysia. Table 2 summarise the work related to the rubber seed on food related application.

2.2.2 Banana

Banana is known as the earliest agricultural crop in the world. The origin which is in India has expanded to South East Asia and the estimated gross production exceeds 139 million tonnes per year. The production of banana in Malaysia is approximately 294,000 metric tonnes from 29,000 ha of plantation (Malaysia Productivity Report 2014). Banana is also known as the second largest fruit crop in the world. Banana is mainly demanded as a fruit but the other parts of banana which are considered as agricultural biomass are also known for variety of usages. Banana is considered as a fruit crop or cash crop alongside other crops such as oil palm, sugarcane, rice and others. Native people tend to consume most parts of banana as their daily food or for other purposes. Few inventions such as renewable energy, textiles and fibre composites, food alternatives and livestock feed were also developed in order to enhance the utilisation of banana biomass.

Figure 3 indicates the agriculture biomass that was produced from a banana tree. The pseudo-stem gave the highest percentage of biomass which is 30.81% followed by leaf and leaf base that is 23.14%. The roots produce 0.97% which is the lowest biomass produced by a banana tree.

Banana biomasses related research are summarise in Table 3.

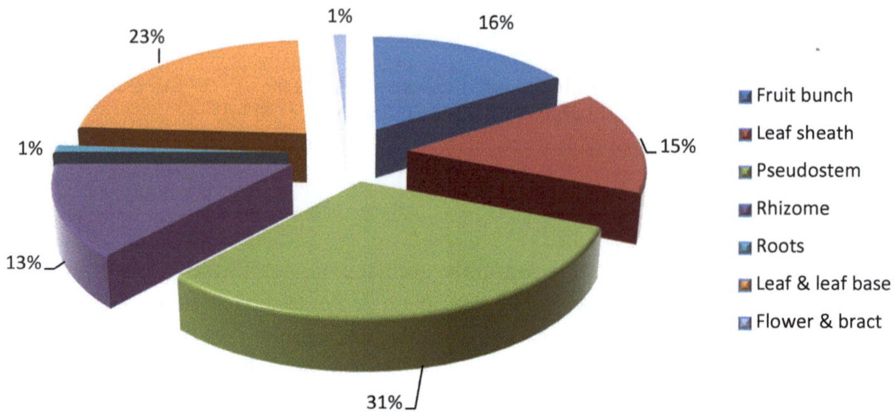

Fig. 3 The agriculture biomass of a banana tree (Saravanan and Aradhya 2011)

Table 3 The utilisation of banana biomass

Biomass	Food ingredients	Summary of research	References
Pseudo-stem	Starch	Food thickeners, gelling agent, reinforcing agent in tablets, fillers	Aziz et al. (2011)
	Flour	Pasting properties of the type of flour partially substituted with banana pseudo-stem flour	Ho et al. (2012)
	Flour	Production of wheat bread incorporated with banana pseudo-stem flour	Ho et al. (2013)
Banana peels	Flour	Production and analysis of physico-chemical properties of banana peel flour	Haslinda et al. (2009)
		Incorporation of banana pulp and peel flour in yellow alkaline noodle	Ramli et al. (2009)
		Production of banana pulp and peel flour	Alkarkhi et al. (2011)

2.2.3 Palm Oil Industries

As of the year 2010, it had been reported that 4.6 million hectares of land in Malaysia were planted with oil palm trees. These plantations produced 16.99 million tonnes of crude palm oil and 2.01 million tonnes of crude palm kernel oil (MPOB 2015). Despite producing a large amount of palm oil, the vast plantation areas require replantation after a stipulated period which generates large amount of biomass waste. It was reported that 40 million tonnes of biomass waste was produced annually which included empty fruit bunch (EFB), oil palm frond (OPF) and oil palm trunk (OPT). This biomass waste has the potential to be consumed as raw sources for bioenergy and new food products (Low 2013). Various investigations had been conducted to utilise this renewable source, thus improving the waste management of palm oil industry.

Empty fruit bunch (EFB) is the leftover of the bunch from which fruits were removed for palm oil production in the palm oil mill (Lee and Ofori-Boateng 2013). In Malaysia, 15.8 million tonnes of EFB were produced annually (Sumathi et al. 2008). The production frequency of EFB is more frequent compared to other types of wastes, as EFB is produced during the harvesting period along with the palm fruits. It was reported that EFB was used as a source of nutrients in the plantation area due to its inherent production of xylose (Rahman et al. 2006).

Locally, the biomass waste of oil palm frond (OPF) was reported to be approximately 54.17 million tonnes in the year of 2010 and 54.24 million tonnes in 2011 (Zahari et al. 2004). Due to a large amount of OPF being produced annually, many research studies were conducted on OPF for the production of binderless board (Laemsak and Okuma 2000), pulp and paper (Wanrosli et al. 2007). The OPF has abundant amount of hemicellulose that is made of polymers of pentoses, hexoses and sugar acids (Sabiha-Hanim et al. 2015).

The oil palm trunk (OPT) is a biomass waste arising from the oil palm replantation process. The OPT is usually high in moisture and starch contents. It has been reported that 13.5 million tonnes of OPT were produced in the year of 2011, with an expected increase of 50% by the year of 2020 (Lee and Ofori-Boateng 2013; MPOB 2012). This oil palm biomass is utilised as particleboard (Hashim et al. 2011), furniture and plywood (Khalil et al. 2010). This shows that most of the OPT applications involve the dry matter (fibre) of the OPT, despite 70% of the trunk weight consists of watery compound (sap). The OPT sap consists of 6.67% sugar and significant amount of amino acids, vitamins and minerals (Kosugi et al. 2010).

Most of the biomass waste is high in hemicellulose and sugars making it an important raw material for food industry. The EFB was used to produce cellulase, an enzyme that is use in food processing (Ariffin et al. 2008). Besides, the utilisation of OPF to produce xylose and xylooligosaccharides opens the opportunity for OPF to be used in food industry (Sabiha-Hanim et al. 2015). In addition, extraction and

Table 4 The utilisation of oil palm biomass waste of food industry

Biomass	Summary of research	References
Empty fruit bunch	Raw material in cellulase production	Ariffin et al. (2008)
Oil palm frond	Hydrothermal treatment and physicochemical properties analysis of hemicellulose extracted from oil palm frond	Fazilah et al. (2009)
	Production of xylose and xylooligosaccharides by autohydrolysis and enzymatic treatment	Sabiha-Hanim et al. (2011)
	Production of hemicellulose fractions impregnated with alkaline or water and treated with steam explosion	Sabiha-Hanim et al. (2011)
Oil palm trunk	Oil palm trunk sap was used as sugar source for lactic acid production	Kosugi et al. (2010)
	Production of alternative sweetener from oil palm trunk sap	Potential
	Incorporation of oil palm trunk syrup as an ingredient in toffee production	Potential

physicochemical properties of hemicellulose from OPF have been studied (Fazilah et al. 2009). Hemicellulose was reported by Sabiha-Hanim et al. (2015) as one of the sources for xylooligosaccharides (Bian et al. 2014; Sabiha-Hanim et al. 2011) and xylitol, which is a valuable additive in food industry (Ping et al. 2013). Table 4 shows the summary of researches on oil palm biomass waste in the food industry of Malaysia.

In addition to that, OPT has the potential to be utilised in food industry as well. The OPT sap can be developed as a food ingredient. High sugar concentrations with the presence of vitamins and minerals give the sap advantage as a potential nutritional food ingredient. The fact that maple syrup and Birch syrup are produced from tree sap eventually makes OPT sap as probable raw material in syrup production.

3 Biomass Related Researches in School of Industrial Technology, Universiti Sains Malaysia

Current work relevant to biomass industry in Malaysia, specifically Universiti Sains Malaysia are summarised in Figs. 4, 5, and 6.

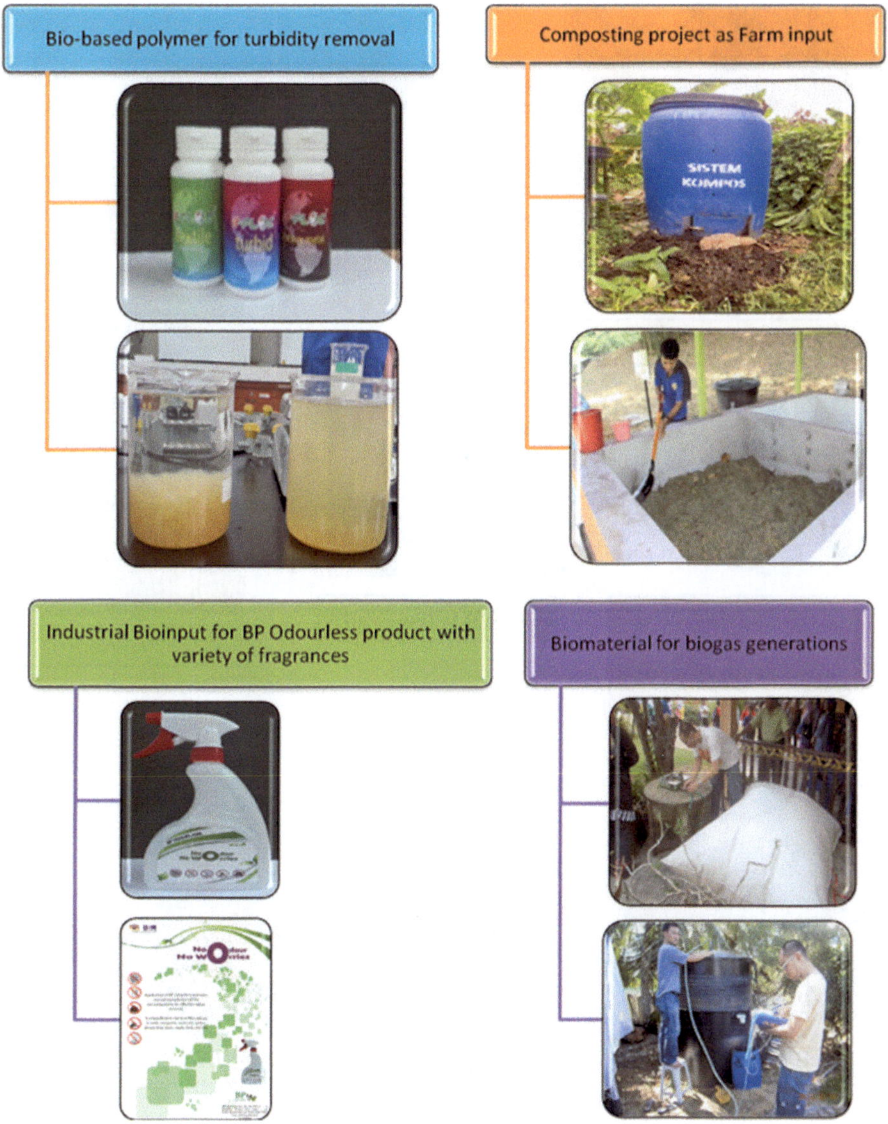

Fig. 4 Utilisation of agriculture biomass for green material

Fig. 5 Summary of BP odourless, bioflocculant and biogas processes

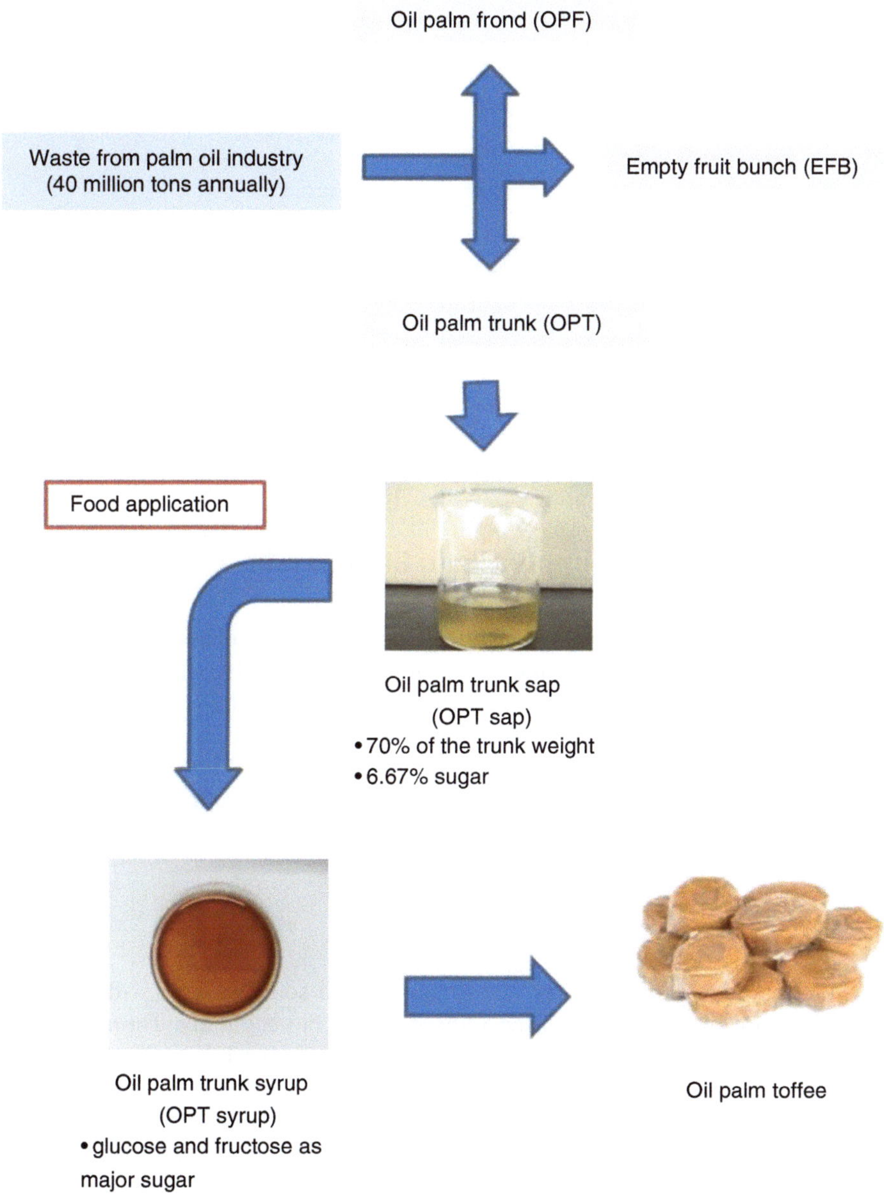

Fig. 6 Utilisation of oil palm trunk sap as potential food ingredients

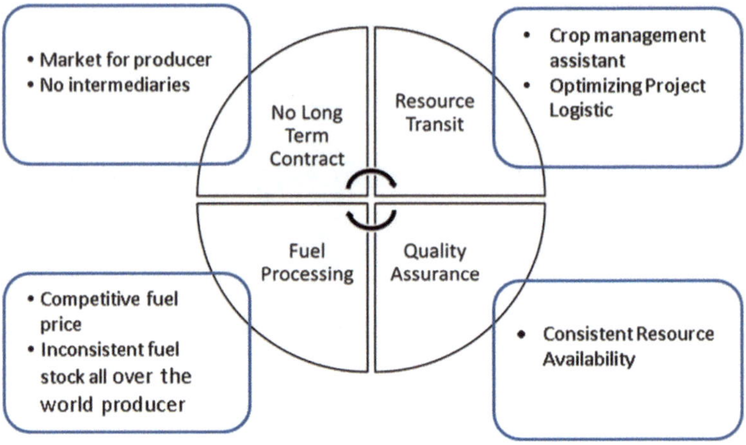

Fig. 7 Issues of agriculture biomass in Malaysia

4 Issue of Biomass Supply Chain in Malaysia

With massive experience in utilising natural resource of biomass for bioeconomy related projects, Malaysia is formalizing a more comprehensive way in ensuring the sustainability of the biomass resources. The complexities in the value chain process will lead to the instability of the supply and demand. There is also some serious underlying issue of a mismatch between supply and demand. Figure 7 illustrates the common factors and way forward for a betterment of the process.

5 Conclusion

As Malaysia is a country that is vast in agricultural sources, there will be plenty of agricultural biomass that can be utilised in the bio-economy. The agriculture biomass is not only suitable for the energy production, but it can also be used as a source for materials and food supply. The aspect of financial investor and the expertise of researchers to broaden the potential of the agriculture biomass are crucial areas which need to be focused on. More promotion and commercialisation by the country and contribution of studies from Malaysia and international researchers need to be done to realise the objectives.

However, to make the target a reality every party such as the authorities, researchers, business community and industrial companies have to collaborate and cooperate in order to make agricultural biomass as one of the potential and main sources for the drive of the bio-economy in Malaysia. Thus, with the enhancement and development of these sources, it is not impossible for the agriculture biomass to become the financial contributor to the Malaysian economy. Indirectly, it may bring positive outcomes not only to the nation but to the world and the people.

Acknowledgement We would like to give our gratitude to the German Academic Exchange service (DAAD), German Federal Ministry of Education and Research and University of Hohenheim for giving us the opportunity to contribute to this book chapter. We hope the information provided will be fruitful to the readers and benefits the target community. At the same time, School of Industrial Technology, Universiti Sains Malaysia welcomes any further discussions and collaborations in order to achieve the target. Last but not least special thanks to Megat, Aisyah, Fariha, Shlrene, Kamilah and Syazana in assisting for the preparation of this book chapter. Thank you.

References

Abdullah, B. M., & Salimon, J. (2009). Physicochemical characteristics of Malaysian rubber (*Hevea brasiliensis*) seed oil. *European Journal of Scientific Research, 31*(3), 437–445.

Abdul-Manan, A. F. N., Baharuddin, A., & Chang, L. W. (2014). A detailed survey of the palm and biodiesel industry landscape in Malaysia. *Energy, 76*, 931–941.

Abu-Jarad, I., Yusof, N. A., & Nikbin, D. (2011). Arrogant businessmen and a challenge in Penang, Malaysia. *Interdisciplinary Journal of Research in Business, 5*(1), 25–33.

Albrecht, K., & Ettling, S. (2014). Bioeconomy strategies across the globe. *Rural, 3*, 10–13.

Alkarkhi, A. F., Ramli, S. B., Yong, Y. S., & Easa, A. M. (2011). Comparing physicochemical properties of banana pulp and peel flours prepared from green and ripe fruits. *Food Chemistry, 129*(2), 312–318.

Annual Report Bioeconomy Malaysia (A.R.B). (2014). *Bioeconmy transformation programme: Enriching the nation, securing the future*. Annual Report 2014.

Ariffin, H., Hassan, M. A., Shah, U. K. M., Abdullah, N., Ghazali, F. M., & Shirai, Y. (2008). Production of bacterial endoglucanase from pretreated oil palm empty fruit bunch by *Bacillus pumilus* EB3. *Journal of Bioscience and Bioengineering, 106*(3), 231–236.

Azali, A., Nasrin, A. B., Choo, Y. M., Adam, N. M., & Sapuan, S. M. (2005). Development of gasification system fuelled with oil palm fibres and shells. *American Journal of Applied Sciences, 1*, 72–75.

Aziz, N. A., Ho, L., Azahari, B., Bhat, R., Cheng, L., & Ibrahim, M. (2011). Chemical and functional properties of the native banana (*Musa acuminata* x *balbisiana Colla cv. Awak*) pseudostem and pseudostem tender core flours. *Food Chemistry, 128*, 748–753.

Bernama. (2010). UPM-Nasmech effort of producing EFB biochar – world's first. *Biochar Malaysia*.

Bian, J., Peng, P., Peng, F., Xiao, X., Xu, F., & Sun, R. C. (2014). Microwave-assisted acid hydrolysis to produce xylooligosaccharides from sugarcane bagasse hemicelluloses. *Food Chemistry, 156*, 7–13.

BiotechCorp. (2015). *Entry point projects*. Kuala Lumpur: BiotechCorp.

Chin, M. J., Poh, P. E., Tey, B. T., Chan, E. S., & Chin, K. L. (2013). Biogas from palm oil mill effluent (POME): Opportunities and challenges from Malaysia's perspective. *Renewable and Sustainable Energy Reviews, 26*, 717–726.

Dit, M. (2007). Palm kernel shell (PKS) is more than biomass for alternative fuel after 2005. In *Proceedings of the PIPOC 2007 International Palm Oil Congress (Chemistry & Technology), Malaysia* (pp. 275–287). Petaling Jaya: Lafarge Malayan Cement Bhd.

Eka, H. D., Aris, Y. T., & Nadiah, W. A. W. (2010). Potential use of Malaysian rubber (*Hevea brasiliensis*) seed as food, feed and biofuel. *International Food Research Journal, 17*, 527–534.

Embrandiri, A., Quaik, S., Rupani, P., Srivastava, V., & Singh, P. (2015). *Waste management: Challenges, threat and opportunities*. New York: Nova Science.

Fazilah, A., Azemi, M. N. M., Karim, A. A., & Norakma, M. N. (2009). Physicochemical properties of hydrothermally treated hemicellulose from oil palm frond. *Journal of Agricultural and Food Chemistry, 57*(4), 1527–1531.

Gurmit, S., Kim, H. L., Teo, L., & Lee, L. D. (1999). Oil palm and the environment a Malaysian perspective. *Malaysian Oil Palm Growers Council, 1*, 1–277.

Hashim, R., Nadhari, W. N. A. W., Sulaiman, O., Kawamura, F., Hiziroglu, S., Sato, M., & Tanaka, R. (2011). Characterization of raw materials and manufactured binderless particleboard from oil palm biomass. *Materials and Design, 32*(1), 246–254.

Haslinda, W. H., Cheng, L. H., Chong, L. C., & Aziah, A. N. (2009). Chemical composition and physicochemical properties of green banana (*Musa acuminata* x *balbisiana Colla cv. Awak*) flour. *International Journal of Food Sciences and Nutrition, 60*, 232–239.

Ho, L. H., Aziah, A. A. N., & Bhat, R. (2012). Mineral composition and pasting properties of banana pseudo-stem flour from *Musa acuminata* x *balbisiana cv. Awak* grown locally in Perak, Malaysia. *International Food Research Journal, 19*(4), 1479–1485.

Ho, L. H., Aziz, N. A. A., & Azahari, B. (2013). Physico-chemical characteristics and sensory evaluation of wheat bread partially substituted with banana (*Musa acuminata* x *balbisiana cv. Awak*) pseudo-stem flour. *Food Chemistry, 139*(1), 532–539.

Ishak, M. A. M., Hassan, S. N. A. M., Ali, S. N., Abdullah, M. F., Ani, A. Y., Kasim, N. N., Jawad, A. H., Ismail, W. I. N. W., & Ismail, K. (2015). Overview of obtaining alternative fuels in the co-liquefaction processes with biomass and coal in Malaysia. In K. Biernat (Ed.), *Biofuels – Status and perspective, Science, technology and medicine* (Vol. 1). Warsaw: InTech.

Kamarudin, S. K., Daud, W. R. W., Yaakub, Z., Misron, Z., Anuar, W., & Yusuf, N. N. A. N. (2009). Synthesis and optimization of future hydrogen energy infrastructure planning in Peninsular Malaysia. *International Journal of Hydrogen Energy, 34*, 2077–2088.

Kelly-Yong, T. L., Lee, K. T., Mohamed, A. R., & Bhatia, S. (2007). Potential of hydrogen from oil palm biomass as a source of renewable energy worldwide. *Energy Policy, 35*, 5692–5701.

Khalil, H. P. S. A., Bhat, A. H., Jawaid, M., Amouzgar, P., Ridzuan, R., & Said, M. R. (2010). Agro-wastes: Mechanical and physical properties of resin impregnated oil palm trunk core lumber. *Polymer Composites, 31*(4), 638–644.

Kong, S. H., Loh, S. K., Bachmann, R. T., Rahim, S. A., & Salimon, J. (2014). Biochar from oil palm biomass: A review of its potential and challenges. *Renewable and Sustainable Energy Reviews, 39*, 729–739.

Kosugi, A., Tanaka, R., Magara, K., Murata, Y., Arai, T., Sulaiman, O., Hashim, R., Hamid, Z. A. A., Yahya, M. K. A., Yusof, M. N. M., Ibrahim, W. A., & Mori, Y. (2010). Ethanol and lactic acid production using sap squeezed from old oil palm trunks felled for replanting. *Journal of Bioscience and Bioengineering, 110*(3), 322–325.

Laemsak, N., & Okuma, M. (2000). Development of boards made from oil palm frond II: Properties of binderless boards from steam-exploded fibers of oil palm frond. *Journal of Wood Science, 46*(4), 322–326.

Lamaming, J., Hashim, R., Sulaiman, O., Sugimoto, T., Sato, M., & Hiziroglu, S. (2014). Measurement of some properties of binderless particleboards made from young and old oil palm trunks. *Measurement, 47*, 813–819.

Lee, K. T., & Ofori-Boateng, C. (2013). *Sustainability of biofuel production from oil palm biomass*. Singapore: Springer.

Lim, K.O. (1986). The energy potential and current utilization of agriculture and logging wastes in Malaysia. *Renewable Energy Review Journal, 8*(2), 57–75.

Low, J. (2013). *Bioeconomy initiative in Malaysia: Lessons learnt and the way forward*. Kuala Lumpur: Roland Berger Strategy Consultants.

Malaysia Productivity Report. (2014). *Connecting lives, changing mindsets*. Malaysia Productivity Report 2014.

Malaysian National News Agency. (2001). *Long term benefit from oil palm biomass*. Kuala Lumpur: Bernama.

MPOB. (2012). *Economics and Industrial Development Division, Malaysian Palm Oil Board*. www.mpob.gov.my

MPOB. (2014). *NKEA: National biogas implementation (EPP5)*. Kajang: Malaysia Palm Oil Board.

MPOB. (2015). *Economics and industrial development division*. Malaysian Palm Oil Board. http://bepi.mpob.gov.my/index.php/en/statistics/production/159-production-2015.html

Neef, H. J. (2009). International overview of hydrogen and fuel cell research. *Energy, 34*, 327–333.

Ping, Y., Ling, H. Z., Song, G., & Ge, J. P. (2013). Xylitol production from non-detoxified corncob hemicellulose acid hydrolysate by *Candida tropicalis*. *Biochemical Engineering Journal, 75*, 86–91.

Prasertsan, S., & Prasertsan, P. (1996). Biomass residues from palm oil mills in Thailand: An overview on quantity and potential usage. *Biomass and Bioenergy, 11*, 387–395.

Rahman, S. H. A., Choudhury, J. P., & Ahmad, A. L. (2006). Production of xylose from oil palm empty fruit bunch fiber using sulfuric acid. *Biochemical Engineering Journal, 30*(1), 97–103.

Ramli, S., Alkarkhi, A. F., Yong, Y. S., Min-Tze, L., & Easa, A. M. (2009). Effect of banana pulp and peel flour on physicochemical properties and in vitro starch digestibility of yellow alkaline noodles. *International Journal of Food Sciences and Nutrition, 60*, 326–340.

Ratnasingam, J., Ramasamy, G., Wai, L. T., Senin, A. L., & Muttiah, N. (2015). The prospects of rubberwood biomass energy production in Malaysia. *BioResources.com, 10*(2), 2526–2548.

Sabiha-Hanim, S., Noor, M. A. M., & Rosma, A. (2011). Effect of autohydrolysis and enzymatic treatment on oil palm (*Elaeis guineensis Jacq.*) frond fibres for xylose and xylooligosaccharides production. *Bioresource Technology, 2*(102), 1234–1239.

Sabiha-Hanim, S., Noor, M. A. M., & Rosma, A. (2015). Fractionation of oil palm frond hemicelluloses by water or alkaline impregnation and steam explosion. *Carbohydrate Polymers, 115*, 533–539.

Saleh, M., Jasmani, L., Yusoff, M. N. M., Ibrahim, R., & Pillay, K. (2009). *Commercial production of pulp and paper from oil palm empty fruit bunches (EFB)* (p. 22). Seminar on Biomass for Biofuels and Value-added Products.

Saravanan, K., & Aradhya, S. (2011). Polyphenols of pseudostem of different banana cultivars and their antioxidant activities. *Journal of Agricultural and Food Chemistry, 59*(8), 3613–3623.

Subramaniam, V., Ma, A. N., Choo, Y. M., & Meriam, N. S. N. (2008). Life cycle inventory of the production of cruide palm oil – A gate to gate case study of 12 palm oil mills. *Journal of Oil Palm Research, 20*, 484–494.

Sulaiman, F., Abdullah, N., Gerhauser, H., & Shariff, A. (2011). An outlook of Malaysian energy, oil palm industry and its utilization of wastes as useful resources. *Biomass and Bioenergy, 35*(9), 3775–3786.

Sumathi, S., Chai, S. P., & Mohamed, A. R. (2008). Utilization of oil palm as a source of renewable energy in Malaysia. *Renewable and Sustainable Energy Reviews, 12*(9), 2404–2421.

Wanrosli, W. D., Zainuddin, Z., Law, K. N., & Asro, R. (2007). Pulp from oil palm fronds by chemical processes. *Industrial Crops and Products, 25*(1), 89–94.

Yin, C. Y., Kadir, S. A. S. A., Lim, Y. P., Syed-Ariffin, S. N., & Zamzuri, Z. (2008). An investigation into physicochemical characteristics of ash produced from combustion of oil palm biomass wastein a boiler. *Fuel Processing Technology, 89*, 693–696.

Ying, T. Y., Teong, L. K., Abdullah, W. N. W., & Peng, L. C. (2014). The effect of various pretreatment methods on oil palm empty fruit bunch (EFB) and kenaf core fibers for sugar production. *Procedia Environmental Sciences, 20*, 328–335.

Zafar, S. (2015). *Agricultural biomass in Malaysia*. Bienergy Consult. http://www.bioenergyconsult.com/agricultural-biomass-in-malaysia/

Zahari, M. W., Sato, J., Furuichi, S., Sukri, I. M., Bakar, C. A., & Yunus, I. (2004). *Recent development on the processing and utilization of complete feed based on oil palm fronds (OPF) for ruminant feeding in Malaysia*. 3rd USM-JIRCAS Joint International Symposium, Penang, Malaysia. JIRCAS.

University-Industry Relationships in the Bioeconomy Innovation System of Denmark

Keld Ejdrup Markedal, Jens Christian Sørensen, and Susanne Sørensen

Abstract Creation of new innovative processes and products within the high frequency of small and medium size enterprises in collaboration with academia can unfold a large potential that can diminish some of the consequences of the four major crises comprising the environmental crises, the food crises, the energy crises, and the economic crises. To unfold this potential the gap between SME's and academia must be bridged. Combining university research and industrial knowhow in an effort to develop holistic, environmentally friendly and economically feasible technologies for optimal processing of agricultural products may result in sustainable production of high value food and feed ingredients as well as bioenergy and non-food products. The technologies must ensure optimal use of natural resources as well as having focus on quality in all parts of the supply chain and thereby increasing the overall economic feasibility of the process. The challenge thereby appears that the processing has to be defined by numerous quality standards which needs to be prioritized by considering parameters like environmental impact, minimizing the formation of waste and optimizing the product portfolio and the profitability. Involving SME's in industrial collaboration ecosystems facilitated by academia, where the residual product of one enterprise is used as a resource by another may serve as a potential solution for utilizing the competences of these companies and the public research in a sustainable bio-economy.

1 Introduction

The annual Danish budget for research and development (R&D) is according to the Danish Ministry of Higher Education and Science around 6.5 billion € including the public funding of approx. 1% of the gross domestic product (GDP) making Denmark one of few European countries that fulfill the common European goals of investing a total of 3% of GDP in R&D (Concern Statistics and Analysis 2012) as defined in the Lisbon strategy (European Commission 2005).

K.E. Markedal (✉) • J.C. Sørensen • S. Sørensen
Faculty of Science, Department of Food Science, University of Copenhagen, Rolighedsvej 30, 1958 Frederiksberg C, Denmark
e-mail: kea@food.ku.dk; jchs@food.ku.dk; sus@food.ku.dk

S. Dabbert et al. (eds.), *Knowledge-Driven Developments in the Bioeconomy*, Economic Complexity and Evolution, DOI 10.1007/978-3-319-58374-7_9

The R&D activities can be divided into subclasses i.e. basic research, applied research and experimental development (Organisation for Economic Co-operation and Development 2002). The Danish universities and other public research institutions have traditionally focused their main research activities on basic research whereas the research activities in the private sector generally are directed towards applied research and experimental development. This picture is very simplified because also private enterprises conduct basic research, aiming at creation of a broad foundation of knowledge in the company (Concern Statistics and Analysis 2012).

The main focus of the private enterprises is business development whereas the public research institutions have a scope of expanding the general knowledge for the benefit of the society and provide the basis for research based teaching and consultancy assignments for the public sector. This is not always conducted in timeframes suitable for private business or with a scope of creating business potential, but may on the other hand have a strength in being shared, tested and further developed by a potentially large group of scientists and may then show potential for commercial utilization in a broader range of industrial sectors. In the early nineties the Danish Government launched a new strategy for funding initiatives to support co-operation between industry and academia and increase knowledge transfer for the benefit of the industry. The timespan for industrial commercialization of academic research therefore seems to decrease (Concern Statistics and Analysis 2012) indicating a tendency towards increased research collaboration between public institutions and private companies or a sharpened focus from the academia research on applied research. Regardless of this trend the majority of the research conducted within the two sectors are still financed internally and only app. 3.2% of the accumulated Danish research budget is transferred across the sectors (Concern Statistics and Analysis 2012).

2 Strategic Focus on Research Collaborations

Governments in several European countries have during the last decade created strategies to improve compatibility and growth in the national private sector. In December 2012 the Danish Minister of Higher Educations and Science published the strategy 'Denmark—Land of Solutions; Strengthened cooperation and better framework for business innovation' in collaboration with Minister for Business and Growth. The aim is to initiate innovative solutions to major societal challenges that will translate into growth and jobs (Denmark and The Government 2012).

The three focus areas that are described in the strategy are:

1. Social challenges must drive innovation: Demand for solutions to specific challenges in society must be given priority in the public innovation efforts.
2. More knowledge should translate into value: Focus on mutual knowledge between businesses and knowledge institutions and effective innovation systems.
3. Programs must increase innovation capacity: a change of culture in the education system with more focus on innovation.

Even though the overall growth of the European economy in 2015 seems to be recovering after the financial crisis this is mostly due to the increasing demands from the domestic markets and accelerated exports (World Bank Group 2016), whereas the social challenges mentioned in the Danish 2012 strategy are still valid. Furthermore The World Bank estimates that there is a risk of spillover from a decline in the GDP growth in Brazil, India, China and especially Russia making innovative growth even more necessary. The challenges described in the 2012 strategy that particularly calls for new innovative solutions are within areas such as sustainable energy supply, reducing environmental and climate impacts, securing healthy and safe food supply, ensuring clean water supply, creation of cheaper and better health and welfare solutions and the creation of a more efficient public sector (Denmark and The Government 2012).

The main drivers in creating the innovative solutions needed to assist solving these global challenges and to generate increased growth and employment are according to the strategy utilization of strong business competences in combination with strong research capabilities. Innovation and knowledge-driven development focusing on solving the global challenges is foreseen to have a large potential as an area to help the recovery of the Danish economy and the consequences inflicted by the recent financial crises.

The high frequency of small and medium size enterprises (SME's) in Denmark and a hypothesis that large enterprises are often more innovative than minor companies (The Ministry of Economic and Business Affairs 2011) has resulted in a strategy that through a change in the public research support system aims at 25 specific focus areas (Denmark and The Government 2012) to support more:

- Professional clusters and networks,
- Establish a combined program for knowledge-based innovation in SME's
- Prioritize research and development that support Danish production as well as put specific focus on innovation in knowledge institutions.

3 Global Challenges in Relation to Danish Innovation

Sustainable energy supply, reducing environmental and climate impacts, healthy and safe food supply and ensuring clean water supply are closely linked (Danish Ethics Committee 2012).

The world population has in 2015 reached 7.3 billion, which indicates a linear growth of 1 billion people in 13 years since 1990, a tendency that according to United Nations will result in a world population of 8.5 billion people in 2030 and between 9.5 and 13.3 billion in 2100. The vast majority of population growth is expected to be in countries with a high fertility rate or with large populations; these countries include India, Nigeria, Pakistan, Democratic Republic of the Congo, Ethiopia, United Republic of Tanzania, United States of America, Indonesia and Uganda. At the same time Europe is expected to experience a decline in the total population (United Nations 2015).

A decrease in the population growth rate is expected to result in a stagnation of the world population over time but as long as the world population is increasing the demand for both food and energy will grow. The demand for agricultural products is according to World Health Organization (WHO) still expected to increase with approx. 1.5% annually over the next 30 years which is less than half of the rate seen over the past 30 years (WHO 2016) but still at a level that calls for attention.

A shift in the dietary composition is furthermore expected as average income increases particularly in developing countries. Staples like cereals, roots and tubers will decrease their share while commodities like meat, dairy products and oil crops will increase. In the developing countries in particular the demand for dairy products and meat is predicted to grow faster than the production. Therefore additional speed in the production growth or increased imports is needed unless a change in the consumer behavior is achieved (WHO 2016).

Increased food production and especially meat with special emphasis on cattle meat production will have a negative impact on water use and water quality, greenhouse gas emission and other factors (Eshel et al. 2014) that may affect the local environment as well as the global climate. Apart from this the productions will also require an energy input that further adds to the energy crises inflicted by an increased global population.

The Danish Ethics Committee has discussed how to address the four major global crises regarding food supply, energy supply, conservation/protection of nature and climate changes, and how to accommodate the apparent need of 50% more food, 45% more energy and 30% more water in 2030 compared to 2012 without destroying or hurting the environment (Danish Ethics Committee 2012). Some ethical questions discussed by the committee concerns how to prioritize. Are humans the sole factor and should a global or a local perspective be applied. Does the effect on nature, environment and animals also conduct an ethical issue that must be addressed and to which extend should this be prioritized compared to the human aspect (Danish Ethics Committee 2012).

One ethical dilemma can be exemplified by the fossil fuel replacement. Fossil fuels are gradually being replaced by renewable energy such as wind, water, and solar power but also from bioenergy sources like biogas, bioethanol and biodiesel. Bioenergy crop production can to a minor extent be performed on marginal lands where growing and harvesting crops like tufted hair grass (*Deschampsia cespitosa*, L., P.Beauv.), rough bluegrass (*Poa trivialis*, L.), couch grass (*Elymus repens*, L., Gould) or common rush (*Juncus effuses*, L.) in wet grassland areas can increase the botanical biodiversity by removing nitrogen, phosphor and potassium from the soil as a side effect to the biomass harvested for biogas production (Berglund et al. 2012). On farm land the most common bioenergy source is wood, wood chips and straws but crops like rapeseed or willow for energy purposes are also being produced (Bentsen 2011). Wet land grass crops or dry land straw and wood chips production for bioenergy use are in terms of land use not in direct competition with the food production if one disregards the use of straw and hay as feed for animal farming. The Danish production of rapeseed on the other hand has been used primarily for bioenergy with 75% of the production being used for biodiesel

(Bentsen 2011). Based on 2015 harvest data ("Statistikbanken," 2014) this equals 145,000 ha of farmland that could have been used for food or feed production. The yield of rapeseed on this land was approx. 615,000 t rapeseeds containing 300,000 t of rapeseed oil that might alternatively have been used for human consumption. The remaining 315,000 t post-extraction rapeseed meal is primarily used in pig or cattle feed but could with the right technology (Andersen et al. 2012) have been converted into food ingredients comprising 100,000 t proteins and 50,000 t dietary fibers and thereby left only 165,000 t for cattle feed with low nutritional value due to low protein and oil content. The incentives to choose either scenario depend on consumer demands or on the policy and subsidies of individual countries all over the world.

According to WHO there are three main sources of growth in crop production: expanding the land area, increasing the frequency at which it is cropped (often through irrigation), and boosting yields. It has been suggested that growth in crop production may be approaching the ceiling of what is possible in respect to all three sources (WHO 2016).

Ethical considerations combined with the actual political and financial situations and the available production engineering technologies are factors that must be evaluated when setting up strategies for developing a society based on bio-economy. An alternative strategy that could supplement the three WHO scenarios mentioned is development of new technologies for more intelligent use of materials already grown and harvested in primary production today.

4 Reduction of Waste

Foods and most feeds are produced with the intention of being used at the end of the supply chain as source of essential nutrients and energy for humans. Not only the product but all the resources used for its manufacturing, storing, transportation etc. are wasted when this intention is not met. On a daily basis huge amounts of foods are discarded on the journey from primary production to final consumption at the end-users. The annual loss of food in the US has been calculated based on 1995 data where a total production of around 43,500 million kg (Heller and Keoleian 2000) resulted in a usage of 73% of the produced foods and a loss of 27% equal to 11,700 million kg primarily in the foodservice chain or as consumer losses. Similar data has been shown for combined figures for USA, Canada, Australia and New Zealand (Gunders 2012). This study also includes seafood and divides the losses between food categories. Here, the post-harvest handling and storage as well as processing and packaging have the lowest incidences of loss in the food supply chain. The main part of the losses was found to be related to consumer losses including out-of-home consumption.

In Europe a directive has defined a waste hierarchy that shall apply as a priority order in waste prevention and management legislation and policy where preventing waste has the highest priority followed by preparation for re-use, recycling, other recovery, e.g. energy recovery, and finally disposal (Organisation for Economic Co-operation and Development 2002). This indicates that energy recovery from materials that could otherwise be used for food or feed production is politically unaccepted and a strategy for producing foods prior to energy has to be developed.

5 Trends in Food Demands

For adult women exercising moderate physical activity USDA recommend an energy intake of 7500 kJ/d whereas for men the recommendation is 8400–10,000 kJ/d. It is also recommended that 10–35% of the energy should originate from protein which equals a daily protein intake of 45–63 g for women and 50–210 g for men (U.S. Department of Agriculture and U.S. Department of Health and Human Service 2010).

In Scandinavia the Nordic nutrition recommendations estimates a daily energy intake based not only on gender but also takes height and weight into consideration (Nordic Council of Ministers 2008), with an average Danish man being 179 cm high and weighing 74 kg and an average Danish woman being 166 cm and weighing 64 kg [Body mass index (BMI) of 23]. With a low to moderate activity level the average intake of energy can then be estimated to 9200 kJ/d for women and 11,550 kJ/d for men, thus being slightly higher than the USDA recommendations. The Nordic recommendations state that 10–20% of the daily energy intake should originate from proteins resulting in a recommended daily protein intake of 55–112 g and 69–139 g for the two genders, respectively.

In the human body the muscles serve as the only storage capacity for proteins and amino acids and that protein has to be built directly into the skeletal muscles; surplus proteins will face the fate of being degraded in the liver. Recent research has thus stressed that an optimal utilization of the dietary protein requires that the protein intake is done throughout the day, and it is therefore suggested that an average of 25–30 g high quality protein should be included in each major meal, and that this will help prevent sarcopenia (Paddon-Jones and Rasmussen 2009) which is noteworthy as the average age of the world population is expected to increase (World Health Organization 1998).

This indicates that the average daily protein intake should be around 75–90 g and that the protein should be of high quality, meaning having a high biological value (Food and Agriculture Organization of The United Nations et al. 1981), and that it must be consumed evenly throughout the day.

Over the last four decades a globally increasing consumption of total energy, meat, vegetable oils and wheat has occurred. This tendency is predicted to continue at least 25 years and will show a shift from high prevalence of under-nutrition to a situation where too high and unhealthy food consumption may cause overweight, obesity and other diet-related illnesses in middle-to-low-income countries (Kearney 2010).

At the same time the availability of fats of animal origin, particularly in many developed countries are decreasing (Food and Agriculture Organization of the

United Nations and World Health Organization 1994). In Europe some trends indicate that consumers strives towards a healthier lifestyle increasing the demand for oil with at least assumed health-promoting properties such as having high concentration of ω-3 fatty acids, lipid soluble vitamins and antioxidants. European food industry expects to decrease their use of oils and gradually shift towards using lesser amounts and replacing the used oils with types of higher quality with respect to functionality in the products. Thereby health driven consumer demands can push the industrial development of functional oils (Pierrot 2015).

The tendency of industrial replacement of oil and fat could furthermore indicate a decrease in the overall energy intake preventing overweight and obesity. An American study however showed that a decrease in oil intake was accompanied by an increased carbohydrate intake resulting in an overall increase of the energy consumption and an increase in obesity prevalence, and it is suggested that substituting protein intake in place of carbohydrate and fat intake could be part of a solution (Austin et al. 2011).

The question is though, how increased demand for high quality proteins, healthy oils and functional food ingredients comply with securing healthy and safe food supply, ensuring clean water supply, decreasing non-communicable diseases as well as reducing environmental and climate impacts?

6 Biofractionation for Food and Feed Ingredients

Vegetables like asparagus, leeks, lettuce, spinach, spring onions, etc. are examples of crops where the entire plant material is harvested and fully usable in the food industry and in theory they do not create any losses or spillover to the feed, energy or fertilizer sector except perhaps the small part of the root system that is left in the field and not used in the future growth season. Other crops are, however, only partly used in the food sector since only part of the plant comprise the edible part that is harvested or because the harvested part is further processed before it is used for foods. Most berries, fruits and vegetables belong to the first group whereas cereals, rice, corn, and oil crops like rapeseeds and sunflower seeds belong to the second group. A fraction of the crops are in particular being processed industrially as alternative to domestic processing e.g. soybeans and lupine seeds used for oil production with the subsequent extraction residue being used as food protein ingredient, as protein supplements in feed applications or for further fractionation into protein concentrates or isolates. Side-streams after processing also follows use of fresh fruits for e.g. juice and wine production leaving a pressed residue for animal feeds, or for further processing, e.g. for extraction of pectin, sugar, acid and/or antioxidant extraction.

The part of the raw material that is not entering the food chain thus faces different possible fates. The cabbage plant, potato plants, soybean plants, carrot tops, straws, roots and stems from cereals etc. may be further processed into foods or food ingredients or is downgraded but still represent a certain value, e.g. as feed or feed ingredient or even as fertilizer after biological conversion and degradation.

Several factors determine how our raw materials are processed and our food produced. Politics, legislations, subsidies, market structures, culture, and traditions are examples of parameters influencing which and how foods are produced. Certain traditions may define which foods you prefer and how you want your foods and groceries presented; it also defines what we find attractive to eat and what we do not. Traditions and preferences shifts, new technologies and food types are developed and inspirations from other parts of the world influence our preferences and mixes with our traditional foods. But how should the ideal food production be designed?

7 Holistic Approach to Biofractionation

The optimal way for product and production optimization is ideally to focus on quality in all parts of the supply chain as a mean to generate overall economic feasibility driven by an increased willingness to pay for the products. Optimal quality from primary producer to end-user is essential for production and consumption of high quality, healthy, nutritious and appealing food.

Defining good quality of a product usually refers to a product having a high standard for the intended use. A product showing low quality in one application may be considered a very high quality product in another. This implies that the specifications and requirements linked to high standard and good quality in the final application must be evaluated and incorporated into every step of the production of a certain product. The challenge thereby appears that most primary products are starting materials for production of various compounds used in several applications and the processing thereby has to be defined by numerous quality standards which do not necessary comply.

Prioritizing between quality standards can be done through the development of a system that through production of a crop and the related harvest, storage, transport, and processing systems ensures the highest possible quality for all ingredients in all sectors and applications. Furthermore the system must contain a hierarchy that defines which products and applications are prioritized over others in the production chain.

Requirements to be considered when defining such a product hierarchy include the environmental impact e.g. by focusing on limiting the consumption of resources and energy, minimizing the formation of side streams and waste and optimizing the product portfolio. All three parts of this strategy will secure minimal environmental impact.

Another parameter that naturally has to be evaluated is the functionalities of the various products and how they comply with the different applications. A suitable strategy is to aim at preserving the natural, native structure of the components in the plant material by avoiding autolysis reactions, un-controlled enzymatic or non-enzymatic degradation during processing or denaturation due to changes in pH, temperature or other physico-chemical parameters. The assumption behind this is, that the structures created naturally has various functionalities that are difficult to re-create if they are lost unintentionally. If a modification is preferred it can be introduced directly in the primary processing, but if this modification limits the usage of the product, modifications can preferably be done on the whole fraction or part of this after the native product has been produced. Thereby tailored modification of products directed to specific applications can extend the usage of the product and the remaining side streams increasing the overall quality and value.

8 Need for Extended Collaboration

Developing a strategy for optimal use of the natural resources increasing the product portfolio and product quality is only part of the challenges in shifting towards a more bio-based economy. The production needs to be implemented and the products need to be introduced to the market and customers need to buy them.

Optimization of product portfolio and product quality is not always compatible with the market and business strategy for a single company. Optimization of earnings is crucial for the compatibility and survival of a company and developing and producing products that are outside the focus area of the company may lead to increased costs and losses negatively influencing the overall economy of the production. Introduction of new production methods and new products must be done in a pace that complies with the development plan of the company and that can be handled by the employees. If the tasks are to complex or to challenging for the company they will not be introduced. The Danish philosopher Søren Kierkegaard has stated that "*If one is truly to succeed in leading a person to a specific place, one must first and foremost take care to find him where he is and begin there*" (Kierkegaard 1859; "Søren Kierkegaard Forskningscenteret—Citater," 2015). Some of the psychological aspects of business development can be exemplified by the theory of zone of proximal development introduced by the Russian psychologist Lev Semyonovich Vygotsky. This concept relates to the difference between what a person can achieve independently and what he/she can achieve with guidance and encouragement from a skilled partner ("Zone of Proximal Development—Scaffolding | Simply Psychology," 2012). If the new concept or new products are outside the development zone it most likely will not succeed.

Leaps in technology therefore require the right timing. Making a differentiated introduction of the individual parts of a new production method or a new product range over a period of time according to technological development in the production, financial surplus or market acceptance may ripen the whole concept both within the company and in the market. Otherwise a single company, especially if SME's are involved, will not be able to handle all aspects of the idea. In such a case collaborations between several companies may be a solution thus introducing generation of a value chain between companies. Through extended collaborations different parts of the developed bio-economical sustainable concept can be utilized by various companies with different business strategies working in different markets.

The Danish agricultural sector is based on a long tradition of cooperative farm marketing and farm supply organizations collaboration primarily based on the principles of consumer cooperatives developed by The Rochdale Pioneers as described in the paper by Zeuli and Cropp [26]. For more than 150 years there has been a history of supply collaboration within the Danish food sector covering areas like creameries and slaughter houses generally between partners whose interests and social, political and cultural position are characterized by high uniformity and who does not appear as being in direct competition (Bjørn 2014).

Cooperative businesses are still successful in Denmark but other types of collaborations may also be beneficial. A food producer may e.g. introduce new production methods increasing the quality and value of a product by removing parts of the original product that decreases the functionality. This extracted material is in its pure form unsuitable for inclusion in foods but may be sold as high quality raw material to a non-food or bio-energy company. Each company then carries a small part of the investment for the whole concept, but is on the other hand dependent on the performance and deliverables of the companies upstream in the production line. These types of industrial collaborations are running in selected areas in Denmark. One example is The Kalundborg Symbiosis between large companies like Novo Nordic, Novozymes, Gyproc, Novoren and the municipality of Kalundborg conducting a collaboration described as an industrial ecosystem, where the residual product of one enterprise is used as a resource by another enterprise, in a closed cycle ("Kalundborg Symbiosis Center," 2015).

Involving SME's in similar collaboration with private and public enterprises are still not widespread in Denmark but entails large potential for utilizing the competences of these companies in a sustainable bio-economy.

9 Collaborations Between Academia and Private Enterprises

The innovative capability of a company and thereby the knowledge gain achieved by companies in network research projects with academia, depends on the 'absorptive capacity' of the company defined by Cohen and Levithal as the '*ability to recognize the value of new, external information, assimilate it, and apply it to commercial ends*' (Cohen and Levinthal 1990; Easterby-Smith et al. 2008). The absorptive capacity refers to the ability to acquire and assimilate new information from the external source but also to the capability to transfer the knowledge across the internal sections in the company and onto the relevant people or divisions that can utilize the knowledge (Cohen and Levinthal 1990).

The company structure may therefore have significant influence on the absorptive capacity. Large companies that are or have been involved in R&D projects normally have an organization that is capable of handling new innovative knowledge and can easily acquire new knowledge from external partners. On the other hand the complex structure of large companies may be an obstacle for the internal transfer of knowledge and thereby may decrease the absorptive capacity. The size of

SME's makes the company structure less complex and generally there is a short distance between the individual company sections making knowledge transfer less complicated. The drawback of the smaller company structure is that a limited number of employees reduces the likelihood of presence of employees within the company that have the capability to receive and recognize the value of the new information. This is especially difficult in SMEs lacking R&D experience or experience with research collaborations.

Collaborations between SME's and academia must aside from creating scientific knowledge also focus on identifying, absorbing and distributing the knowledge and the potential benefits of this throughout the relevant departments in the SME and sometimes even beyond the sales and marketing division to the customers or end-users to identify and illustrate the value. This type of collaboration requires an open and easy flow of knowledge between the different partners. Insight to business plans and company strategies are highly valuable for the development of tailored solutions suitable for the individual partners. Building a tight network between core industrial partners significantly increases the innovative capability of the companies and increases the likelihood to access useful knowledge generated in the project (Tsai 2001) and at the same time provides the industrial partners with the best position to utilize and benefit from the collaboration also with academia.

10 Critical Success Factors for New Product Development

The commercial success rate of industrial innovation ideas depends to a certain extent on the type of industry in which it is generated. In general industrial innovations have <1% likelihood to reach the market as a commercial success even after an initial self-screening process performed by the person or team responsible for the idea (Stevens and Burley 1997). Increasing the success rate of collaboration projects may find inspiration through comparison to intercorporate success parameters. Cooper and Kleinschmidt (2007) have pinpointed four key drivers to profitable new product development. The existence of a high-quality, rigorous new product process including preliminary market assessment, preliminary technical assessment, detailed market studies, detailed technical assessment and financial and business analysis as well as a sharp and early product definition was found to be the strongest driver. Secondly having a defined new product strategy with clearly defined areas of strategic focus, setting goals or objectives for the total new product effort as well as communicating of the role of new product development in achieving business goals is also a key driver. Furthermore the needed resources in terms of allocated time and money as well as overall commitment to the product development have to be devoted to the project if the success rate should improve (Cooper and Kleinschmidt 2007).

SME's can have difficulties handling the above mentioned requirements by themselves, and collaborations with external partners can increase the success rate of an innovative project idea. Therefore we have developed a model for initial screening

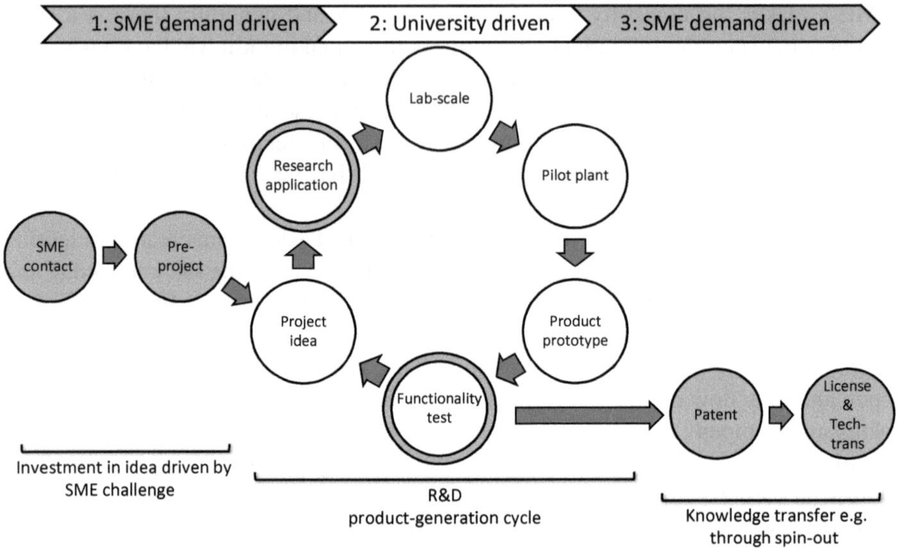

Fig. 1 Project life circle model for collaboration between academia and SME's. The initial pre-project phase is aiming at making preliminary market assessment, preliminary technical assessment, initial market studies, initial technical assessment and preliminary financial and business analysis as well as attempting an early product definition. This creates the basis for identifying the technology and developmental needs to define a project idea and setting a relevant and motivated project team from the initial project partner networks both within academia, larger private companies as well as other SME's

of potential product specifications, market potentials, technical requirements and needed resources in terms of time, manpower, expert knowledge and money. These preliminary data are used for preparing and filing a research project application for performing the more thorough research and product development. A scheme of the developed method is shown in Fig. 1.

One important feature of this project process is the production of prototypes. Developing prototypes may for some applications be possible from laboratorial scale trials, but within the food ingredient and especially the feed ingredient market larger quantities are required for performance and functionality testing. For this reason pilot plant productions are essential for several reasons. First of all product quality and application performance can be tested and modified to fulfill requirements defined by the end-users, but larger scale trials also gives possibilities of estimating mass balances of the various products, energy consumption etc., and finally calculations of expected investments and variable production costs can be done based on results from the pilot plant trials. All of this contributes to increasing the likelihood of getting the optimal processing technology developed producing high quality and high value products and getting the developed products successfully introduced to the right markets.

11 Final Remarks

SME's have large potential for expanding the Danish economy and creating new innovative processes and products that can mitigate some of the consequences of the four major crises comprising the environmental crises, the food crises, the energy crises, and the economic crises. To unfold this potential the gap between SME's and academia must be bridged. Thereby university research and the industrial knowhow from the private companies can be combined in an effort to develop holistic, environmentally friendly and economically feasible technologies for optimal processing of agricultural products, producing food and feed ingredients as well as bioenergy and non-food products.

Acknowledgement Funding for the research projects, of which the developed collaboration model is based, is highly acknowledged. This include Ministry of Environment and Food: GUDP J.no. 34009-13-0773 NyProFOOD—Purification of new plant proteins for food applications; GUDP J.no. 34009-12-0549 Value optimized Danish grown rapeseed proteins and fibers; MUDP MST-141-00671 Technology clarification by purification of high value ingredients from side-stream. InnovationsFonden—Denmark—HTF 058-2012-01 HITFOOD—High quality ingredients for the food industry. The Danish Agency for Science, Technology and Innovation HAPFAM—Healthy and affordable protein rich foods for African markets.

References

Andersen, K. E., Sørensen, J. C., Sørensen, H., & Sørensen, A. D. (2012). *A process for the manufacture of a product from a plant material*. WO/2012/116703.

Austin, G. L., Ogden, L. G., & Hill, J. O. (2011). Trends in carbohydrate, fat, and protein intakes and association with energy intake in normal-weight, overweight, and obese individuals: 1971-2006. *The American Journal of Clinical Nutrition, 93*, 836–843. doi:10.3945/ajcn.110.000141.

Bentsen, N. S. (2011). *Bioenergi: Udvikling, anvendelse og miljømæssige forhold – Bilag til det etiske råds redegørelse: Bioenergi, fødevarer og etik i en global verden*. Det Etiske Råd, Skov & Landskab, Det Biovidenskabelige Fakultet, Københavns Universitet.

Berglund, K., Berglund, M., Berglund, Ö., Grissén, C., Hald, A. B., Hedberg-Fält, P., Helander, I., Larsen, S. U., Lundgrén, J., Lærke, P. E., Nielsen, L., Stefanek, K., Stubsgaard, A., Svensson, S. E., Jørgensen, T. V., & Wennerberg, P. (2012). Evalueringsrapport Marginale Jorder och Odlingssystem. *BioM Bæredygtig bioenergi*.

Bjørn, C. (2014). *Cooperative movement* [WWW Document]. Accessed February 16, 2016, from http://denstoredanske.dk/Erhverv,_karriere_og_ledelse/Erhvervsliv/Andelsbev%C3%A6gelsen/andelsbev%C3%A6gelse

Cohen, W. M., & Levinthal, D. A. (1990). Absorptive capacity: A new perspective on learning and innovation. *Administrative Science Quarterly, 35*, 128. doi:10.2307/2393553.

Concern Statistics and Analysis. (2012). *Public research: Effects on innovation and economic growth*. Ministry of Higher Education and Sciences.

Cooper, R. G., & Kleinschmidt, E. J. (2007). Winning businesses in product development: The critical success factors. *Research Technology Management, 50*, 52–66.

Danish Ethics Committee. (2012). *Bioenergy, foods and etichs in a globalized world*. Det Etiske Råd.

Denmark, The Government. (2012). *Denmark – Land of solutions: Enhanced cooperation and a better framework for business innovation*. Copenhagen: Regeringen.

Easterby-Smith, M., Graça, M., Antonacopoulou, E., & Ferdinand, J. (2008). Absorptive capacity: A process perspective. *Management Learning, 39*, 483–501. doi:10.1177/1350507608096037.

Eshel, G., Shepon, A., Makov, T., & Milo, R. (2014). Land, irrigation water, greenhouse gas, and reactive nitrogen burdens of meat, eggs, and dairy production in the United States. *Proceedings of the National Academy of Science, 111*, 11996–12001. doi:10.1073/pnas.1402183111.

European Commission. (2005). *European Commission, 2010: Investing in European research* [WWW Document]. ec.europa.eu. Accessed February 23, 2016, from http://ec.europa.eu/invest-in-research/index_en.htm

Food and Agriculture Organization of the United Nations, World Health Organization (Eds.). (1994). *Fats and oils in human nutrition: Report of a joint expert consultation*, Rome, 19–26 October 1993. FAO food and nutrition paper. Rome: World Health Organization, FAO.

Food and Agriculture Organization of The United Nations, World Health Organization, The United Nations University. (1981). *The use of biological value of protein in evaluation its quality for human requirements* [WWW Document]. Accessed March 1, 2016, from http://www.fao.org/3/contents/fafed3e7-a4b8-59c4-afb1-fd16ec21d91d/M2835E00.HTM

Gunders, D. (2012). Wasted: How America is losing up to 40 percent of its food from farm to fork to landfill. *Natural Resources Defense Council, 26*.

Heller, M. C., & Keoleian, G. A. (2000). *Life cycle-based sustainability indicators for assessment of the US food system* (No. CSS00-04) (Report of the Center for Sustainable Systems). Ann Arbor, MI: Citeseer, The Center for Sustainable Systems.

Kalundborg Symbiosis Center. (2015). *Symbiosis Center Denmark, 2015: Kalundborg Symbiosis Center* [WWW Document]. Accessed June 3, 2016, from http://www.symbiosecenter.dk/en

Kearney, J. (2010). Food consumption trends and drivers. *Philosophical Transactions of The Royal Society B Biological Sciences, 365*, 2793–2807. doi:10.1098/rstb.2010.0149.

Kierkegaard, S. (1859). *Synspunktet for min forfatter-virksomhed: en ligefrem meddelelse, rapport til historien*. C.A. Reitzels Forlag.

Nordic Council of Ministers, N.C. of M. (2008). Nordic nutrition recommendations 2012. *Nordic Nutrition Recommendations, 5*, 1–3. doi:10.6027/Nord2014-002.

Organisation for Economic Co-operation and Development. (2002). *Frascati manual 2002: Proposed standard practice for surveys on research and experimental development: The measurement of scientific and technological activities*. Paris: Organisation for Economic Co-operation and Development.

Paddon-Jones, D., & Rasmussen, B. B. (2009). Dietary protein recommendations and the prevention of sarcopenia. *Current Opinion in Clinical Nutrition and Metabolic Care, 12*, 86–90. doi:10.1097/MCO.0b013e32831cef8b.

Pierrot, J. (2015). *Vegetable oils in Europe (CBI trends)*. Hague: CBI Ministry of Foreign Affairs.

Søren Kierkegaard Forskning… (2015). *Søren Kierkegaard Forskningscenteret, 2013: Citater* [WWW Document]. Accessed June 3, 2016, from http://www.sk.ku.dk/citater.asp#SFV.s96

Statistikbanken. (2014). *Danmarks Statistik, 2015: Statistikbanken* [WWW Document]. Dan. Stat. Accessed February 29, 2016, from http://www.statistikbanken.dk/10474

Stevens, G. A., & Burley, J. (1997). 3,000 raw ideas = 1 commercial success! *Research Technology Management, 40*, 16.

The Ministry of Economic and Business Affairs. (2011). *Economic analysis – Growth through innovation* (No. 3).

Tsai, W. (2001). Knowledge transfer in intraorganizational networks: Effects of network position and absorptive capacity on business unit innovation and performance. *The Academy of Management Journal, 44*, 996–1004.

U.S. Department of Agriculture, U.S. Department of Health and Human Service. (2010). *Dietary guidlines for Americans* (7th ed.). Washington, DC: U.S. Government Printing Office.

United Nations. (2015). *World population prospects: The 2015 revision, key findings and advance tables* (No. ESA/P/WP.241). New York: United Nations, Department of Economic and Social Affairs, Population Division.

WHO. (2016). *Global and regional food consumption patterns and trends* [WWW Document]. World Health Organ. Accessed February 16, 2016, from http://www.who.int/nutrition/topics/3_foodconsumption/en/print.html

World Bank Group. (2016). *Global economic prospects, spillovers amid weak growth* (A World Bank Group Flagship Report).

World Health Organization. (1998). *WHO | 50 Facts: Global health situation and trends 1955-2025* [WWW Document]. WHO. Accessed February 29, 2016, from http://www.who.int/whr/1998/media_centre/50facts/en/

Zeuli, K.A., & Cropp, R. (2004). *Cooperatives: Principles and practices in the 21st century (A1457) – A1457.pdf* [WWW Document]. Accessed February 15, 2016, from http://www.learningstore.uwex.edu/assets/pdfs/A1457.pdf

Zone of Proximal Development. (2012). *McLeod, 2012: Zone of proximal development – Scaffolding | Simply psychology* [WWW Document]. Accessed June 3, 2016, from http://www.simplypsychology.org/Zone-of-Proximal-Development.html

[illegible]

Part III
Resources of the Bioeconomy: Sustainable Biomass Supply

Increasing Biomass Production to Sustain the Bioeconomy

Iris Lewandowski

Abstract The bioeconomy builds on biomass as a resource base. Increased demand for biomass in a growing bioeconomy will lead to increased competition for this resource. However, current bioeconomy strategies are not sufficient to ensure the additional biomass demand is met sustainably. This contribution describes the criteria for a sustainable production and supply of biomass and suggests approaches for increasing the availability of sustainably produced biomass. In this context, the concept of sustainable agricultural intensification is elaborated by showing how breeding and more efficient cropping and land use systems can contribute to increasing biomass production. The participation and empowerment of farmers is addressed as a prerequisite for the implementation of sustainable biomass production. It is concluded that sustainable intensification on available agricultural land has large potential for increasing biomass supply and appears a more promising strategy than mobilizing additional land resources, mainly marginal land. In this strategy, the integration of land use functions, biomass production systems and biomass uses offer an encouraging method for avoiding competition for biomass and land for its production. A biomass supply strategy is envisioned, which makes use of all forms of biomass in an integrated approach and takes account of the interactions between them through the conceptualization of bioeconomic networks. The implementation of efficient strategies for the securing of sustainable biomass production and supply will require both continued research and political support.

1 Biomass Resources

The bioeconomy builds on biomass as a resource base. The term "biomass" refers to all organic material originating from plants, animals or microorganisms. This includes edible biomass, such as starch, sugar and oil-rich biomass, and non-edible lignocellulosic biomass from dedicated crop production, residues and wastes (Lewandowski 2015). Today, "biomass" is most frequently used to refer to organic

I. Lewandowski (✉)
Department of Biobased Products and Energy Crops, Institute of Crop Science (340b), University of Hohenheim, Fruwirthstr. 23, 70593 Stuttgart, Germany
e-mail: Iris_Lewandowski@uni-hohenheim.de

S. Dabbert et al. (eds.), *Knowledge-Driven Developments in the Bioeconomy*, Economic Complexity and Evolution, DOI 10.1007/978-3-319-58374-7_10

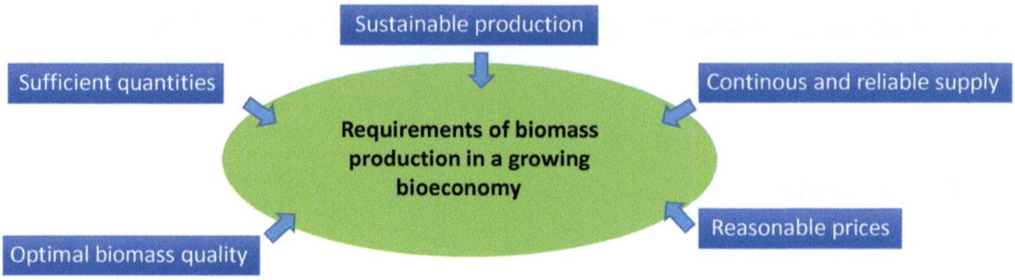

Fig. 1 What is required of biomass production and supply in the bioeconomy?

material utilised for energy production and other non-food applications such as the production of biogenic materials and chemicals. In the following text we use the more general definition of biomass, which includes edible as well as non-edible organic material.

The envisioned "ideal" bioeconomy will pay heed to ecological, social and health considerations and at the same time be internationally competitive. It will enable the use of biomass as a resource for food, feed, materials and energy through the development of new technologies and biotechnological processes (Staffas et al. 2013). It is thus apparent that the transition to a bioeconomy will be dependent on a sufficient supply of biomass of adequate quality for its intended application and that this biomass should derive from reliable and sustainable production (see also van Dam et al. 2005) (see Fig. 1).

The lessons learnt from the development of bioenergy have shown that concerns about the security of feedstock supply, sustainability of biomass production and impact of increased biomass use on food security are major bottlenecks for the further expansion of bioenergy production (Lewandowski 2015). It is estimated that by 2020 an additional 33 million tonnes of biomass will be required for the growing bioeconomy in Europe.

This leaves us with the main challenge to the sustainable growth of the bioeconomy: the sustainable production and supply of biomass. Therefore the objectives of this chapter are, firstly, to discuss how sustainability of biomass production and supply can be defined and, secondly, how the production of biomass can be increased in a sustainable way.

2 What Is Sustainable Biomass Supply?

There are various approaches to defining the sustainability of biomass supply, including legislative ones such as the EU directive 2009/28/EC (EC 2009). The main aspects of sustainable biomass supply addressed in this directive are (1) achieving minimum greenhouse gas (GHG) emission savings and (2) banning the use of land with high biodiversity (e.g. primary forest and highly biodiverse grasslands) or high carbon stocks (e.g. wetlands, continuously forested areas and peatlands) for biomass production. It also regulates agricultural production according to EU Cross Compliance rules and the traceability of biomass origin (EC 2009). However, the EC definition ignores the social and economic criteria involved. To date, the most comprehensive approach to the definition of sustainable biomass production and supply has been developed by roundtables, such as the Roundtable on Sustainable Palm Oil (RSPO) http://www.rspo.org/, Roundtable Responsible Soy (RTRS) http://www.responsiblesoy.org/?lang=en, Bonsucro (for sugar cane) http://bonsucro.com/ and the Roundtable on Sustainable Biomaterials (RSB) http://rsb.org/. Roundtables are multi-stakeholder initiatives that develop sets of principles, criteria and indicators for sustainable or "responsible" biomass production and supply, mostly for a specific feedstock and for the purpose of providing the basis for certification schemes (Lewandowski 2015). Table 1 gives an overview of criteria used for sustainable biomass supply addressed by those roundtables. In this study, we refer to "biomass supply" as the process of biomass production, harvesting, pre-treatment and transport to the plant gate. This broader scope is chosen because the sustainable production of biomass alone cannot guarantee that this biomass is available to consumers and industry, should processing facilities to prepare the biomass for transport or transport infrastructure be missing.

These criteria for sustainable biomass supply were developed in response to the increasing demand for biomass in the context of biofuel mandates and biofuel consumption, especially in the EU and the US. However, despite the efforts to define sustainable biomass supply for bioenergy, public perception of a non-sustainable production of biomass for bioenergy is one of the main barriers to the introduction and expansion of modern bioenergy (Alasti 2011). Clearly, the wide diversity of opinions and positions on sustainability of biomass supply is grounded in the different and often contradictory interests and perceptions of the stakeholders involved (Wicke et al. 2014). There is as yet no consensus on the best approach to a more sustainable biomass supply due to the manifold criteria applied (see Table 1) and the numerous trade-offs between productivity and environmental performance (Pretty 2008). Therefore the development of a common understanding of sustainability is a prerequisite for sustainable biomass supply in a growing bioeconomy (Ros 2014; Schnabel 2014).

What can the development of bioenergy tell us about the sustainability of biomass supply?

1. The theoretical potential for sustainable biomass supply is very high (Smeets et al. 2007; van Vuuren et al. 2009; Dornburg et al. 2010), but the actual implementation is currently extremely limited.
2. We are aware of the demand for and expectations of sustainable biomass supply (see Table 1 for an overview of criteria), but we are struggling with their simultaneous implementation and with the trade-offs between them (Pretty 2008; The World Bank 2009; Dornburg et al. 2010; Pfau et al. 2014; Lewandowski 2015). The question of how to conduct agricultural production in such a way that productivity, farmers' income and environmental aspects (such as nitrogen

Table 1 Summary of criteria for sustainable biomass supply, compiled from the sustainability studies of the Roundtable on Sustainable Palm Oil (RSPO), Roundtable Responsible Soy (RTRS), Bonsucro and the Roundtable on Sustainable Biomaterials (RSB) (from Lewandowski 2015)

Social criteria
Respect of human and labour rights
– No child labour
– Consultation/stakeholder involvement
– Payment/fair salary
– No discrimination (sex, race)
– Freedom of association
– Health and safety plans
– Respect of customary rights and indigenous people
Smallholders' rights
Responsible community relations
Socio-economic development
Well-being
Ecological criteria
Protection of biodiversity/wildlife/HCV areas
Environmental responsibility
– Minimization of waste
– Reduction of GHG
– Efficient use of energy
– Responsible use of fire
Soil degradation
Water resources/quality
Air pollution
Use of best practice/responsible agricultural practices
– Responsible use of agrochemicals
– Training of employees
Responsible development of infrastructure and new areas of cultivation/plantations
– Impact assessment prior to establishment
– No replacement of HCV areas after year X
– No establishment on fragile soils
– Restoration of degraded land
– Compensation of local people, informed consent
– Maintenance of sites with high carbon soil content
General and economic criteria
Commitment to continuous improvement
Wise use of biotechnology
Climate change and GHG mitigation
Food security
Use of by-products
Traceability
Transparency
Legality
Responsible business practices
Respect for land-use rights

leaching and soil erosion) are balanced has been on the agenda of agricultural research for decades (Pretty et al. 2010; Wolters et al. 2014). At the same time we need to address sustainability criteria on different levels. These range from aspects of soil conservation—relevant at field level—through maintenance of biodiversity—relevant at landscape level—to food security—relevant at local to global level. Therefore, we will need to think of new approaches to defining sustainable biomass production. One approach could include the ranking and prioritizing of sustainability criteria. A priority list could help to reduce the complexity and support the implementation ability of sustainable biomass production and supply.

3. Of all the sustainability criteria for biomass, food security has emerged as one of the most relevant in public perception (Solomon 2010; Tait and Barker 2011; Pfau et al. 2014; Lewandowski 2015). Tait and Barker (2011) even suggests an ethical standard for biofuels which puts food security in first place. At the same time its implementation has turned out to be rather complex. Whereas certification schemes can offer concrete implementation strategies for most criteria for sustainable biomass supply, as listed in Table 1, there are no effective strategies available for dealing with the competing claims on biomass or on land used for its production. These competing claims include the "food-versus-fuel" as well as the "biomass-production-versus-biodiversity" conflicts (Tilman et al. 2009; Pfau et al. 2014).

In the context of biomass production there are two types of competition which need to be addressed:

(a) Competition for biomass *per se*
(b) Competition for land on which biomass is grown

The discussion on the competition for biomass has been most controversially led in the context of road transportation biofuels (Solomon 2010). So-called "first generation" biofuels, which make up the majority of biofuels consumed today, are produced from sugars and starch (ethanol) or vegetable oils (biodiesel). Two examples of the direct effects of the expansion of transportation biofuels on the supply and costs of foodstuffs are the so-called "tortilla crisis" in Mexico and the increase in palm oil prices in Asia:

- From 2008 onwards the US used about half of its corn supplies for ethanol production instead of exporting them to Mexico. Until then, Mexico had imported corn at dumping prices from the US. Mexican farmers had previously produced enough corn for the total population but, as a consequence of being forced to open their markets to heavily subsidized imports from the US, they struggled to compete and eventually gave up production (Levitt 2011). However when the US started to increase its production of ethanol from corn, exports to Mexico decreased and corn and tortilla prices in Mexico tripled.
- Palm oil, which makes up 35% of global vegetable oil production (USDA 2015), is increasingly being used for biofuel production. In 2011, 68% of palm oil was used for food, 27% for industrial and 5% for energetic purposes (FNR 2014).

Additional demand for palm oil for food (especially by poorer people as palm oil is the cheapest of all vegetable oils) and for fuel purposes has been predicted for the future and has already led to price increases (OECD-FAO 2012). Increasing food prices impact low-income households in particular as these often spend up to 75% of their income on food (Naylor et al. 2007).

These examples show that the energetic or material use of edible biomass in particular should be carefully considered on account of its direct effect on food prices and availability. However competing claims not only impact on food prices and availability. The rapid expansion of palm oil plantations in Indonesia has been and still is at the cost of deforestation of tropical rainforest, which often stands on carbon-rich peatlands (Laurance et al. 2010; Wilcove and Koh 2010). This has led to losses in biodiversity and high GHG emissions (Wicke et al. 2008). Although there is an estimated potential of 8–30 million hectares of degraded imperata grasslands in Asia which could be used for oil palm plantations (Kosonen et al. 1997; Otsamo et al. 1995), none of this grassland is currently employed for this purpose. This is because the high establishment costs of oil palm plantations are financed by the sale of wood from the rainforest and the exploitation of imperata grasslands incurs additional costs.

Another example of competing claims on land can be seen in Madagascar. In 2009 foreign companies tried to gain access to land here to establish jatropha plantations. The state government was interested in granting access to land because it wanted to attract investment into the country and create new jobs. But local villagers use the land as pasture and claimed a right to the management of the area (Burnod et al. 2013). In this example there was competition for land control between the central and local governments and the conflict centred around the question of whether it is more important to create new jobs or to maintain traditional land-use rights (Burnod et al. 2013).

The following section looks at approaches to increasing the sustainable production and supply of biomass for a growing bioeconomy in ways which avoid competition for biomass and agricultural land.

3 How Can Biomass Production and Supply be Increased Sustainably?

In the context of the bioeconomy, three approaches to increasing biomass production are currently being discussed: (1) Sustainable agricultural intensification; (2) Use of fallow or marginal land; and (3) Reducing losses and increasing efficiency of biomass utilization.

3.1 Sustainable Agricultural Intensification

The term "sustainable intensification" dates back to the 1990s and was first discussed in the context of African agricultural development (Reardon et al. 1997). It became common following the UK Royal Society's report "Reaping the benefits" of 2009, which explored the future of crop production (Garnett and Godfray 2012). Thoughts on sustainable intensification resulted from debates about increasing yields, mainly of arable crops, in the face of resource scarcity and environmental challenges (Garnett et al. 2013). The Royal Society (2009) defined sustainable intensification as a form of agricultural production whereby "yields are increased without adverse environmental impact and without the cultivation of more land". More recently, Pretty et al. (2011) extended this definition of sustainable agricultural intensification to "producing more output from the same area of land while reducing the negative environmental impacts and at the same time increasing contributions to natural capital and the flow of environmental services."

Sustainable intensification of agriculture clearly targets "land sparing" (high-yield farming combined with protection of natural habitats from conversion to agricultural land) and "land sharing" (integration of several functions on the same land) as opposed to increasing the area of agricultural land, because the latter carries high environmental costs (Garnett et al. 2013). According to Garnett et al. (2013), for sustainable intensification to be successful, the following questions need to be addressed:

- How can land sharing deliver sufficiently high yields and also multiple ecosystem services?
- How can trade-offs between yields and the various environmental benefits be quantified?
- How can these trade-offs be best resolved on different spatial scales?

Approaches to sustainable intensification include: increasing yield per hectare; increasing cropping intensity (i.e. two or more crops per unit of land) or other inputs (water); and changing from low-value to higher-value crops (Pretty et al. 2011). According to Pretty et al. (2011), the intention is to achieve productive and sustainable agricultural systems that make the best use of both crop varieties and livestock breeds and also their agro-ecological and agronomic management. Strategies include:

- Utilizing crop varieties and livestock breeds with a high productivity ratio;
- Efficient use of external inputs and avoidance of any unnecessary use;
- Harnessing agro-ecological processes such as nutrient cycling, biological nitrogen fixation, allelopathy, predation and parasitism;
- Minimizing the use of technologies or practices that have adverse impacts on the environment and human health:
- Productive use of human capital (knowledge and capacity to adapt) and innovative and social capital to resolve common landscape-scale problems.

Clearly, sustainable intensification in agriculture strives to make best use of new breeds, modern technologies and efficient novel systems to increase productivity. These are discussed in the following sections.

3.1.1 Breeding

The improvement of biomass production and land-use efficiency by increasing yields has been at the forefront of breeding and agronomic research for decades. The combined efforts of plant breeding and crop management improvement in the period 1961–2007 led to annual yield increases of 2.1% for wheat, 0.7% for sugar cane, 2.5% for rapeseed and 3.1% for palm oil (Lywood et al. 2009). The potential for yield increases in "new" biomass crops currently under development, such as miscanthus, willow and jatropha, is estimated to be particularly high because these crops have not yet been bred and selected intensively and they exhibit a large genetic variability which can be exploited. However, because investment in the development of these crops is relatively low, the rate of yield increase is likely to be lower than that achieved for cereals in the past. In addition, perennial crops have long breeding timelines, the yield response of perennial grasses to fertilizer is particularly low and the manipulation of the harvest index is not applicable (Searle and Malins 2014).

Important goals for the development of improved crop varieties are: higher yields; optimal quality of the biomass for its intended use; tolerance and adaptability to biotic and abiotic stresses such as pests and drought; and an improvement of resource use efficiency, especially water and nutrients. Sugar cane is the most productive crop for ethanol production in Brazil, as is maize for biogas production in Germany. Both are so-called "C4" crops which have a more efficient photosynthetic pathway than so-called "C3" crops, such as wheat and rape seed. The theoretical maximal photosynthetic energy conversion efficiency is 4.6% for C3 and 6% for C4 plants (Zhu et al. 2010). Therefore, the development of C4 crops could be one method of achieving higher yields. These are however only more productive than C3 crops in temperate to warm climates. Zhu et al. (2010) discuss several possibilities of increasing photosynthetic efficiency, including physiological approaches such as plant architecture modifications, and biochemical approaches such as carbon metabolism engineering. Another crop production mechanism for optimizing resource-use efficiency is perennial growth. All "new" biomass crops, such as miscanthus and willow, are perennials, which are planted only once in a plantation lifetime of 20–30 years. They use nutrients very efficiently because they recycle a large proportion of them back from the leaves into underground rhizomes (perennial grasses) or stems and roots (short rotation coppice trees) during the growth period, from where newly emerging shoots and leaves are nourished in the spring (Jørgensen 2011; Lewandowski 2013).

There is large variation in "crop:residue ratios", for example 0.8–1.6 for wheat, 0.9–1.2 for maize and 1.4–2.0 for rape (Scarlat et al. 2010). This is due to differences in cultivation practices for the same crop, adverse field conditions over the years, climatic conditions, variation in soil quality and the use of different varieties (Lindenẽt al. 2000; Graham et al. 2007; Scarlat et al. 2010). This indicates that there is potential to increase residue yields through proper cultivation practices and the use of improved varieties. The use of residues has only recently become relevant in the context of increasing biomass demand for bioenergy. In the past decades, breeding has concentrated on high grain yields and these were partly achieved at the expense of straw yields. Future breeding programmes for bioeconomy crops should strive to optimize the overall crop yield and maximize biomass yields also from the "residual" part of the crop.

Presently, around 60% of globally available biomass is used for animal feed (von Braun 2014). Considering the upward trend in global meat consumption, it is unrealistic to assume that biomass potentials can be mobilized through a reduction in feed demand by changing human diets. Therefore the use of feed biomass should be optimized by the improvement of feed-use efficiency in animal husbandry. Animal breeding can significantly contribute to this. The magnitude of the differences in feed-use efficiency can be seen by comparing the amount of fodder required in different regions. The production of 1 kg bovine meat requires 99 kg feed in Sub-Saharan Africa and 72 kg feed in East Asia, but only 24 and 26 kg feed in West Europa and the US, respectively (Bouwman et al. 2005; FAO 2003).

Biomass resources other than land-based crops and animals, for example algae, are presently being sought. The production of algae in open or closed systems is not dependent on agricultural land and could therefore be performed in areas with unfertile soil, such as deserts, that are close to the sea or a water supply. Because their production in closed systems is very expensive and technically challenging, algae biomass is currently only used for high-value niche products, such as carotinoids (Wijffels and Barbosa 2010). Algae production in open raceways faces the challenge of contamination issues and is not yet cost-competitive (Gallagher 2011). The search for algae strains suitable for large-scale, commercialised production that can concentrate the target substances, such as oils, in sufficient quantities is still at the research stage (Rodolfi et al. 2009). Other challenges, such as the high water content of the algae suspension, difficulties in separating the algae from the suspension, and the target substances from the algae biomass, have altogether led to a situation in which it has not yet been possible to establish commercial algae production plants (Reijnders 2012).

3.1.2 More Efficient Cropping and Farming Systems

At the level of cropping systems, agricultural research investigates the optimal combination of crop and variety choice, soil cultivation, crop establishment, fertilization and crop protection regimes. The aim is to maximize yield, ensuring biomass quality and at the same time balancing ecological impacts (Wolters et al. 2014). The highest potentials for increasing yields by improving cropping systems are seen in approaches targeted at closing the gap between obtainable and actually harvested yields. In many regions of Africa, Latin America and Eastern Europe this yield gap averages up to 55% (FAO 2003) and as much as 77% for specific crops, such as sweet potato in Cameroon (Yengoh and Ardö 2014). This is often not a problem of the biophysical suitability of the site or "site x crop combination", but of insufficient agronomical practices and policy support (Yengoh and Ardö 2014). Strategies for closing the yield gap are: use of improved varieties; use of fertilizers or improved fertilizer application regimes; crop protection measures and weed control; and efficient harvest technology (Yengoh and Ardö 2014). Avoiding losses to pests is an important contribution to closing the yield gap, as this is currently in the order of 30–40% depending on the type of crop (Oerke 2006). While increasing crop yields will remain an important objective of agricultural research, the future challenge lies in the combination of increased productivity and environmentally beneficial performance of biomass production (Pretty et al. 2010; Wolters et al. 2014). For this purpose, agricultural research focuses on efficient cropping systems, which on the one hand improve yields and biomass qualities by choosing efficient crops and varieties and on the other hand minimize inputs, especially of agrochemicals with a high potential environmental impact such as nitrogen fertilizers. The range of management and technical solutions for improving the efficiency of cropping systems is very broad and includes, for example:

- Optimizing crop rotation to allow maximum exploitation of the productive season and of nutrients in the system and to reduce pressure from pests and diseases (see e.g. Krupinsky et al. 2002);
- Optimizing fertilization amount, timing and application technology, especially nitrogen fertilizer, and use of designer fertilizers adapted to specific crop needs in order to reduce nutrient losses (see e.g. Weisz et al. 2001; Adegbidi et al. 2003; Lewandowski and Kauter 2003);
- Optimizing the use of organic material for fertilization and soil carbon maintenance (see e.g. Adegbidi et al. 2003; Triberti et al. 2008);
- Optimizing soil cultivation regimes by appropriate soil cultivation timing and technologies to help avoid erosion and soil degradation and simultaneously prepare optimal conditions for crop establishment and growth. In this context, tillage systems with reduced soil cultivation intensities, such as low- or no-till soil cultivation or strip-till techniques are currently being investigated and no-till systems are already widely applied in the US (see e.g. Lal 2004; Fu et al. 2006);
- Reclaiming contaminated and degraded soils by means such as phytoremediation (see e.g. Dickinson et al. 2010) or increasing soil fertility and carbon contents by additives, for example biochar (Biederman and Harpole 2013);
- Optimizing crop protection regimes by applying preventive measures, biological crop protection, thresholds for intervention decisions and effective application technologies. These technologies prevail in ecological and integrated farming systems (see e.g. Doyle 1997; Letourneau and van Bruggen 2006);
- Optimizing irrigation regimes (if required) by applying efficient irrigation technologies which use water sparingly (see e.g. Howell 2001);
- Optimizing harvest regimes by reducing pre-harvest and harvest losses through optimal timing of harvest and efficient, well maintained harvest technologies (see e.g. Meehan et al. 2013);
- Applying precision farming technologies including sub-area-specific application of agrochemicals (Lencsés and Takács-György 2014).

Integration of functions at the farming system level is an important approach to reducing competing claims on biomass. The integration of production for the 4 Fs (Food-Feed-Fibre-Fuel) and the integration of the agricultural sectors (Crop-Animal-Bioenergy) are addressed here. There are many examples of modern bioenergy uses where the production of bioenergy is integrated with crop and animal production. In Germany, the number of biogas plants quadrupled in the last decade up to 7772 in the year 2013 and biogas now contributes 7% of electricity production. Over 40% of the biogas substrate used here is animal slurry (FNR 2014). This use of slurry in biogas plants can fulfil several functions. First, renewable energy is produced from a waste product. Secondly, gaseous N emissions from slurry are reduced. Thirdly, the accumulation and oversupply of nutrients in areas with high livestock densities is diminished because the nutrients are channelled into feedstock (slurry) for biogas plants and consequently into the biogas digestate. They can be processed to transportable and storable fertilizer, e.g. by separating and/or drying of the digestates or by extracting the nutrients (Fraunhofer IGB 2014). Other examples of integrated food and fuel production are the combustion of wheat straw in combined heat and power plants in Denmark (CADETT 1998) and sugar cane production in Brazil. In the former, the straw is a by-product of wheat grain production. In the latter, 30% of the by-product bagasse is used to fuel the process of sugar extraction and concentration, and 70% is used to produce electricity. The fibrous bagasse can also be used for material applications, such as biocomposites.

Multi-product use of crops is an important part of integrated systems, as can be seen clearly through the example of cassava. Currently, the starch-rich cassava roots are used for food or ethanol production and the leaves are left as residues in the field. However the protein content of the leaves (approximately 17.7–38.1% of DM) could also be used as protein and the biomass remaining after extraction could serve energetic purposes, such as biogas production (Latif and Müller 2014). Another multi-product example is the use of wheat grain for bread making and wheat straw as feedstock for CHP or second generation ethanol production. Biorefinery concepts are currently being developed to support the full exploitation of crops and their residues for various uses. In order to avoid the use of crop residues in these systems leading to soil carbon depletion (see e.g. Lal 2005), the residues from the processing or energetic use of biomass, such as biogas digestates, can be returned to the field (Möller et al. 2009).

Urban farming—the integration of food production into urban areas—can improve the food supply situation of the local population and decrease the demand for food imports from rural areas. While some urban agriculture concepts, such as skyfarming (Sauerborn 2010) are still theoretical, other forms, for example commercial farms in first-ring suburbs and rooftop gardens, are already in place in major cities such as New York and Berlin (Hodgson et al. 2011). Urban farming is seen as an approach to improving the supply of fresh food (e.g. vegetables), significantly reducing food transportation, creating jobs and income and making a positive contribution to the urban environment (Ackerman et al. 2014).

3.1.3 Improved and Novel Land-Use Systems

An important approach to the reduction of competition for land is the combination of different land-use functions in multiple land-use systems (MLU). Nature conservation, extraction of abiotic renewable resources, waste treatment and disposal, buffer zones and infrastructural improvements are among the land-use functions that can be combined with biomass production (Lewandowski et al. 2004). Many of these functions, e.g. soil erosion prevention, carbon sequestration, providing habitats for wildlife and soil decontamination, can best be combined with the production of perennial crops, mainly due to the long periods of soil rest and lower input demands (Freibauer et al. 2004; Jørgensen 2011; Lewandowski 2013). A practical example for an MLU biomass production system can be seen in Sweden, where willows are produced to supply the CHP Enköping with biomass. The willow plantations are irrigated with municipal waste water and benefit from the nutrients in this water. At the same time the plantations clean the waste water and therefore serve as a replacement for sewage plants (Grelle et al. 2007). This example shows how MLU systems can increase the value generation of and income from biomass production systems. Another example is described by Lewandowski et al. (2006) for the remediation of heavy-metal-contaminated soils by willow. However, most land-use functions are ecological functions of the agro-ecosystem for which farmers are not remunerated, but which are well suited to balancing productivity with ecological needs. The fulfilment of some of these functions creates additional workload for farmers or can compromise the productivity of the biomass production system, for example when limiting the amount of fertilizer or pesticide leads to lower yields. For such cases in particular, mechanisms or policies need to be developed which provide incentives for the implementation of ecological land-use functions. The EU agricultural policy (CAP) is an example of this. Direct payments are coupled with environmental measures, such as greening or environmentally benign fertilization regimes, through cross compliance (CC) rules.

Another land-use system where biomass production and environmental functions can be combined is grassland. Estimates of the proportion of the earth's land area covered by grassland vary between 20% and 40%, depending on the definition (FAO 2014). In 2000 there were an estimated 3.5 billion hectares of pasture and fodder crops globally, representing 26% of the earth's land area and 70% of global agricultural land (FAO 2014). The nature of grassland varies widely and ranges from natural to intensively managed. In the EU, the conversion of grassland to arable land is limited and in some countries it is prohibited altogether because grassland, especially natural grassland, is of high biodiversity value (Peeters et al. 2014) and important for soil carbon storage (Freibauer et al. 2004). While the existence of natural grasslands is not related to human activities, other grasslands are dependent on management regimes (Peeters et al. 2014). The conservation of biodiversity in semi-natural grasslands requires continuous management (Heinsoo et al. 2010). When managed carefully (e.g. by appropriate cutting frequency and timing; adapted fertilization strategies), the harvesting of grassland biomass can help maintain typical grassland communities and landscapes and at the same time potentially increase biomass supply (Rösch et al. 2009; Buhle et al. 2010). On account of the decreasing demand for roughage fodder in Europe, a surplus grassland area of 9.2–14.9 × 10^6 ha has been estimated in the European Union for the year 2020, which represents about 13–22% of permanent grassland (Prochnow et al. 2009). The yields and properties of the harvested biomass and hence the options for its use vary widely depending on the type and intensity of the management of these grasslands (Thumm et al. 2014). The use of biomass from natural to semi-natural grasslands may be uneconomical due to the low yields harvested and infrastructural limitations. However, on the other hand biomass use from these grasslands could create a win-win situation if it helps maintain the grassland community and simultaneously provides additional income for farmers. If the selling price for the biomass does not meet the extra costs that farmers incur to harvest it, the question arises of how the "service" of landscape maintenance should be paid for and who should pay for it.

3.1.4 Empowerment of Farmers

Sustainable intensification requires farmers to have greater knowledge and skills because they need to able to apply the often more complex management systems and techniques involved (Pretty et al. 2011).

The provision of technical solutions for the improvement of cropping systems alone will however not be sufficient to close the yield gaps. Farmers must also be willing to adopt these solutions and see an advantage in their application. For example, modern improved crop varieties often do not always possess the qualities of traditional cultivars preferred by farmers and consumers (Nhamo et al. 2014). Farmers also need to have the financial means to acquire the necessary agricultural inputs and knowledge and skills to apply them. This could be promoted through credit services and training programmes for farmers and by facilitating access to markets (Nhamo et al. 2014). The direct involvement of farmers in the development of improved varieties and agricultural systems using participatory approaches would serve to enhance their acceptance (Hilger et al. 2015; Nhamo et al. 2014; Pretty et al. 2011). This would also support the design of locally specific approaches which are necessary for successful yield improvement, especially in smallholder farmer systems (Nhamo et al. 2014; Krawinkel 2012).

Also, land-use planning for biomass production should involve smallholders and any farmers that actually use the land. In Madagascar, regional governments granted licenses to international companies for the establishment of jatropha plantations without involving the local farmers. These then sabotaged the project (Burnod et al. 2013). There are many other examples. To avoid local farmers' interests and rights being neglected and to prepare the ground for their support and co-operation in land-use projects, participatory approaches such as IREPA (Integrated Renewable Energy Potential Assessment) should be applied (Hilger et al. 2015). IREPA provides a people-centred, bottom-up approach to assessing the implementation potential of renewable energy technologies in smallholder agricultural systems (Winkler et al. 2016). The key to assessing the sustainable biomass potential is the ranking of sustainability criteria in a participatory approach. The search for operational methods to define the sustainable biomass potential often fails due to trade-offs between sustainability goals and the manifold criteria which are expected to address ecological, economic as well as social aspects. In this respect, natural science approaches alone are not sufficient to identify "the" way to sustainability. Social science approaches also need to be applied in the identification of the "best compromise" for sustainability.

3.2 Use of Fallow and Marginal Land

Marginal land is often seen as an untapped resource, in this case as land that could be used to improve sustainable biomass supply (Dauber et al. 2012). Potential analyses estimate that about 250–1580 Mha of marginal, degraded or abandoned agricultural land could be available for biomass production (Smeets et al. 2007; van Vuuren et al. 2009; Dauber et al. 2012). The figures depend on the definition of the term "marginal land" and which categories (waste land, degraded land, abandoned land etc.) it includes (Dauber et al. 2012; Lewis and Kelly 2014). Marginal production conditions can be defined in economic and biophysical terms (Dauber et al. 2012). If marginal land is defined as land that does not support economically viable agricultural (food) production, the status of marginality will depend on and change with land-use and biomass prices. This can be seen by the example of Germany, where land-use prices in some areas increased sharply in the vicinity of emerging biogas plants that require a supply of local biomass (Theuvsen et al. 2014). One of the consequences was that grassland was converted to biogas maize production (Bundesamt für Naturschutz BfN 2014). Another caveat to the use of land which is marginal in economic terms is the fact that the production of whatever biomasses, be it for food or energetic or material uses, on this land will result in low profit (van Dam et al. 2009; Dornburg et al. 2010; Dauber et al. 2012). It is to be expected that the GHG reduction potential of biomass production on marginal land will also be lower than on good quality land because the same level of inputs (energy, fertilizer etc.) results in lower yields (Dauber et al. 2012). Another socio-economic consideration is that the low yields harvested may lead to only low levels of income which require cheap labour and generate only low-value employment (Giller et al. 2007). Therefore the potential of economically marginal land to contribute to sustainable biomass supply should not be overestimated.

The effect of marginal land use on biodiversity should be assessed carefully before projects are implemented. The relatively low yields to be expected may require highly automated systems and large, monoculture plantations (Dauber et al. 2012). Another point of consideration is that land susceptible to agricultural marginalization is often in extensive farming regions where small-scale farming systems prevail. These are considered important for the preservation of agricultural biodiversity (Baldock et al. 1996). From a biodiversity perspective, the vegetation type dictates whether the use of marginal land has a positive or negative effect. There are many examples where biodiversity comes under pressure due to the expansion of energy crop plantations, for example the forests of the Chaco in Brazil (Beringer et al. 2011, Raghu et al. 2011), and where careful ex-ante assessment of potential impacts on biodiversity is advisable (Dale et al. 2010). An assessment should not only be performed with regard to impacts on biodiversity but should also consider the actual use and rights of use of the marginal land in order to avoid displacement of smallholder farmers or people who use the land for cattle production (Beringer et al. 2011).

Biophysical marginal land includes areas that are contaminated, degraded or have biophysical constraints such as steep slopes. Such production conditions can best be dealt with using perennial cropping systems, which can contribute to the amelioration of the land, as discussed above. These production systems mainly supply lignocellulosic biomass. There are also promising perennial systems that produce vegetable oils under biophysical marginal conditions, such as the oil-bearing shrub jatropha and the oil-bearing palm acrocomia (Poetsch et al. 2012; Becker et al. 2013). However, these crops and their management systems require further development before commercial production can be recommended.

3.3 Reducing Losses and Improving Biomass Use Efficiency

3.3.1 Harvest, Pre-treatment, Storage and Transportation

A large proportion of the biomass produced agriculturally does not reach the consumer due to losses in the supply chain. According to FAO estimates, approximately 1.3 billion tonnes of food are lost or wasted globally per year (Gustavsson et al. 2011). Food losses vary considerably depending on commodity and region. Taking the whole value chain into consideration, these can be as much as 25–35% for cereals and more than 50% for perishable products such as roots, tubers, fruits and vegetables (Aulakh and Regmi 2013). The reasons for these losses are manifold and mainly dependent on the three factors weather, infrastructure and the availability of technology (Aulakh and Regmi 2013). In less developed countries with less mechanized supply chains the largest losses occur during drying, storage, processing and transportation (Aulakh and Regmi 2013). More efficient farming and management systems in more developed countries, including crop protection, better transport, storage and processing facilities, lead to considerably lower food losses in the middle stages of the value chain. Here they occur mainly at the consumer level (Hodges et al. 2011). Losses also occur in biomass production for feed and energy chains. These are mainly due to inefficient harvesting and drying procedures and storage problems, for example if the biomass is too wet (Jirjis 1995).

Reduction of food and biomass losses is an effective way of increasing biomass availability without necessitating additional production resources. Relevant measures for reducing losses include:

- Improved training of farmers, enabling them to make better management decisions
- More efficient harvest technologies
- Providing farmers access to credit
- Drying of biomass
- Improved storage facilities and conditions
- Protection from insect pests, mites, rodents and birds
- Transportation infrastructure
- Adequate and efficient market systems
- Improved and more efficient processing technologies
- Use of by-products

The best approaches to the reduction of post-harvest losses vary depending on region. In developed countries, where most food is wasted at the consumer level, consumer education would be a promising strategy. In the less developed world the most important measures include better training of farmers, improved infrastructure to connect smallholders to markets, provision of sufficient financial incentives at the producer level and improved technologies supported by easier access to microcredits (Aulakh and Regmi 2013). Finally, in these regions large investments will be required to enable an effective and cost-efficient transportation of biomass. This is a prerequisite for making the biomass available to the market. Here the question remains as to who is able and willing to invest and whether the sharing of these investments by the public and private sector is a feasible option.

3.3.2 Biomass Conversion

As biomass is a limited resource, its use needs to be efficient. The energetic use of biomass in particular exhibits large variation in efficiency, ranging from 10% in traditional biomass for cooking in open fires to 60–90% for combined heat and power production in modern combustion plants (IEA 2007). A comparison of the efficiencies of biofuel production pathways by Huang and Zahng (2011) revealed a biomass-to-fuel (BtF) efficiency of only 36.5% for sugar-based ester diesel and up to 81.3% for methane production from maize via fermentation. The efficiency of biomass conversion depends both on the conversion technology and the properties of the biomass, such as water and target component contents [see e.g. van Loo and Koppejan (2008) for combustion]. Similar considerations hold true for the conversion of biomass into food and feed products. Therefore the optimization of biomass properties in line with the demands of the conversion technology along the supply chain is an important prerequisite for efficient biomass conversion.

Biorefineries are also seen as a promising approach to the efficient exploitation of biomass resources (Parajuli et al. 2015). Biorefining can be defined as "the sustainable processing of biomass into a spectrum of marketable products and energy" (IEA Bioenergy 2009). Part of the biorefinery approach is that all biomass components and by-products are made use of and materials are recycled during the process as much as possible. A simple example of a biorefinery concept is the biogas digester. Here the biomass is first used energetically and the remaining biodigestate is then made available for other purposes. The main application of biodigestates is as fertilizers because digestates contain significant amounts of plant nutrients, such as N, P and K, and can fully replace mineral fertilizer (Möller et al. 2009). Pulp and paper factories are an example of wood-based biorefineries. Green biorefining technologies are used to exploit fresh grass or silage-based feedstocks and convert them into a number of materials for feed, material and energetic uses (Mandl 2010). There are currently several green biorefinery activities being pursued, for example in the Netherlands, Denmark, Austria and Germany. However, most of these are in the pilot phase and not yet economically viable (Cherubini 2010).

In addition to the physical integration of material and energetic applications of biomass in biorefineries, the sequential integration of its use in cascade systems has been proposed as an approach to increasing biomass use efficiency and maximizing greenhouse gas (GHG) emission reduction potentials. Cascading concepts foresee biomass being utilised for various functions throughout its life cycle. These functions successively follow on one after the other starting with material use, followed by recycling and energy recovery (Dornburg and Faaij 2005). It is anticipated that the cascade use has a higher GHG emission reduction potential compared to a single energetic use of biomass because carbon is sequestered in the material use phase and not released until sometime afterwards when the biomass is used energetically (Dornburg and Faaij 2005). However, this is only the case if the material is recycled efficiently and incinerated instead of being disposed of in landfill. Recently, the term "cascade use" is also applied to the combined and simultaneous energetic and material use of different biomass components.

3.3.3 Optimizing Biomass Use and Allocation

The efficient use of biomass and biogenic products can reduce competition for biomass by lowering the demand for and consumption of these products. The relevance of reducing food losses at the consumer level has been discussed in previous sections. It is estimated that in Germany and the US about a third of all food losses or approximately 100 kg per capita and year of food are wasted at the consumer level (Gustavsson et al. 2011). Therefore a large proportion of the additional food necessary to feed undernourished people could be fulfilled by reducing losses elsewhere—if the distribution issue is neglected in this theoretical consideration.

Generally we need to make wise choices about the use of whatever biomass is available. There is wide agreement that the energetic use of biomass should not compromise food security. For energetic use, the pathway with the highest potential for reducing GHG emissions is often recommended as the optimum. The

European biofuel policies were initiated with the main aim of greenhouse gas emission reduction. The 2009 European Directive on Renewable Energy set a target of 35% GHG emission reduction (compared to fossil fuels) for current biofuel production and 50 or 60% (for old and new conversion plants respectively) GHG emission reduction for future biofuel production. However there are other biomass applications that actually reduce GHG emissions by much more than liquid biofuels. For instance, combined heat and power production by combustion of lignocellulosic biomass can achieve GHG emission reductions of more than 90% (Giuntoli et al. 2014). Another example is biochemicals, where the GHG reduction potential per hectare can be seven times that of sugar-based ethanol, provided they are produced from lignocellulose crops (Barry et al. 2011).

Often the criteria for the most efficient are not the same as those for the most sustainable biomass use. This is further complicated by the varying perceptions of what should be considered sustainable. Therefore the definition of sustainability criteria for biomass use should be specified in a participatory societal process and the optimal allocation of biomass to the various uses should be supported by political measures (Kidane et al. 2006). However, political measures can become counterproductive. For instance, the EU policy incentivizing the production of biodiesel from vegetable oils is often criticised as being in competition with food production (Grethe et al. 2013). To avoid the support of one bioeconomic value chain threatening sustainable development, the impacts of each value chain should be analysed in the context of bioeconomic value chain networks. These networks integrate value chains into the diverse sectors and activities of the bioeconomy and show their interrelations (Lewandowski 2015).

4 Conclusions

Increased demand for biomass in a growing bioeconomy will increase competition for this resource, or at least reinforce the perception of competing claims on biomass and land for its production. Current bioeconomy strategies are not sufficient to secure this additional biomass supply.

The bottlenecks to a sustainable biomass supply become apparent from the analysis of the development of bioenergy. Here, political programmes mainly had the effect of increasing demand for biomass by setting targets for bioenergy use and by incentivizing biofuel and bioelectricity production. However they failed to support the mobilization of the available biomass potentials. This study suggests that this mobilization should start at the level of farmers, with their direct participation in the development of solutions from the beginning and with training schemes. Both should focus on supporting smallholder farmers in particular, as these supply the largest proportion of globally-produced agricultural biomass. The prerequisite for mobilizing the biomass potential and securing sustainable biomass supply is the establishment of a legal and political framework for efficient and sustainable production, use and allocation of biobased raw materials.

There should also be a renewed effort to define "sustainability" in biomass production. Due to the numerous trade-offs involved, demanding the fulfilment of all relevant sustainability criteria at the same time and to the same extent will become a show-stopper in bioeconomy development. Therefore the ranking of sustainability criteria by stakeholder participation in a local to regional context should be applied in order to find the best compromise.

The "food-versus-fuel" discussion needs to be conducted in a more informed and less biased manner. In this context it will doubtlessly become clear that solutions to food security need to start at a local to regional level and that an impact assessment on food security should be performed first before considering any biomass production or land-use change strategies.

There is currently a large untapped potential capable of increasing sustainable biomass supply by technical means and improvements. The more efficient use of biomass resources and avoidance of biomass losses along the value chain up to the consumer are the most promising approaches in quantitative terms to increasing security in biomass supply and avoiding food shortages and competition for biomass. They are also the easiest to implement.

Sustainable intensification on available agricultural land has a large potential for increasing biomass supply and appears a more promising strategy than mobilizing additional land resources, mainly marginal land. This is due to the fact that marginal land is often characterized by low productivity and high production costs on the one hand and high ecological value, which may be compromised by agricultural use, on the other.

The integration of land-use functions, biomass production systems and biomass uses seems a promising approach to avoiding competition for biomass and land for its production. A bioeconomy which makes use of all forms of biomass in an integrated approach and takes account of the interactions in the context of bioeconomic networks, therefore seems more promising than the isolated development of bioenergy. For the implementation of efficient strategies, both research and political support are necessary to secure sustainable biomass production and supply. With the impact of climate change, the sustainable increase of biomass production in sufficient quantity to feed the growing population and provide biomass to replace depleting fossil resources will become even more challenging.

References

Ackerman, K., Conard, M., Cullgigan, P., Plunz, R., Sutto, M.-P., & Whittinghill, P. (2014). Sustainable food systems for future cities: the potential of urban agriculture. *The Economic and Social Review, 45*, 189–206.

Adegbidi, H., Briggs, R. D., Volk, T. A., White, E. H., & Abrahamson, L. P. (2003). Effect of organic amendments and slow-release nitrogen fertilizer on willow biomass production and soil chemical characteristics. *Biomass and Bioenergy, 25*, 389–398.

Alasti, E. (2011). *Social acceptance of bioenergy in Europe*. IIIEE Theses 2011:13. http://www.syntropolis.net/media/library/Social_Acceptance_of_Bioenergy_in_Europe.pdf

Aulakh, J., & Regmi, A. (2013). Post-harvest food losses estimation – Development of consistent methodology. http://www.fao.org/fileadmin/templates/ess/documents/meetings_and_workshops/GS_SAC_2013/Improving_methods_for_estimating_post_harvest_losses/Final_PHLs_Estimation_6-13-13.pdf

Baldock, D., Beaufoy, G., Brouwer, F., & Godeschalk, F. (1996). *Farming at the margins: Abandonment of redeployment of agricultural land in Europe* (pp. 1–182). London/The Hague: Institute for European Environmental Policy (IEEP)/Agricultural Economics Research Institute (LEI-DLO).

Barry, M.-L., Steyn, H., & Brent, A. (2011). Selection of renewable energy technologies for Africa: Eight case studies in Rwanda, Tanzania and Malawi. *Renewable Energy, 36*, 2845–2852.

Becker, K., Wulfmeyer, V., Berger, T., Gebel, J., & Münch, W. (2013). Carbon farming in hot, dry coastal areas: An option for climate change mitigation. *Earth System Dynamics, 4*, 237–251.

Beringer, T., Lucht, W., & Schaphoff, S. (2011). Bioenergy production potential of global biomass plantations under environmental and agricultural constraints. *GCB Bioenergy, 3*, 299–312.

Biederman, L. A., & Harpole, W. S. (2013). Biochar and its effects on plant productivity and nutrient cycling: A meta-analysis. *GCB Bioenergy, Special Issue: Biochar, 5*, 202–214.

Bouwman, A. F., Van der Hoek, K. W., Eickhout, B., & Soenario, I. (2005). Exploring changes in world ruminant production systems. *Agricultural Systems, 84*, 121–153.

Buhle, L., Donnison, I., Heinsoo, K., Sudekum, K.-H., & Wachendorf, M. (2010). PROGRASS – Erhalt von Naturschutzgrünland durch eine dezentrale energetische Verwertung. *Mitteilungen der Gesellschaft für Pflanzenbauwissenschaften, 22*, 275–276.

Bundesamt für Naturschutz (BfN). (2014). *Grünland-Report. Alles im grünen Bereich?* Bundesamt für Naturschutz (BfN), Referat Presse und Öffentlichkeitsarbeit, Konstantinstr. 110, 53179 Bonn. Accessed December 22, 2014, from www.bfn.de

Burnod, P., Gingembre, M., & Andrianirina Ratsialonana, R. (2013). Competition over authority and access: International land deals in madagascar. *Development and Change, 44*, 357–379.

CADETT. (1998). Straw fired Plants in Denmark. *CADDET Renewable Energy Newsletter*, March 1997, 10–12. Accessed December 22, 2015, from http://www.caddet-re.org/assets/1-97art3.pdf

Cherubini, F. (2010). The biorefinery concept: Using biomass instead of oil for producing energy and chemicals. *Energy Conversion and Management, 51*, 1412–1421.

Dale, V. H., Kline, K. L., Wiens, J., & Fargione, J. (2010). Biofuels: Implications for land use and biodiversity. *Biofuels and Sustainability Reports*, 1–13. Accessed December 2, 2014, from http://www.esa.org/biofuelsreports/

Dauber, J., Brown, C., Fernando, A. L., et al. (2012). Bioenergy from "surplus" land: Environmental and socio-economic implications. *BioRisk, 7*, 5–50.

Dickinson, N. M., Baker, A. J. M., Doronila, A., Roger, S., & Reeves, D. (2010). Phytoremediation of inorganics: Realism and synergies. *International Journal of Phytoremediation, 11*, 97–114.

Dornburg, V., & Faaij, A. (2005). Cost and CO_2-emission reduction of biomass cascading – Methodological aspects and case study of SRF poplar. *Climatic Change, 71*, 373–408.

Dornburg, V., van Vuuren, D., van de Ven, G., et al. (2010). Bioenergy revisited: Key factors in global potentials of bioenergy. *Energy and Environmental Science, 3*, 258–267.

Doyle, C. J. (1997). A review of the use of models of weed control in integrated crop protection. *Agriculture, Ecosystems and Environment, 64*, 165–172.

EC. (2009). Directive 2009/28/EC of the European parliament and of the council of 23 April 2009 on the promotion of the use of energy from renewable sources and amending and subsequently repealing directives 2001/77/EC and 2003/30/EC.

FAO. (2003). *World agriculture: Towards 2015/2030. An FAO perspective, food agricultural organisation*. London, UK: Earthscan Publications Ltd.

FAO. (2014). *Are grasslands under threat*? Accessed December 4, 2014, from http://www.fao.org/ag/agp/agpc/doc/grass_stats/grass-stats.htm

FNR. (2014). Accessed December 10, 2014, from www.fnr.de

Fraunhofer IGB. (2014). *Rückgewinnung von Nährstoffen zur Herstellung von Düngemitteln*. Accessed December 22, 2014, from http://www.igb.fraunhofer.de/content/dam/igb/de/documents/broschueren/naehrstoffrueckgewinnung.pdf

Freibauer, A., Rounsevell, M., Smith, P., & Verhagen, J. (2004). Carbon sequestration in the agricultural soils of Europe. *Geoderma, 122*, 1–23.
Fu, G., Chen, S., & McCool, D. K. (2006). Modelling the impacts of no-till practice on soil erosion and sediment yield with RUSLE, SEDD, and ArcView GIS. *Soil and Tillage Research, 85*, 38–49.
Gallagher, B. J. (2011). The economics of producing biodiesel from algae. *Renewable Energy, 36*, 158–162.
Garnett, T., & Godfray, C. (2012). *Sustainable intensification in agriculture. Navigating a course through competing food system priorities*. Food Climate Research Network and the Oxford Martin Programme on the Future of Food, University of Oxford, UK.
Garnett, T., Appleby, M. C., Balmford, A., Bateman, I. J., Benton, T. G., Bloomer, P., Burlingame, B., Dawkins, M., Dolan, L., Fraser, D., Herrero, M., Hoffmann, I., Smith, P., Thornton, P. K., Toulmin, C., Vermeulen, S. J., & Godfray, H. C. J. (2013). Sutainable intensification in agriculture: Premises and policies. *Science, 341*(5), 33–34.
Giller, K., van den Veen, G., & van Ittersum, M. (2007). Competing claims on natural resources: Food, fuel, fibre or forest? In A. Haverkort, P. Bindraban, & H. Bos (Eds.), *Food, fuel or forest? Opportunities, threats and knowledge gaps of feedstock production for bio-energy* (pp. 37–41). Wageningen: PlantResearch International B.V. (Report 142).
Giuntoli, J., Agostini, A., Edwards, R., & Marelli, L. (2014). *Solid and gaseous bioenergy pathways: Input values and GHG emissions* (JRC Science and Policy reports Report EUR 26696 EN).
Graham, R. L., Nelson, R., Sheehan, J., Perlack, R. D., & Wright, L. L. (2007). Current and potential US corn stover supplies. *Agronomy Journal, 99*, 1–11.
Grelle, A., Aronsson, P., Weslien, P., Klemedtsson, L., & Lindroth, A. (2007). Large carbon-sink potential by Kyoto forests in Sweden – A case study on willow plantations. *Tellus, 59*, 910–918.
Grethe, H., Deppermann, A., & Marquardt, S. (2013). *Biofuels: Effects on global agricultural prices and climate change*. Study for Heinrich Böll Stiftung and Oxfam Germany. Institute of Agrcultural Policy and Markets, University of Hohenheim, Stuttgart.
Gustavsson, J., Cederberg, C., Sonesson, U., van Otterdijk, R., & Meybeck, A. (2011). *Global food losses and food waste: Extent, causes and prevention*. Rome: Food and Agriculture Organization (FAO) of the United Nations.
Heinsoo, K., Melts, I., Sammul, M., & Holm, B. (2010). The potential of estonian semi-natural grasslands for bioenergy production. *Agriculture, Ecosystems and Environment, 137*, 86–92.
Hilger, T., Lewandowski, I., Winkler, B., Ramsperger, B., Kageyama P., & Colombo, C. (2015). Seeds of change—Plant genetic resources and people's livelihoods, agroecology. In V. Pilipavicius (Ed.). InTech. doi:10.5772/59984. Available from https://www.intechopen.com/books/agroecology/seeds-of-change-plant-genetic-resources-and-people-s-livelihoods
Hodges, R. J., Buzby, J. C., & Bennett, B. (2011). Postharvest losses and waste in developed and less developed countries: Opportunities to improve resource use. *Journal of Agricultural Science, 149*, 37–45.
Hodgson, K., Campbell, M. C., & Bailkey, M. (2011). *Investing in healthy, sustainable places through urban agriculture*. American Planning Association, Planning Advisory Service Report Number 563, January 2011. http://www.fundersnetwork.org/files/learn/Investing_in_Urban_Agriculture_Final_110713.pdf
Howell, T. A. (2001). Enhancing water use efficiency in irrigated agriculture. *Agronomy Journal, 93*, 281–289.
Huang, W.-D., & Zahng Y-H, D. (2011). Energy efficiency analysis: Biomass-to-wheel efficiency related with biofuels production, fuel distribution, and powertrain systems. *PloS One, 6*(e22113), 1–10.
IEA. (2007).*IEA energy technology essentials. Biomass for Power Generation and CHP*. ETE03. http://www.iea.org/textbase/techno/essentials.htm
IEA Bioenergy . (2009). *Biorefineries: Adding value to the sustainable utilisation of biomass*. Accessed March 14, 2014, from http://www.ieabioenergy.com/wp-content/uploads/2013/10/Task-42-Booklet.pdf

Jirjis, R. (1995). Storage and drying of wood fuel. *Biomass and Bioenergy, 9*, 181–190.

Jørgensen, U. (2011). Benefits versus risks of growing biofuel crops: The case of *Miscanthus*. *Current Opinion in Environmental Sustainability, 3*, 24–30.

Kidane, W., Maetz, M., & Dardel, P. (2006). *Food security and agricultural development in sub-Saharan Africa: Building a case for more public support*. Sub-regional Office for Southern and East Africa (Harare), Food and Agriculture Organisation of United Nations (FAO), Rome.

Kosonen, M., Otsamo, A., & Kuusiplao, J. (1997). Financial, economic and environmental profitability of reforestation of Imperata grasslands in Indonesia. *Forestry Ecology and Management, 99*, 247–259.

Krawinkel, M. B. (2012). Overcoming undernutrition with local resources in Africa, Asia and Latin America. *Journal of the Science of Food and Agriculture, 92*, 2757–2759.

Krupinsky, J. M., Bailey, K. L., McMullen, M. P., Gossen, B. D., & Turkington, T. K. (2002). Managing plant disease risk in diversified cropping systems. *Agronomy Journal, 94*, 198–209.

Lal, R. (2004). Soil carbon sequestration impacts on global climate change and food security. *Science, 304*, 1623–1627.

Lal, R. (2005). World crop residues production and implication of its use as a biofuel. *Environment International, 31*, 575–586.

Latif, S., & Müller, J. (2014). Cassava – How to explore the “all-sufficient”. *Ruralia, 21*(3), 30–31.

Laurance, W., Koh, L. P., Butler, R., Sodhi, N. S., Bradshaw, C. J. A., Neidel, J. D., Consunji, H., & Vega, J. M. (2010). Improving the performance of the roundtable on sustainable palm oil for nature conservation. *Conservation Biology, 24*, 377–381.

Lencsés, E. I., & Takács-György, K. (2014). Farmers’ perception of precision farming technology among Hungarian farmers. *Sustainability, 6*, 8452–8465.

Letourneau, D., & van Bruggen, A. (2006). Crop protection in organic agriculture. In P. Kristiansen, A. Taji, & J. Reganold (Eds.), *Organic agriculture: A global perspective* (pp. 93–114). Collingwood, VIC: CSIRO Publishing.

Levitt, T. (2011, September 13). Mexico’s poor suffer as food speculation fuels tortilla crisis. *Ecologist*. Accessed August 26, 2014, http://www.theecologist.org/trial_investigations/1051194/mexicos_poor_suffer_as_food_speculation_fuels_tortilla_crisis.html

Lewandowski, I. (2013). Soil carbon and biofuels. In R. Lal, R. Hüttel, B. U. Schneider, & J. von Braun (Eds.), *Multifunctionality of ecosystem services*. Ecosystem services and carbon sequestration in the biosphere (pp. 333–356). Dordrecht: Springer Science+Business Media.

Lewandowski, I. (2015). Securing a sustainable biomass supply in a growing bioeconomy. *Global Food Security, 6*, 34–42.

Lewandowski, I., & Kauter, D. (2003). The influence of nitrogen fertilizer on the yield and combustion quality of whole grain crops for solid fuel use. *Industrial Crops and Products, 17*, 103–117.

Lewandowski, I., Londo, M., Schmidt, U., & Faaij, A. (2004). Biomass production in multiple land use systems: Categorization of feasible land use functions and their quantification by the example of phytoremediation. In *Proceedings of the 2nd World Conference and Technology Exhibition on Biomass for Energy, Industry and Climate Protection* (pp. 54–57). Rome, Italy.

Lewandowski, I., Schmidt, U., Londo, M., & Faaij, A. (2006). The economic value of the phytoremediation function. *Agricultural Systems, 89*, 68–89.

Lewis, S. M., & Kelly, M. (2014). Mapping the potential for biofuel production on marginal lands: Differences in definitions, data and models across scales. *ISPRS International Journal of Geo-Information, 3*, 430–459.

Linden, D. R., Clapp, C. E., & Dowdy, R. H. (2000). Long-term corn grain and stover yields as a function of tillage and residue removal in east central Minnesota. *Soil and Tillage Research, 56*, 167–174.

Lywood, W., Pinkney, J., & Cockerill, S. (2009). The relative contributions of changes in yield and land area to increasing crop output. *GCB Bioenergy, 1*, 360–369.

Mandl, M. G. (2010). Status of green biorefining in Europe. *Biofuels, Bioproducts and Biorefining, 4*, 268–274.

Meehan, P. G., McDonnell, K. P., & Finnan, J. M. (2013). An assessment of the effect of harvest time and harvest method on biomass loss for *Miscanthus × giganteus*. *GCB Bioenergy, 5*, 400–407.

Möller, K., Schulz, R., & Müller, T. (2009). *Mit Gärresten richtig düngen*. Stuttgart: Aktuelle Informationen für Berater. Institut für Pflanzenernährung der Universität Hohenheim.

Naylor, R. L., Liska, A. J., Burke, M. B., Falcon, W. P., Gaskell, J. C., Rozelle, S. D., & Cassman, K. G. (2007). The ripple effect: Biofuels, food security, and the environment. *Environment, 49*, 30–43.

Nhamo, N., Rodenburg, J., Zenna, N., Makombe, G., & Luzi-Kihupi, A. (2014). Narrowing the rice yield gap in East and Southern Africa: Using and adapting existing technologies. *Agricultural Systems, 131*, 45–55.

OECD-FAO. (2012). *Agricultural Outlook* 2012. Accessed December 22, 2014, from http://www.fao.org/fileadmin/templates/est/COMM_MARKETS_MONITORING/Oilcrops/Documents/OECD_Reports/Ch5StatAnnex.pdf

Oerke, E.-C. (2006). Crop losses to pests. *Journal of Agricultural Science, 144*, 31–43.

Otsamo, A., Adjers, G., Hadi, T. S., Kuusipalo, J., Tuomeal, K., & Vuokko, R. (1995). Effect of site preparation and initial fertilization on the establishment and growth of four plantation tree species used in reforestation of Imperata cylindrical (L.) Beauv. Dominated grasslands. *Forest Ecology and Management, 73*, 271–277.

Parajuli, R., Dalgaard, T., Jørgensen, U., Adamsen, A. P. S., Tydeman Knudsen, M., Birkved, M., Gylling, M., & Schjørring, J. K. (2015). Biorefining in the prevailing energy and material crisis: A review of sustainable pathways for biorefining value chains and sustainability assessment methodologies. *Renewable and Sustainable Energy Reviews, 43*, 244–263.

Peeters, A., Beaufoy, G., Canals, R. M., De Vliegher, A., et al. (2014). Grassland term definitions and classifications adapted to the diversity of European grassland-based systems. *Grassland Science in Europe, 19*, 743–750.

Pfau, S. F., Hagens, J. E., Dankbaar, B., & Smits, A. J. M. (2014). Visions of sustainability in bioeconomy research. *Sustainability, 6*, 1222–1249.

Poetsch, J., Haupenthal, D., Lewandowski, I., Oberländer, D., & Hilger, T. (2012). *Acrocomia aculeata* – A sustainable oil crop. *Ruralia, 21*(3), 41–44.

Pretty, J. (2008). Agricultural sustainability: Concepts, principles and evidence. *Philosophical Transactions of the Royal Society B: Biological Sciences, 363*, 447–465.

Pretty, J., Sutherland, W. J., Ashby, J., et al. (2010). The top 100 questions of importance to the future of global agriculture. *International Journal of Agricultural Sustainability, 8*, 219–236.

Pretty, J., Toulmin, C., & Williams, S. (2011). Sustainable intensification in African agriculture. *International Journal of Agricultural Sustainability, 9*(1), 5–24.

Prochnow, A., Heiermann, M., Plöchl, M., Linke, B., Idler, C., Amon, T., & Hobbs, P. J. (2009). Bioenergy from permanent grassland – A review: 1. Biogas. *Bioresource Technology, 100*, 4931–4944.

Raghu, S., Spencer, J. L., Davis, A. S., & Wiedenmann, R. N. (2011). Ecological considerations in the sustainable development of terrestrial biofuel crops. *Current Opinion in Environmental Sustainability, 3*, 15–23.

Reardon, T., Kelly, V., Crawford, E., Diagana, B., Dioné, J., Savadogo, K., & Boughton, D. (1997). Promoting sustainable intensification and productivity growth in Sahel agriculture after macroeconomic policy reform. *Food Policy, 22*(4), 317–327.

Reijnders, L. (2012). Lipid-based liquid biofuels from autotrophic microalgae: Energetic and environmental performance. *WIREs Energy and Environment, 2*, 73–85.

Rodolfi, L., Zittelli, G. C., Bassi, N., Padovani, G., Biondi, N., Bonini, G., & Tredici, M. R. (2009). Microalgae for oil: Strain selection, induction of lipid synthesis and outdoor mass cultivation in a low-cost photobioreactor. *Biotechnology and Bioengineering, 102*, 100–112.

Ros, J. (2014). Biomass Potential, Oral presentation at the OECD Workshop on "Sustainable Biomass drives the next Bioeconomy: A new industrial revolution?". 10./11.6.2014. Paris: OECD Headquarters.

Rösch, C., Skarka, J., Raab, K., & Stelzer, V. (2009). Energy production from grassland – Assessing the sustainability of different process chains under German conditions. *Biomass and Bioenergy, 33*, 689–700.

Royal Society. (2009). Reaping the benefit, science and the sustainable intensification of global agriculture. *The Royal Society*. https://royalsociety.org/~/media/Royal_Society_Content/policy/publications/2009/4294967719.pdf

Sauerborn, J. (2010, Dezember). Mein Dach ist mein Acker. *dlz-agrarmagazin*, 116–120.

Scarlat, N., Martinov, M., & Dallemand, J. F. (2010). Assessment of the availability of agricultural crop residues in the European Union: Potential and limitations for bioenergy use. *Waste Management, 30*, 1889–1897.

Schnabel, D. (2014). Head of Unit Bioeconomy/Biotechnology, Ministry of Innovation, Science & Research of the German State of North-Rhine Westfalia, Düsseldorf, Germany; oral presentation on the conference Bioökonomie – Nachhaltige Alternative zur fossilen Wirtschaft? 3./4.11.2014 in Berlin, Germany.

Searle, S. Y., & Malins, C. J. (2014). Will energy crop yields meet expectations? *Biomass and Bioenergy, 65*, 3–12.

Smeets, E., Faaij, A., Lewandowski, J., & Turkenburg, W. (2007). A bottom-up assessment and review of global bioenergy potentials to 2050. *Energy and Combustion Science, 33*, 56–106.

Solomon, B. D. (2010). Biofuels and sustainability. *Annals of the New York Academy of Sciences, 1185*, 119–134.

Staffas, L., Gustavsson, M., & McCormick, K. (2013). Strategies and policies for the bioeconomy and bio-based economy: An analysis of official national approaches. *Sustainability, 5*, 2751–2769.

Tait, J., & Barker, G. (2011). Global food security and the governance of modern biotechnologies. *EMBO Reports, 12*, 763–768.

The World Bank. (2009). *Awakening Africa's sleeping giant prospects for commercial agriculture in the Guinea Savannah Zone and beyond*. Washington, DC. doi:10.1596/978-0-8213-7941-7.

Theuvsen, L., Plumeyer, C.-H., & Emmann, C. H. (2014). *Endbericht zum Projekt: Einfluss der Biogasproduktion auf den Landpachtmarkt in Niedersachsen*. Georg-August-Universität Göttingen, Department für Agrarökonomie und Rurale Entwicklung, Betriebswirtschaftslehre des Agribusiness, Platz der Göttinger Sieben 5, 37073 Göttingen.

Thumm, U., Raufer, B., & Lewandowski, I. (2014). Novel products from grassland (bioenergy and biorefinery). *Grassland Science in Europe, 19*, 429–437.

Tilman, D., Socolow, R., Foley, J. A., Hill, J., Larson, E., Lynd, L., Pacala, S., Reilly, J., Searchinger, T., Somerville, C., et al. (2009). Beneficial biofuels – The food, energy, and environment trilemma. *Science, 325*, 270–271.

Triberti, L., Nastri, A., Giordani, A., Comellini, F., Baldoni, G., & Toderi, G. (2008). Can mineral and organic fertilization help sequestrate carbon dioxide in cropland? *European Journal of Agronomy, 29*, 13–20.

USDA. (2015). US Department of Agriculture, Statistics 2015. Accessed January 08, 2015, from http://www.statista.com/statistics/263937/vegetable-oils-global-consumption/

van Dam, J. E. G., de Klerk-Engels, B., Struik, P. C., & Rabbinge, R. (2005). Securing renewable resource supplies for changing market demands in a bio-based economy. *Industrial Crops and Products, 21*(1), 129–144.

van Dam, J., Faaij, A. P. C., Hilber, J., Petruzzi, H., & Turkenburg, W. C. (2009). Large scale bioenergy production from soybeans and switchgrass in Argentina Part B. Environmental and socioeconomic impacts on a regional level. *Renewable and Sustainable Energy Reviews, 13*, 1697–1709.

van Loo, S., & Koppejan, J. (2008). *The handbook of biomass combustion and co-firing*. London: Earthscan

van Vuuren, D. P., van Vliet, J., & Stehfest, E. (2009). Future bio-energy potential under various natural constraints. *Energy Policy, 37*, 4220–4230.

von Braun, J. (2014). Bioeconomy and sustainable development – Dimensions. *Ruralia, 21*(3), 6–9.

Weisz, R., Crozier, C. R., & Heiniger, R. W. (2001). Optimizing nitrogen application timing in no-till soft red winter wheat. *Agronomy Journal, 93*, 435–442.

Wicke, B., Dornburg, V., Junginger, M., & Faaij, A. (2008). Different palm oil production systems for energy purposes and their greenhouse gas implications. *Biomass and Bioenergy, 32*, 1322–1337.

Wicke, B., van der Hilst, F., & Daioglou, V. (2014, April 29). Model collaboration for the improved assessment of biomass supply, demand, and impacts. *GCB Bioenergy*, Article first published online. doi:10.1111/gcbb.12176.

Wijffels, R. H., & Barbosa, M. J. (2010). An outlook on microalgal biofuels. *Science, 13*, 796–799.

Wilcove, D. S., & Koh, L. P. (2010). Addressing the threats to biodiversity from oil palm agriculture. *Biodiversity and Conservation, 19*, 999–1007.

Winkler, B., Lemke, S., Ritter, J., & Lewandowski, I. (2016). Integrated assessment of renewable energy potential: Approach and application in rural South Africa. *Environmental Innovation and Societal Transitions*. doi:10.1016/j.eist.2016.10.002.

Wolters, V., Isselstein, J., Stützel, H., et al. (2014). Nachhaltige ressourceneffiziente Erhöhung der Flächenproduktivität: Zukunftsoptionen der deutschen Agrarökosystemforschung. Grundsatzpapier der DFG Senatskommission für Agrarökosystemforschung. *Journal für Kulturpflanzen, 6*, 225–236.

Yengoh, G. T., & Ardö, J. (2014). Crop yield gaps in cameroon. *Ambio, 43*, 175–190.

Zhu, X.-G., Long, S. P., & Ort, D. R. (2010). Improving photosynthetic efficiency for greater yield. *Annual Review of Plant Biology, 61*, 235–261.

Importance of Sugarcane in Brazilian and World Bioeconomy

Reges Heinrichs, Rafael Otto, Aline Magalhães, and Guilherme Constantino Meirelles

Abstract The area of sugarcane (*Saccharum* spp.) cultivation totaled 27 million hectares in the world and 10 million hectares in Brazil. Sugarcane is a valuable crop considering the potential to produce sugar, ethanol, biodegradable products, energy generation and food for animal production. In tropical conditions, high biomass production in the range of 150–300 Mg ha^{-1} $year^{-1}$ can be achieved, depending on the management and production system employed. Due to great adaptation to different types of soil and environment, sugarcane could be produced in over 100 countries to supply biofuel and food to the world. Improvement in the production process adopted in Brazil in the last decade, including mechanical planting and harvesting, new methods of sugarcane planting, control of pests, diseases, nutrition and fertilization, has increased sugarcane yield in Brazil while improving work-conditions and social aspects of sugarcane cultivation. Therefore, the high potential production of sugarcane, its varied uses and its ability to be cultivated in regions with low economic and social development indicates that sugarcane cultivation could become a key source of income and improve life-quality in many regions. However, political and governmental organization is required to achieve this goal.

R. Heinrichs (✉)
College of Technology and Agricultural Sciences, São Paulo State University (UNESP), Dracena, São Paulo, Brazil
e-mail: reges@dracena.unesp.br

R. Otto
"Luiz de Queiroz" College of Agriculture, University of São Paulo (USP), Piracicaba, São Paulo, Brazil
e-mail: rotto@usp.br

A. Magalhães • G.C. Meirelles
Graduate Program in Animal Science and Technology, UNESP, Dracena, São Paulo, Brazil
e-mail: aline.magalhaes@live.com; gui_meirelles2312@hotmail.com

S. Dabbert et al. (eds.), *Knowledge-Driven Developments in the Bioeconomy*, Economic Complexity and Evolution, DOI 10.1007/978-3-319-58374-7_11

1 Introduction

Sugarcane is a key crop in the Brazilian economy. Brazil has the world's largest cultivated area and is the world's largest producer of sugar and ethanol, being a world leader in the international market with the use of biofuels as an energy alternative. Brazilian production accounts for more than half the sugar traded in the world and production is estimated to increase 3.25% per year until 2018/2019. This corresponds to an increase of 14.6 million tons as compared to 2007/2008. Exports are expected to reach a volume of 32.6 million tons of sugar by 2019. Ethanol production should reach 58.8 billion liters in 2019, more than double the amount produced in 2008. Due to the increase in domestic consumption, internal consumption is projected at 50 billion liters and exports at 8.8 billion liters (MAPA 2010).

The success of Brazilian ethanol production started in 1975 with the creation of Proálcool to support the production of ethanol in Brazil. This was a good example of public policy for the development of biofuels, allowing Brazil to reach second position in ethanol production in 2008, and also to have the lowest production costs (Amaral et al. 2008).

Production of sugarcane in Brazil is estimated to increase by 3.3% per year until 2024, rising 884 Mt, i.e. 42% higher than production obtained in 2008, mainly due to the increase in cultivated areas. In the same period, the total cultivated area is estimated to increase by 2.9% per year. In contrast, the average yield fell between 2010 and 2014 due to climatic and management constraints, but should moderately increase during this projection (FAO 2015).

Energy production from sugarcane also plays an important role for the Brazilian economy. There are currently around 408 sugarcane mills in Brazil, and all of them are self-sufficient in the production process through burning bagasse. This results in a significant cost reduction. Some mills also present the cogeneration of electricity, allowing them to sell on the surplus energy, increasing income and reducing dependence on other sources of energy (thermal, hydroelectric, etc.).

The agroindustrial system of sugarcane is complex. The sector depends on suppliers of raw materials and high capital investment for sugar, ethanol and energy production (Neves and Conejero 2007). After industrialization, ethanol, sugar and energy are transferred to fuel distributors, electrical power systems, food industry, wholesale and retail, and export trading companies. The byproducts generated, such as filter cake, vinasse, and residual water, are used as bio-fertilizer in the production process, thus reducing expenses with synthetic fertilizers.

2 Planted Area and Production of Sugarcane

2.1 *Brazilian Planted Area and Production*

The Brazilian Agro-Energy Statistical Yearbook 2014, consolidating data from the agroenergetic chain of the Ministry of Agriculture, Livestock and Supply (MAPA 2015), presents the areas planted and harvested in the country during the period 2002–2013 (Table 1). The planted area increased more than 80% from 2002 to 2012.

Sugarcane production in Brazil in the 2014/2015 season reached 630 million tons, of which 575 million tons were grown in the South Central region, 48 million tons in the Northeast and 7 tons in the North (MAPA 2015). From this amount 35 million tons of sugar and 29 million cubic meters of ethanol were produced. The State of Sao Paulo, located in the South Central region of the country, accounts for 60% of total production of sugarcane in Brazil.

2.2 *World Planted Area and Production*

The total area cultivated with sugarcane in the world increased from 20.5 million hectares in 2002 to 26.1 million hectares in 2012. During this period, Brazil took first place among the main producing countries with a total area of 8.5 million hectares, followed by India (5.1 million hectares), China (1.8 million hectares) and Thailand (1.3 million hectares) in 2012 (Table 2).

Worldwide sugarcane production reached 1.8 billion tons in 2012. From that amount, Brazil reached 594 million tons, followed by India (348 million tons) and China (134 million tons) (Table 3).

Considering the expansion in area and production of sugarcane in major countries (Tables 2 and 3), it can be seen that from 2003 to 2012 the four major producing countries (Brazil, India, China and Thailand) consistently increased the area and

Table 1 Planted area and harvested area of sugarcane in Brazil

Year	Planted area (ha)	Harvested area (ha)
2002	5,206,656	5,100,405
2003	5,377,216	5,371,020
2004	5,633,700	5,631,741
2005	5,815,151	5,805,518
2006	6,390,474	6,355,498
2007	7,086,851	7,080,920
2008	8,210,877	8,140,089
2009	8,845,833	8,617,555
2010	9,164,756	9,076,706
2011	9,616,615	9,535,194
2012	9,424,615	9,407,078

Table 2 Sugarcane area harvested of main producing countries, in millions hectares

Country	2003	2004	2005	2006	2007	2008	2009	2010	2011	2012
Brazil	5.4	5.6	5.8	6.1	6.9	7.0	7.4	8.0	8.4	8.5
India	4.5	3.9	3.7	4.2	5.1	5.0	4.4	4.2	4.9	5.1
China	1.4	1.4	1.4	1.4	1.4	1.6	1.7	1.7	1.7	1.8
Thailand	1.1	1.1	1.0	0.9	1.0	1.0	0.9	1.0	1.3	1.3
Pakistan	1.1	1.1	1.0	0.9	1.0	1.2	1.0	0.9	1.0	1.0
Mexico	0.6	0.6	0.7	0.7	0.7	0.7	0.7	0.7	0.7	0.7
Indonesia	0.3	0.4	0.4	0.4	0.4	0.4	0.4	0.4	0.4	0.5
Philippines	0.4	0.4	0.4	0.4	0.4	0.4	0.4	0.4	0.4	0.4
USA	0.4	0.4	0.4	0.4	0.4	0.4	0.4	0.4	0.4	0.4
Argentina	0.3	0.3	0.3	0.3	0.3	0.4	0.3	0.4	0.4	0.4
Colombia	0.4	0.4	0.4	0.4	0.4	0.4	0.4	0.3	0.4	0.4
Australia	0.4	0.4	0.4	0.4	0.4	0.4	0.4	0.4	0.3	0.3
South Africa	0.3	0.3	0.3	0.3	0.3	0.3	0.3	0.3	0.3	0.3
Vietnam	0.3	0.3	0.3	0.3	0.3	0.3	0.3	0.3	0.3	0.3
Guatemala	0.2	0.2	0.3	0.2	0.2	0.2	0.2	0.2	0.2	0.3
Egypt	137.5	135.3	135.0	137.3	140.8	135.9	133.0	134.5	136.7	143.5
World	**20, 517.6**	**20, 154.5**	**19, 714.7**	**20, 611.5**	**22, 684.4**	**24, 085.4**	**23, 693.6**	**23, 784.1**	**25, 581.2**	**26, 088.6**

Source: FAO (2015); MAPA (2015)

Table 3 Sugarcane production of main producing countries, in million tons

Country	2003	2004	2005	2006	2007	2008	2009	2010	2011	2012
Brazil	349.6	374.7	382.3	428.3	495.5	552.8	622.6	627.3	565.8	594.3
India	287.4	233.9	237.1	281.2	355.5	348.2	285	292.3	342.4	347.9
China	90.2	89.8	86.6	92.6	113	124.2	115.6	110.8	114.4	123.5
Thailand	74.3	65.0	49.6	47.7	64.4	73.5	66.8	68.8	96.0	96.5
Pakistan	52.1	53.8	47.2	44.7	54.7	63.9	50.0	49.4	55.3	58.4
Mexico	47.5	48.7	51.6	50.7	52.1	51.1	49.5	50.4	49.7	50.9
Philippines	31.0	33.5	31.4	31.6	32.0	34.0	32.5	28.0	30.0	30.0
USA	33.9	29.0	26.6	29.8	27.8	25.0	27.6	24.8	26.7	27.9
Australia	37.0	37.0	37.8	37.1	36.4	32.6	30.3	31.5	25.2	26.0
Argentina	22.1	20.9	24.4	26.5	24.0	27.0	27.0	26.2	27.1	25.0
Indonesia	24.5	26.8	29.3	29.2	25.2	25.6	26.4	26.6	24.0	26.3
Colombia	39.0	40.0	39.8	38.5	38.5	38.5	43.0	37.0	42.0	38.0
Guatemala	17.4	20.0	18.0	17.6	20.3	20.3	21.5	22.3	20.6	21.8
Vietnam	16.9	15.6	14.9	16.7	17.4	16.1	15.6	16.2	17.5	19.0
South Africa	20.4	19.1	21.3	20.3	19.7	19.3	18.7	16.0	16.8	17.3
Egypt	16.2	16.2	16.3	16.7	17.0	16.5	15.5	15.7	15.8	16.5
World	**1, 378.6**	**1, 340.9**	**1, 316.4**	**1, 421.9**	**1, 618.5**	**1, 7535**	**1, 693.5**	**1, 707.9**	**1, 819.4**	**1, 832.5**

Source: FAO (2015); MAPA (2015)

Table 4 Variation in area and production of sugarcane from 2003 to 2012

Country	Increase or reduction in area (%)	Increase or reduction in production (%)
Brazil	58.0	70.0
India	12.6	21.1
China	25.9	36.9
Thailand	17.6	29.9
Pakistan	–5.0	12.1
Mexico	14.1	7.2
Philippines	12.9	–3.2
USA	–7.9	–17.7
Australia	–24.4	–29.7
Argentina	18.6	13.1
Indonesia	36.0	7.3
Colombia	–12.2	–2.6
Guatemala	30.2	25.3
Vietnam	–5.0	12.4
South Africa	–2.1	–15.2
Egypt	4.4	1.9
World	**27.2**	**32.9**

Source: Adapted FAO (2015) and MAPA (2015)

production of sugarcane (Table 4). In Pakistan there was a reduction in planted area, although production increased, reflecting gains in productivity. On the other hand, countries with large potential for expansion of the crop, such as the USA, Australia, Colombia and South Africa reduced area and production; results that can be linked to specific agricultural policies of each country. However, these policies can change as demonstrated during the Paris Climate Change Conference (November, 2015) during which leaders of 195 countries pledged to reduce emissions of greenhouse gases. As a result, sugarcane is a crop with great potential for expansion due to its production of clean, renewable energy.

It is noteworthy that Brazil and India account for 52% of the total harvested area of sugarcane. The sum of the six highest sugarcane producing countries accounts for 70% of the total world area of sugar cane. Similar results are observed in the production of stems, with Brazil and India accounting for 51% of world production. With the addition of China, Thailand, Pakistan and Mexico this figure reaches 72% of world production of sugar cane (Fig. 1). Through analysis of these data it is apparent that sugarcane production is concentrated in a small number of countries.

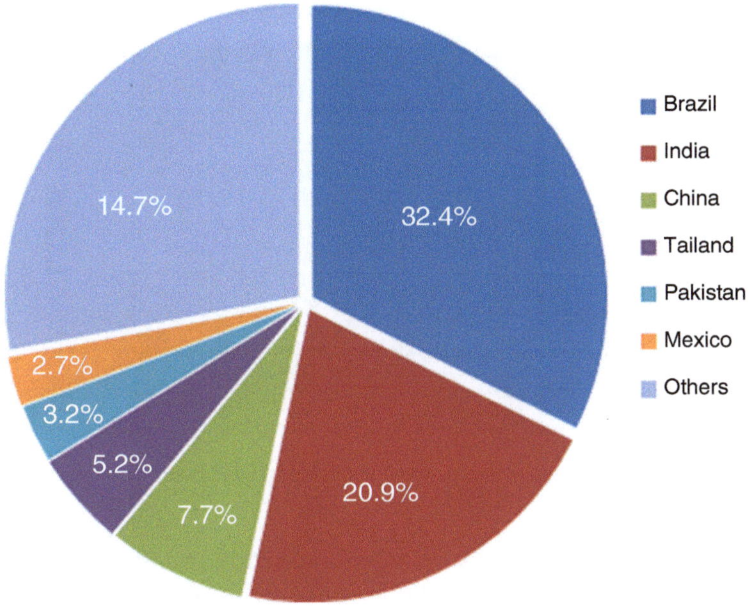

Fig. 1 Major producing countries of sugarcane. Source: Adapted by FAO (2015) and MAPA (2015)

However, there is a good possibility of expanding the area cultivated with sugar cane, by finding favorable climatic and soil conditions between latitudes 36.7° N and 31° S (Humbert 1968), covering the tropical and subtropical regions of the world. In this scenario, more than 100 countries could supply biofuels to 200 nations. Comparatively, currently only 20 producers provide fossil fuels to the rest of the world.

3 Technological Evolution in Cultivation and Productivity

Besides an increase in area, sugarcane production in Brazil also has the potential to increase due to development of technologies in planting and harvesting systems. This would allow an increase in production of bioenergy, technological advancement and also discovery of new products such as biobutanol, cellulosic ethanol and bioplastics, causing major changes in the industry's structure (Viana and Perez 2013).

The main advance in sugarcane production systems in Brazil in the last decade was the elimination of burning before harvest, with social, economic and environmental benefits. In the system without burning, the harvest is performed mechanically, leaving the crop residues (straw) on the soil surface. This conser-

Fig. 2 Pre-sprouted seedling (*left*) and planting in the field (*right*). Source: Authors

vational system has advantages such as soil protection against erosion and water loss, increase in carbon storage (Leal et al. 2013), improvement in soil fertility and reduction in CO_2 emission to the atmosphere.

New technologies such as no tillage and manure application have also been incorporated into the sugarcane system in Brazil. The no-tillage system, as compared to the conventional system, has increased productivity and improvement in soil conservation (Duarte Junior and Coelho 2008).

The use of vinasse provides higher concentrations of potassium in the soil, increasing the potential productivity, especially in sandy soil. The filter cake (press mud) provides better soil fertility by providing macronutrients and micronutrients, lower levels of aluminum, by acting as a corrective of acidity, providing higher levels of phosphorus and nitrogen in the plant. Sugarcane fields receiving either vinasse or filter cake display an increase in yields compared to non-amended fields, reduction in fertilizers usage and a decrease in costs.

Another developing technology is the planting of pre-sprouted seedlings, in order to reduce the amount of cane-bullets used during field establishment, as well as improved control of diseases (Fig. 2). Changes in plant spacing has also been evaluated, such as planting alternating double-rows (1.5 × 0.9 m) which allow controlled traffic, reduction in row compaction and thus increases in yield and longevity.

The removal of straw from fields for energy or second-generation ethanol production is in full expansion in many producing regions of sugarcane in Brazil (Cantarella et al. 2013). However, despite the economic appeal of this practice, sustainability issues need to be clarified, given the positive effect of straw in maintaining soil moisture, increases in C and N stocks and sugarcane productivity (Leal et al. 2013), especially in regions subjected to high temperatures and limited rainfall.

4 Employment in Sugarcane Production System

Historically, the Brazilian sugarcane industry was associated with poor working conditions, especially for manual harvesting of sugarcane. Currently, with the advancement of mechanized harvesting, which has already reached 85% of the cultivated area in the South Central region, working conditions have improved significantly.

However, there is concern related to unemployment that may be caused by mechanization. In this view, despite the reduction in jobs caused by mechanization, an increase in demand for better-qualified manual labour has been recorded. The mechanization process is creating opportunities for tractor drivers, truck drivers, mechanics, combine harvester drivers and electronics technicians, among others. As a result, this reduces the demand for low qualified hand-labor (Moraes and Momenti 2006).

The economic development of the regions varies due to different local or national actions, which could determine growth rates and can generate socio-economic inequality between regions. Some regions have advantages in structure and higher productivity, favoring the development of the sugar and ethanol industry and creating jobs.

5 Strategies to Increase Productivity and Sustainability

Brazil has a large area available for growing sugarcane, without causing damage to the production of other foods, as well as having production structures and distribution of technological products. The country incorporates the whole cycle of ethanol production, beginning in fields with high yields all the way to installation of equipment for the providers of this biofuel industry (MAPA 2015).

The expansion in sugarcane fields often takes the place of degraded areas with grains or pastures due to economic reasons, i.e. the availability of areas with low production efficiency transformed into productive areas with sugarcane cultivation. Sugarcane expansion is not occurring over areas of native vegetation. For recovery of degraded pasture, for example, areas can be used to plant soybean for one or more years to improve soil conditions for implementation of the sugarcane crop (Macedo and Seabra 2008).

In order to guide the sustainable expansion of sugarcane in Brazil, the federal government launched a policy based on environmental, economic and social factors. The *Agro-Ecological Zoning of Sugarcane* defined areas suitable for planting the crop considering climate types, soil, biomass, land slope and need of irrigation, among other characteristics (MAPA 2015). The study revealed an available area to expansion of sugarcane or other crops up to 65 million hectares, without the need for deforestation or encroaching upon protected areas such as the Amazon or Pantanal.

However, the increase in sugarcane production can not be based exclusively on an increase in the cultivated area but, in contrast, should be driven by increases in productivity. The sugarcane yield in Brazil increased from 43 Mg ha^{-1} in 1961 to 75 Mg ha^{-1} in 2013 (FAOSTAT 2014). The increased productivity of sugarcane comes from the improvement in varieties, plant protection treatments, changes in cultural practices, correct use of fertilizers, choice of regions with favorable climate and soil production and better control of weeds, pests and diseases.

6 Green Energy from Sugarcane

The production of ethanol from sugarcane in Brazil is a model that is well accepted because it is renewable and from biomass stocks for which the world has more sustainable agricultural production. The recent growth of the sugarcane industry is essentially due to the development of new technologies for the production of duel-fuel vehicles, or flex fuel, that is, vehicles capable of using both ethanol and gasoline, or even a mixture of both. The final aim is to increase the use of clean energy sources in order to reduce carbon monoxide emissions to meet the requirements of the Kyoto Protocol (Souza and Miziara 2010).

Internally in Brazil, sugarcane mills signed the "Environmental Protocol of the Sugarcane Sector" in order to conserve soil and water resources, protecting forests, recovering river basins, reducing the emission of greenhouse gases and increasing the efficiency of fertilizer use and agrochemicals products (Amaral et al. 2008). Reducing water in the industrial process is a requirement for sustainable ethanol production. Re-use of water in a closed circuit in the processing stage can reduce 90% of water usage (Salazar et al. 2013).

The cogeneration of energy has been an option for sugar and alcohol companies, due to the fluctuation of energy production by hydropower and the variation in rainfall, so cogeneration is a safe, cheap and environmental friendly option.

Competition for bagasse for energy generation, genetic and physiological improvement of sugarcane, and requirements for pre-hydrolysis of bagasse are variables that can affect second-generation ethanol (Raele et al. 2014).

Different levels of integration between first and second-generation ethanol are possible, using technologies for hydrolysis and fermentation of pentoses, resulting in great benefits due to higher ethanol production and better economic results. Second-generation ethanol has higher production of cane processed per ton, between 200 and 400 L ton^{-1} of dry matter in the fermentation and degradation of pentoses (Dias et al. 2012a, b).

The integrated system, based on the use of biomass and the combined cycle gasification used in sugar mills, provides better generation and power exportation than the direct consumption of bagasse combustion in the high-pressure steam cycle. This is beginning to be used in the sugar industry (Deshmukh et al. 2013). The increased volume of bagasse in the last few years represented 19.3% of Brazil's energy matrix in 2010, and all renewable energy sources accounted for 47.6%. Meanwhile, on the global scale renewable sources reached approximately 15.6% (Hofsetz and Silva 2012).

7 Byproducts of the Sugar and Ethanol Manufacturing Process

In the sugar and ethanol production process, the byproducts generated are reused in the industrial or agricultural processes, reducing production costs and environmental impacts. Bagasse is a fibrous residue from the extraction of the juice by the mills and the amount produced depends on the processed sugarcane fiber. Bagasse can be used as a source of fuel (energy) for boiler, pulp production and cattle confined feed.

Filter cake (or press mud) is a byproduct generated mainly in the production of sugar, in rates varying from 5 to 30 kg Mg^{-1} of sugarcane processed, depending on the extraction process of the juice. It contains around 75% of humidity and composting processes have been adopted to reduce humidity, dosages and quality of application to the fields. Filter cake is applied to the fields as a source of organic matter, phosphorus, nitrogen, calcium, sulfur and other nutrients, in planting or to the ratoon, in rates varying from 5 to 20 Mg ha^{-1} of dry matter.

The vinasse is the main byproduct of ethanol production. It is produced at a rate of 13 L per 1 L of ethanol produced and presents considerable amounts of potassium. This liquid byproduct is applied to sugarcane in the form of irrigation, supplying the whole amount of potassium required by sugarcane, as well as a portion of sulfur and nitrogen. Environmental legislation has advanced greatly in recent years and, currently, the industrial plants must draw up a vinasse implementation plan to be submitted annually to the environmental agencies to permit the milling. One strategy to adjust the vinasse application to environmental requirements refers to the concentration of vinasse. Equipment is available in Brazil for installation annexed to the mills, allowing concentration of between 8 and 10 times the vinasse. In this system, the dosages of conventional vinasse normally used (between 60 and 150 m^3 $year^{-1}$) are reduced to 6–12 m^3 ha^{-1} $year^{-1}$, with operational, economic and environmental benefits.

8 Animal Feed

Livestock activity is expensive and the production sector seeks lower costs of alternative food sources. Grinding sugarcane for feeding is a potential strategy, since the sugarcane production coincides with the period of lowest forage production (winter), when a loss of weight of animals is often observed (Murta et al. 2011). The sugarcane can also be used as silage, usually available in feedlots, being effective as forage for beef cattle. Cattle farmers have sought alternatives to reduce their production costs with feeding, since the confinement is a high-risk economic activity (Pinto et al. 2010). Sugar cane is commonly used in feedlots in the form of roughage food, being an economically viable alternative to substitute the silage when properly supplemented (Barros et al. 2010).

The sugarcane bagasse as a form of animal feed can be implemented with the use of treatments to improve digestibility, such as alkalizing agents that can been used for hydrolysis (Murta et al. 2011). Another technique that can be used is the ammonization of bagasse with urea to improve nutritional characteristics, by increasing the digestibility of fiber and crude protein content (Pires et al. 2004).

Final remarks

The sugarcane cultivated area is expanding rapidly in Brazil, but without replacing areas of native forest or other protected areas. Sugarcane yield has also increased consistently in the last decades in Brazil. Sugarcane is a renewable alternative to the production of sugar, ethanol and electricity. Second-generation ethanol requires further development, in order to allow ethanol production through enzymatic hydrolysis of bagasse or straw.

Brazil is the largest producer of sugarcane in the world and better farming practices are being developed to increase productivity. These include mechanical harvesting without burning, mechanical planting, soil amendments and fertilization practices, changes in the form of planting and the development of varieties adapted to the soil and climate of the expanding areas. Increased mechanization of planting and harvesting has offset the reduction of jobs by increasing requirements of qualified labor. The mechanization of harvest process has reduced historical problems of labor relations and improved social aspects of biofuel production in Brazil.

An international public policy to expand the sugarcane planted areas in the world is required in order to make sugarcane a commodity to meet the global energy demand. Brazil, due to its technical and scientific knowledge of the production chain, can contribute to the diffusion of agricultural and industrial technology needed for high productivity in a variety of soil and climatic environments, which would significantly advance sustainable energy production.

References

Amaral, W. A. N., Marinho, J. P., Tarasantchi, R., Beber, A., & Giuliani, E. (2008). Environmental sustainability of sugarcane ethanol in Brazil. In P. Zuubier & J. V. Vooren (Eds.), *Sugarcane ethanol: Contributions to climate change mitigation and the environment* (1st ed., pp. 29–62). Wageningen: Wageningen Academic.

Barros, R. C., Rocha Júnior, V. R., Souza, A. S., Franco, M. O., Oliveira, T. S., Mendes, G. A., Pires, D. A. A., Sales, E. C. J., & Caldeira, L. A. (2010). Economic viability of substitution of sorghum silage by sugarcane or sugarcane bagasse ammoniated with urea for cattle in feedlot. *Brazilian Journal of Animal Health and Production, 11*(3), 555–569.

Cantarella, H., Cerri, C. E. P., Carvalho, J. L. N., & Magalhães, P. S. G. (2013). How much sugarcane trash should be left on the soil? (Editorial). *Scientia Agricola, 70*(5).

Deshmukh, R., Jacobson, A., Chamberlin, C., & Kammen, D. (2013). Thermal gasification or direct combustion? Comparison of advanced cogeneration systems in the sugarcane industry. *Biomass and Bioenergy, 55*, 163–174.

Dias, M. O. S., Junqueira, T., Cavalett, O., Cunha, M. P., Jesus, C. D. F., Rossell, C. E. V., Maciel Filho, R., & Bonomi, A. (2012a). Integrated versus stand-alone second generation ethanol production from sugarcane bagasse and trash. *Bioresource Technology, 103*(1), 152–161.

Dias, M. O. S., Junqueira, T., Rossell, C. E. V., Cunha, M. P., Maciel Filho, R., & Bonomi, A. (2012b). Evaluation of process configurations for second generation integrated with first generation bioethanol production from sugarcane. *Fuel Processing Technology, 30*, 172–176.

Duarte Junior, J. B., & Coelho, F. C. (2008). Sugarcane in a no-tillage system compared to the conventional system with, and without, manuring. *Agriambi Journal, Campina Grande, 12*(6), 576–583.

FAO. (2015). Organização das Nações Unidas para Alimentação e Agricultura. OCDE-FAO PERSPECTIVAS AGR´iCOLAS 2015-2024. 54 p.

FAOSTAT. (2014). Food and Agriculture Organization of the United Nations. Statistics Division. Available at: http://faostat3.fao.org/home/E. Accessed 18 Apr 2015.

Hofsetz, K., & Silva, M. A. (2012). Brazilian sugarcane bagasse: Energy and non-energy consumption. *Biomass and Bioenergy, 46*, 564–573.

Humbert, R. P. (1968). *The growing of sugar cane* (779p). New York: Elsevier.

Leal, M. R. L. V., Galdos, M. V., Scarpare, F. V., Seabra, J. E. A., Walter, A., & Oliveira, C. O. F. (2013). Sugarcane straw availability, quality, recovery and energy use: A literature review. *Biomass and Bioenergy, 53*, 11–19.

Macedo, I. C., & Seabra, J. E. A. (2008). Mitigation of GHG emissions using sugarcane bioethanol. In P. Zuubier & J. V. Vooren (Eds.), *Sugarcane ethanol: Contributions to climate change mitigation and the environment* (1st ed., pp. 95–112). Wageningen: Wageningen Academic.

MAPA. (2010). BRASIL. Ministério da Agricultura, Pecuária e Abastecimento. Secretaria de Desenvolvimento Agropecuário e Cooperativismo. Boletim técnico: biotecnologia agropecuária (pp. 1–72). Brasília: MAPA.

MAPA. (2015). BRASIL. Ministério da Agricultura, Pecuária e Abastecimento. Secretaria de Produção e Agroenergia. Anuário Estatístico de Agroenergia de 2014 (pp. 1–24). Brasília.

Moraes, M. A. F. D., & Momenti, M. P. (2006). *Setor Sucroalcooleiro. Latin American Trade Network (LATN)* (Working Paper, n. 68).

Murta, R. M., Chaves, M. A., Pires, A. J. V., Veloso, C. M., Silva, F. F., Rocha Neto, A. L., Eustaquio Filho, A., & Santos, P. E. F. (2011). Performance and nutrients apparent digestibility in sheep fed diets containing sugar cane bagasse treated with calcium oxide. *Brazilian Journal of Animal Science, 40*(6), 1325–1332.

Neves, M. F., & Conejero, M. A. (2007). Sistema Agroindustrial da cana: cenários e agenda estratégica. *Economia Aplicada, São Paulo, 11*(4), 587–604. Available at: http://www.scielo.br/pdf/ecoa/v11n4/07.pdf. Accessed 29 Jan 2009.

Pinto, A. P., Abrahão, J. J. S., Marques, J. A., Nascimento, W. J., Perotto, D., & Lugão, S. M. B. (2010). Performance and carcass characteristics of crossbred young bulls finished in a feedlot on diets with sorghum silage substituted by sugar cane. *Brazilian Journal of Animal Science, 39*(1), 198–203.

Pires, A. J. V., Garcia, R., Valadares Filho, S. C., Pereira, O. G., Cecon, P. R., Silva, F. F., Silva, P. A., & ´itavo, L. C. V. (2004). Diets with of sugarcane bagasse treated with anhydrous ammonia and/or sodium sulfate fed heifers. *Brazilian Journal of Animal Science, 33*(4), 1078–1085.

Raele, R., Boaventura, J. M. G., Fischmann, A. A., & Sarturi, G. (2014). Scenarios for the second generation ethanol in Brazil. *Technological Forecasting and Social Change, 87*, 205–223.

Salazar, K. J. M., Bereche, R. P., Rodríguez, M. C., Seabra, J., & Nebra, S. A. (2013). Reduction of water consumption in an integrated first- and second-generation ethanol plant. *Energy for Sustainable Development, 17*(5), 531–535.

Souza, C. B., & Miziara, F. (2010). Políticas de financiamento à expansão do setor sucroalcooleiro em Goiás versus políticas ambientais. In *Congresso: VIII Congresso Latino Americano de Sociologia Rural, Porto de Galinha* (pp. 1–19).

Viana, K. R. O., & Perez, R. (2013). Survey of sugarcane industry in Minas Gerais, Brazil: Focus on sustainability. *Biomass and Bioenergy, 58*, 149–157.

Economic Evaluation of Short Rotation Eucalyptus Plantation Harvesting System: A Case Study

Saulo Philipe Sebastião Guerra, Guilherme Oguri, Izabel Cristina Takitane, Giulia Lembo Caterina, and Maura Seiko Tsutsui Esperancini

Abstract Recently, the New Holland Company brought to Brazil a forager machine to harvest short rotation coppice (SRC) Eucalyptus plantation focusing on high quantity of low-priced woodchips. There are other harvesting machines available on market and each harvesting system has pros and cons. Since, in general, the harvesting and chipping costs represents the main operational costs, evaluating the economic feasibility of the chosen harvesting system is crucial. Therefore, a case study was conducted to analyse the cost of this new system in Brazil that uses a modified forager harvester and a pulled-tractor silage trailer in SRC Eucalyptus plantation. The cost analysis methodology was adapted from ASABE and the costs obtained were determined in two units: cost per time and quantity harvested in oven-dry ton (odt). The system's effective field productivity and productivity were 0.44 ha h^{-1} and 31.0 odt h^{-1}, respectively. The harvest system's total operational cost was € 258 pmh^{-1} or € 18.9 odt^{-1} and the harvester machine was the largest contributor of total cost with fixed total cost of € 87 pmh^{-1} and € 6.4 odt^{-1}. In spite of high labor charges values and high exchange rates in Brazil, the total estimated cost was cheaper than the ones found in temperate countries. From the total cost, depreciation and fuel consumption were the biggest influences. Thus, the experience levels of the harvester and tractor operators are crucial to this system economy.

S.P.S. Guerra (✉) • I.C. Takitane • G.L. Caterina • M.S.T. Esperancini
Bioenergy Research Institute (IPBEN), School of Agronomical Sciences (FCA), Sao Paulo State University (UNESP), Botucatu, Sao Paulo 18610-337, Brazil
e-mail: ssguerra@fca.unesp.br; takitante@fca.unesp.br; giulcaterina@gmail.com; maura@fca.unesp.br

G. Oguri
Cooperative Program for Forestry Mechanization and Automation (PCMAF), Forestry Science and Research Institute (IPEF), Piracicaba, Sao Paulo 13415-000, Brazil
e-mail: guilherme@ipef.br

S. Dabbert et al. (eds.), *Knowledge-Driven Developments in the Bioeconomy*, Economic Complexity and Evolution, DOI 10.1007/978-3-319-58374-7_12

1 Introduction

In Brazil, Eucalyptus is a promising genus for renewable energy production in short rotation coppice (SRC) (Guerra et al. 2012) given that the surface area covered by this species plantations designed for production was around 5.6 million ha in 2014 (IBA 2015). Conventional commercial plantations (of 3.0 × 2.0 m spacing, harvest at 6–8 years) can reach mean productivities of 45 m^3 ha^{-1} $year^{-1}$ due to genetic improvement and ideal edaphoclimatic conditions (Gonçalves et al. 2013). The key components of Eucalyptus high growth rate are its nutrients capture and use-efficiency and abundant water supply, mainly (Stape et al. 2004). According to Guerra et al. (2014), eucalypt short-rotation plantations, clearcut at 2–3 years, may present superior biomass production in shorter time when compared with the conventional forestry, reaching 120 m^3 ha^{-1} in just 1 year.

Short rotation coppice (SRC) systems are design to increase the population density resulting in high quantity of low-priced final biomass product (Spinelli et al. 2009). In addition, as a fast growth species, eucalypt cultivated in SRC, system characterized by constant removal of biomass, requires an extra fertilization to maintain the soil fertility and high productivity rates (Mitchell et al. 1999). Therefore, this system demands highly efficient operations to active success, especially within harvesting, whereas its cost can account of 50% of total (Spinelli et al. 2009). In addition, as energy source, the final product woodchips should have uniform size (Spinelli et al. 2011) and low moisture content. This latter quality parameter is directly linked to market price (Nyström and Dahlquist 2004) and calorific power (Ergül and Ayrilmis 2014). It is important for power plants that the delivered woodchip has low moisture content variation to avoid combustion problems (Nyström and Dahlquist 2004).

In some European and American countries, the use of modified forager machines for forest harvesting has been the target of studies since the 1990s. Modified foragers have the characteristic of transforming the entire tree into woodchips in one-single pass, mostly known as cut-and-chip system (CAC). Some researchers like, Spinelli and Magagnotti (2010), Schweier and Becker (2012a, b) and Eisenbies et al. (2014) analysed the operating performance of available forage harvesters in the market from the following brands; Jaguar, Class, John Deere, and New Holland. In these studies, the main trees genus used were poplar (*Poplar* spp.) and willow (*Salix* spp.).

Forest mechanized harvesting is normally composed by two different systems in Brazil: cut-to-length (CTL) and whole-tree logging (WTL). In the former case, the machine responsible for cutting the trees, the harvester header cut-base the stems, delimber, cross-cut and bunch in one process, then, the forwarder machine collect the harvested stems and remove from the plot to alongside the road (Nurminen et al. 2006). In the latter case, the operation consists of a feller-buncher, responsible for cutting-base the trees and bunching, a processor, for delimbering, and a grapple skidder, to remove the logs outside the plot (Ghaffariyan et al. 2011). The CTL system is considered more environmental friendly because has higher productivity rate and requires less machines on the field. Both harvesting types need another machine in order to process the logs and cut them into woodchips, known as woodchipper. Chippers can be divided in mobile or stationary (Spinelli and

Fig. 1 Woodchip pile, forager and Eucalyptus SRC

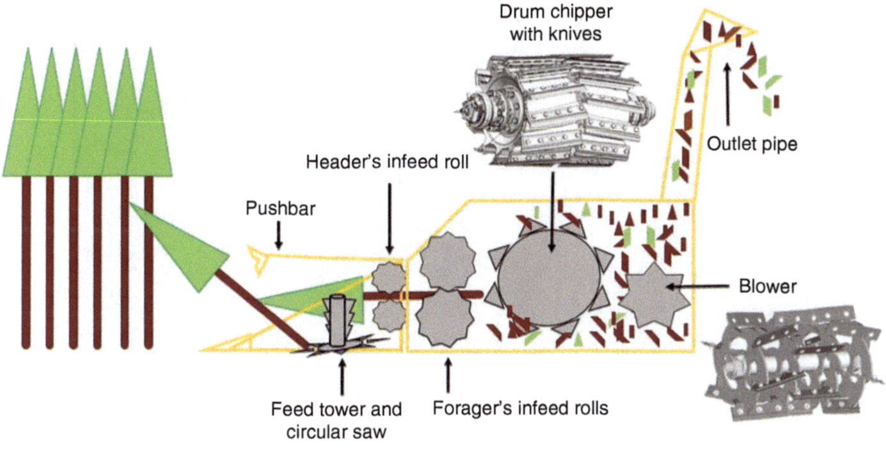

Fig. 2 Main features scheme of adapted forage harvester

Magagnotti 2010) and be classified for its cutting device, drum or discs (Manzone and Balsari 2015). Drum devices are more productive than discs (Spinelli et al. 2013) but produce higher diversification of medium-size chips (Spinelli et al. 2005).

The introduction of modified forager machines in forest systems is quite recent in Brazil. A few years ago, New Holland brought to the country a forager machine (Fig. 1) with the purpose to harvest SRC Eucalyptus plantation with high quantity of low-priced woodchips.

A forager machine and a coppice header (Fig. 2) compose the harvester. The header or cutting head is equipped with two circular saws that cut-base located at

Fig. 3 Woodchips length: 30 mm (*top*) and 5 mm (*bottom*)

the bottom of two feeding towers, one push-bar that push the felled trees towards opposite direction to the machine, one paddle roll that lifts up the trees into two verticals infeed rolls toward inside the forager. The drum chipper with knives is located inside the forager after the forage's infeed roll and metal detector roll. After the chipper, the woodchips are engaged by the blower and discharged through the outlet pipe.

The chip length can be chosen electronically from the cab through increasing or decreasing the feeding speed, giving the option for chip length from 30 to 5 mm (Fig. 3) in 1 mm steps. The largest one is recommended for directly burning and the smaller, can be used for pellet manufacturing.

Fig. 4 Cut quality: low (*top*) and high (*bottom*)

The further concern regarding this kind of harvesting is the stump quality after the coppice (Fig. 4). In order to produce high standard cutting, the circular saw must have the following features: be sharpened, ideal rotation speed and ideal forward-motion speed (forager harvesting speed). Low cutting quality decrease the odds for stumps to produce vigorous resprouts. In other hand, "clean" cutting allows both healthy resprouting (Soler et al. 2013) and increase the biomass productivity

Fig. 5 Eucalypt resprouts on stump few weeks after mechanized harvested

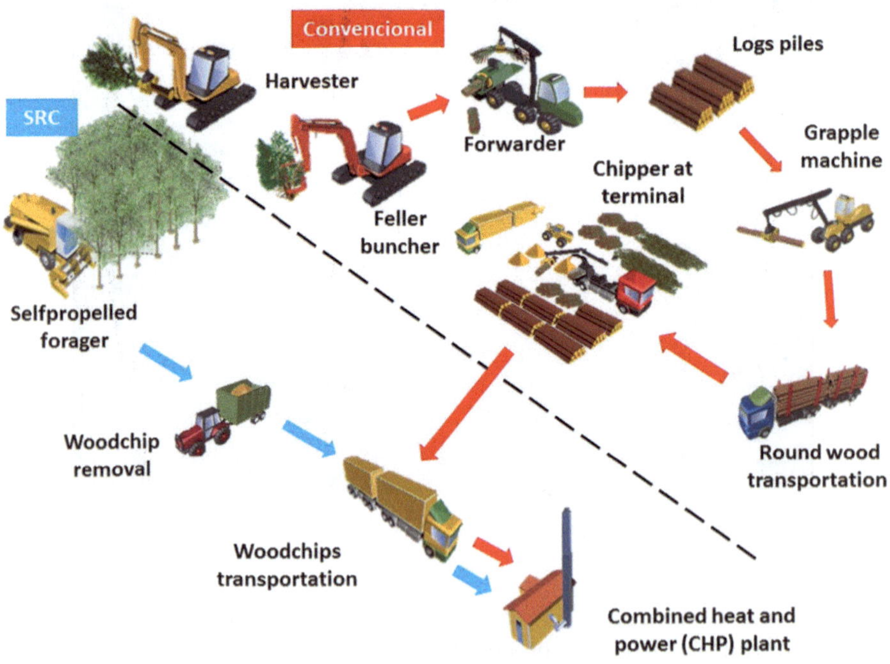

Fig. 6 CTL, WTL and CAC harvesting process (Eufrade Junior et al. 2016)

(Dillen et al. 2013). However, one must have in mind that the regrowth ability varies between species (Fig. 5).

Hence, there is plenty harvesting machines available on market. However, evaluating the economic feasibility of the introduction of eucalypt as energy crop, the harvesting and chipping cost represents the main cost item (Sgroi et al. 2015). For this reason, a wisely choice is necessary for a profit production. Each harvesting system has pros and cons (Culshaw and Stokes 1995). Figure 6 represents each harvesting system discussed above.

Thus, a case study was conducted to analyse the cost of this new system in Brazil using modified forager harvester and pulled-tractor silage trailer in short-rotation eucalypt plantation.

2 Material and Methods

The study was conducted at Sao Paulo State (48°24′43″W, 22°58′10″S), in a 1.7 ha *Eucalyptus grandis* × *Eucalyptus urophylla* hybrid short-rotation plantation. The plantation spacing was 3.0 × 1.0 m (3333 plants ha^{-1}) and at harvest time, trees had 2.8 years and an average base diameter (at 9 cm of height) of 10 cm. The area maximum land slope was 6%. The climate is mesothermic with dry winter and an average temperature of 20°C. The average annual precipitation is 1500 mm with 58% occurring from January to June. The annual potential evapotranspiration is 945 mm with 33% occurring in summer time (Cunha and Martins 2009).

Integrated the harvest system: a New Holland self-propelled forager machine attached to New Holland coppice header, two New Holland tractors and four TMA 24 m^3 silage trailers (Table 1).

Table 1 Equipments description

Forager		
Make		New Holland
Model		FR9060
Power	kW	441
Mass	kg	12,600
Header		
Make		New Holland
Model		130FB
Mass	kg	2000
Tractor		
Make		New Holland
Model		TM7040
Power	kW	132
Mass	kg	6800

The system's cost analysis methodology was adapted from ASABE (2011a) and the costs obtained were determined in two units: cost per time and quantity harvested in oven-dry ton (odt).

Ownership costs or fixed costs (FC) consist of depreciation, interest on fixed assets and other costs (taxes, housing and insurance), which focus on all the equipment of the mechanized harvesting system (harvester, tractor, and silage trailer) and are detailed in the equation:

$$FC = \frac{\{[\frac{V_A+V_R}{2} + i] + [\frac{V_A-V_R}{EL}] + THI\}}{pmh}$$

Where:

FC = fixed cost (€ $hour^{-1}$)
V_A = acquisition value of machinery and equipment (€)
V_R = residual value of machinery and equipment (€)
i = interest rate per year (%)
EL = economic life (years)
THI = taxes, housing and insurance (% of V_A in € $year^{-1}$)
pmh = productive machine hours per year

Operating costs or variable costs (VC) consist of fuel, oil and lubricants, repairs and maintenances, and labor focusing on forage machines and tractors. The average fuel consumption is based on the actual power that is required or on the actual consumption measured in the field. The data of fuel consumption were collected using the machine's on-board computer (*Intelliview*). To collect the tractor's hourly consumption we followed the methodology described by Fiorese et al. (2012). To calculate oil lubricants and greases costs, we used the 15% factor in the cost of fuel.

According to ASABE (2011b), the accumulated costs of repairs and maintenance to a typical field velocity can be determined with the following expression using the repair and maintenance factors RF1 and RF2:

$$CRM = \frac{\left\{V_A \times RF1 \times \left(\frac{h+pmh}{1000}\right)^{RF2}\right\} - \left\{V_A \times RF1 \times \left(\frac{h}{1000}\right)^{RF2}\right\}}{pmh}$$

Where:

CRM = cost of repair and maintenance (€ $hour^{-1}$)
RF1 and RF2 = repair and maintenance factors
V_A = acquisition value of machinery and equipment (€)
h = hours accumulated
pmh = productive machine hour per year

Labor cost was calculated based on the monthly wage and work hours, including a correction factor of 25% due to idle time, in other words, time taken for repairs and supply of machines. Cost spend with employees transportation to the workstation were discarded. One workday consisted by two shifts of 4 h each. Wages and labor charges were based on the database provided by the forestry companies' partners.

Some harvester data were estimated because the product is not considered as commercial machine in Brazil. Prices were acquired in Brazilian currency (R$) and converted to Euros (€) using the average exchange rate for 2015 of R$ 3.16 €$^{-1}$ according to the official website of the Central Bank of Brazil (www.bcb.gov.br). Only repairs and maintenance cost were included on variable costs for silage trailers (Table 2).

In order to estimate the cost per tonne it was used the productivity expressed in dry tonne per hour, once the woodchip moisture was 52%.

The effective field productivity (EFP) was calculated as shown:

$$EFP = \frac{d \times s}{1000 \times t}$$

Where:

EFP = effective field productivity (ha h^{-1})
d = distance (m)
s = space between rows (m)
t = time (h)

Table 2 Input parameter for cost calculation

Description		Harvester[a]	Tractor	Silage trailer
Estimated acquisition value	€	316,500.00	55,380.00	28,500.00
Expected economic life	years	7	9	12
Annual use	pmh	600	1500	n.a.
Residual value	%	10.00	10.00	10.00
Interest rate	%	5.00[b]	5.00	5.00
Housing	%	0.75	0.75	0.75
Insurance	%	0.25	0.25	0.25
Fuel consumption	L h^{-1}	80	25	n.a.
Fuel price	€ L^{-1}	0.80	0.80	n.a.
Repair factor 1		0.03	0.003	0.16
Repair factor 2		2.0	2.0	1.6
Operator wage	€ month^{-1}	633.00	633.00	n.a.
Labor charges	%	125.00	125.00	n.a.

Note:
pmh productive machine hours, *n.a.* not applicable
[a]FR9060 and 130FB
[b]KTBL (2014)

3 Results and Discussion

EFP and productivity were 0.44 ha h^{-1} and 31.0 odt h^{-1}, respectively. Using the same forage machine in a SRC poplar in Germany, an EFP average of 0.90 ha h^{-1} and a productivity of 14.6 odt h^{-1} were obtained (Schweier and Becker 2012a). Tests conducted in the United States, harvesting willow, showed results from 1.8 to 2.3 ha h^{-1} of EFP and a productivity between 23.9 and 24.9 odt h^{-1}, requiring speeds from 8.0 to 10.0 km h^{-1} in which is unrealistic regarding a SRC (Eisenbies et al. 2014). These studies were conducted in temperate regions where the most common source of raw material are *Salix* spp. and *Poplar* spp. Plantations. These species are generally harvested around 3 and 4 years, and the basic wood density average is 350 and 410 kg m^{-3}, respectively (Tharakan et al. 2003). *Eucalyptus* spp. has an average basic wood density of 430 kg m^{-3} (Garcia 2013). This information might justify the result of higher productivity in Brazil, once Eucalyptus's wood density is 4.6% and 18.6% higher than willow and poplar, respectively.

The harvester EFP is related to the working speed in which could be limited by terrain conditions (i.e. slope and soil type), planting condition (i.e. trees diameter, planting spacing and presence of old stumps between planting lines), operator experience level, and forage harvester power.

The harvest system's total operational cost was € 258 pmh^{-1} or € 18.9 odt^{-1} being the harvester the largest contributor of total cost with fixed total cost of € 87 pmh^{-1} and € 6.4 odt^{-1} (Table 3).

Table 3 Operational costs

	Harvester		Tractors[a]		Silage trailers[b]	
Element	€ pmh^{-1}	€ odt^{-1}	€ pmh^{-1}	R$ odt^{-1}	€ pmh^{-1}	€ odt^{-1}
Fixed costs						
Depreciation	67.81	4.97	7.38	0.54	5.70	0.42
Interest rate	14.50	1.06	2.03	0.15	2.09	0.15
Housing	3.96	0.29	0.55	0.02	0.57	0.04
Insurance	1.32	0.10	0.18	0.01	0.19	0.01
FCT	*87.59*	6.42	*10.15*	0.71	*8.54*	0.63
Variable costs						
Fuel	64.05	4.70	40.03	2.93	–	–
Oils and lubricants	9.61	0.70	12.01	0.88	–	–
Repairs and maintenance	5.70	0.42	0.50	0.04	5.81	13.21
Operator	4.94	0.36	9.89	0.73	–	–
VCT	*84.30*	6.18	*62.43*	4.58	*5.81*	*13.21*
Total cost	**258.83**	**18.95**				

Note:
FCT fixed costs total, *VCT* variable costs total
Exchange rate (2015 average value): € 1.00 = R$ 3.16
[a]Two tractors
[b]Four silage trailers

Schweier and Becker (2012a) determined an estimated total cost of € 281 h^{-1} and € 19.70 odt^{-1}. The harvester per unit time individual cost (excluding labor costs) was found by Berhongaray et al. (2013), which was 212.5 € h^{-1}. Despite the high value of labor charges on the operator's wage and the rise of the exchange rate, operational costs are below those found in the literature. This difference is even greater due to the mean annual increment (MAI) of eucalypt in SRC in Brazil and its impact on the productivity generated by the harvester in this system.

The percentage contributions to total cost of each equipment item are listed in Table 4.

Depreciation and fuel are the two factors that most contributed to the total cost of the harvester, justified by the high purchase price and high fuel consumption of this equipment. Operator's experience can be crucial to reduce fuel consumption because it is necessary to adjust the speed according to the forest conditions in order to increase wood chips production while manage operational time and fuel consumption efficiently, without wasting trees and preserving both sets of base cutting disks and cutting knives.

Whereas this forager harvester is non-commercial equipment, lifespan and productive annual hours were estimated, thus, with the adaptations improvement for Brazilian forests conditions both parameters can be even greater reducing the depreciation cost of the harvester.

The greater part of the tractor's cost is related to the consumed diesel and here, again, the experience of the operator is decisive. Proper engine rotation for each harvest stage, proper adjustment of ballast weights, the type of tires and their internal calibration are factors that significantly influence the tractor's performance (Lopes et al. 2003; Filho et al. 2010; Monteiro et al. 2011; Berhongaray et al. 2013).

Suitable fleet sizing is essential to reduce idle time in the harvesting process; however, this analysis demands full time studies and measurement of the maximum

Table 4 Participation in percentage of each element on individual cost

	Harvester	Tractor	Silage trailer
Element	(%)		
Fixed costs			
Depreciation	39.5	10.2	39.7
Interest rate	8.4	2.8	14.5
Housing	2.3	0.8	4.0
Insurance	0.8	0.3	1.3
Variable costs			
Fuel	37.3	55.2	–
Lubricants	5.6	16.5	–
Repairs and maintenance	3.3	0.7	40.5
Operator	2.9	13.6	–
	100	100	100

distance between harvested area and wood chip discharge area in order to optimize the logistics. In this assessment, the required fleet sizing was estimated from authors' experience.

Once the produced biomass destination is bioenergy production, the calorific value contained in this material becomes relevant. Guerra et al. (2014) conducted a study testing the same clone in order to quantify the calorific value of a SRC under different spacing and fertilization levels. At 2 years old, in 2.8 × 1.0 m spacing (3571 plants ha^{-1}) and applying the conventional fertilization dose, this clone reached an average of 20 GJ t^{-1} or 761 GJ ha^{-1}. Converting to megawatt-hour, a productive day of harvesting would be able to produce around 2500 GWh. This energy is sufficient to generate electricity during 3 h in a European city with 542,000 households (Kavgic et al. 2013), or 16 h in a large Brazilian city with 2,000,000 inhabitants (Pereira and de Assis 2013). Clearly, this comparison is both informative and illustrative because it does not include generation neither transmission loss of energy, among other factors.

4 Conclusion

The use of modified forage machines for harvesting eucalypt in SRC, despite harvesting a smaller area per time, could reach a greater amount of harvested material per area compared to consolidated harvesting systems for willow and poplar plantations in temperate countries due to the difference of wood basic density.

Even with high labor charges values and high exchange rates, the total estimated cost is cheaper than those from temperate countries, with depreciation and fuel consumption being the biggest influences of total cost. The experience level of the harvester and tractor operators is crucial to this system economy.

References

American Society of Agricultural and Biological Engineers. (2011a). ASABE Standards 2011. ASABE Standards EP4963—Agricultural machinery management data (p. 6). St. Joseph, MI: ASABE.

American Society of Agricultural and Biological Engineers. (2011b). ASABE Standards 2011. ASABE Standards D4977—Agricultural machinery management data (p. 8). St. Joseph, MI: ASABE.

Berhongaray, G., El Kasmioui, O., & Ceulemans, R. (2013). Comparative analysis of harvesting machines on an operational high-density short rotation woody crop (SRWC) culture: One-process versus two-process harvest operation. *Biomass and Bioenergy, 58*, 333–342.

Culshaw, D., & Stokes, B. (1995). Mechanisation of short rotation forestry. *Biomass and Bioenergy, 9*, 127–140.

Cunha, A. R., & Martins, D. (2009). Classificação climática para os municípios de Botucatu e São Manuel, SP. *Irriga, Botucatu 14*, 1–11 [Portuguese].

Dillen, S., Djomo, S. N., Al Afas, N., Vanbeveren, S., & Ceulemans, R. (2013). Biomass yield and energy balance of a short-rotation poplar coppice with multiple clones on degraded land during 16 years. *Biomass and Bioenergy, 56*, 157–165.

Eisenbies, M. H., Volk, T. A., Posselius, J., Foster, C., Shi, S., & Karapetyan, S. (2014). Evaluation of a single-pass, cut and chip harvest system on commercial-scale, short-rotation shrub willow biomass crops. *Bioenergy Research, 7*, 1506–1518.

Ergül, E., & Ayrilmis, N. (2014). Effect of outdoor storage conditions of wood chip pile on the technological properties of wood-based panel. *Biomass and Bioenergy, 61*, 66–72.

Eufrade Junior, H. J., Melo, R. X., Sartori, M. M. P., Guerra, S. P. S., & Ballarin, A. W. (2016). Sustainable use of Eucalyptus biomass on short rotation coppice for bioenergy. *Biomass and Bioenergy, 90*, 15–21.

Filho, A. G., Lanças, K. P., Leite, F., Acosta, J. J., & Jesuino, P. R. (2010). Desempenho de trator agrícola em três superfícies de solo e quatro velocidades de deslocamento. *Revista Brasileira de Engenharia Agrícola e Ambiental, 14*, 333–339 [Portuguese].

Fiorese, D. A., Dallmeyer, A. U., Romano, L. N., Schlosser, J. F., & Machado, P. R. N. (2012). Desempenho de um motor de trator agrícola em bancada dinamométrica com biodiesel de óleo de frango e misturas binárias com óleo diesel. *Ciência Rural, 42*, 660–666 [Portuguese].

Garcia, E. A. (2013). Qualidade energética da madeira de eucalipto em função do espaçamento, da adubação e da idade conduzida no sistema florestal de curta rotação (p. 57). Botucatu/SP: Universidade Estadual Paulista [Portuguese]

Ghaffariyan, M., Brown, M., Acuna, M., Sessions, J., Kuehmaier, M., & Wiedemann, J. (2011). Biomass harvesting in Eucalyptus plantations in Western Australia. *Southern Forests: A Journal of Forest Science, 73*, 149–154.

Gonçalves, J. L. M., et al. (2013). Integrating genetic and silvicultural strategies to minimize abiotic and biotic constraints in Brazilian eucalypt plantations. *Forest Ecology and Management, 301*, 6–27.

Guerra, S. P. S., Lanças, K. P., Garcia, E. A., & Spinelli, R. (2012). Eucalipto adensado: manejo para florestal energéticas. In E. G. M. Lemos & N. R. Stradiotto (Eds.), *Bioenergia: desenvolvimento, pesquisa e inovação* (p. 1069). São Paulo: Ed. Unesp [Portuguese]

Guerra, S. P. S., Garcia, E. A., Lanças, K. P., Rezende, M. A., & Spinelli, R. (2014). Heating value of eucalypt wood grown on SRC for energy production. *Fuel, 137*, 360–363.

IBA. (2015). Statistics of the Brazilian Tree Industry, Report 2015 (p. 64). Brasília: Brazilian Tree Industry.

Kavgic, M., Mumovic, D., Summerfield, A., Stevanovic, Z., & Ecim-Djuric, O. (2013). Uncertainty and modeling energy consumption: Sensitivity analysis for a city-scale domestic energy model. *Energy and Buildings, 60*, 1–11.

Kuratorium für Technik und Bauwesen in der Landwirtschaft. (2014). KTBL 2014: Datensammlung. Kuratorium für Technik und Bauwesen in der Landwirtschaft. KTBL 2014. Datensammlung: Auflage [German]

Lopes, A., Lanças, K. P., Furlani, C. E., Nagaoka, A. K., Castro Neto, P., & Grotta, D. D. C. (2003). Consumo de combustível de um trator em função do tipo de pneu, da lastragem e da velocidade de trabalho. *Revista Brasileira de Engenharia Agrícola e Ambiental, 7*, 382–386 [Portuguese].

Manzone, M., & Balsari, P. (2015). Productivity and woodchip quality of different chippers during poplar plantation harvesting. *Biomass and Bioenergy, 83*, 278–283.

Mitchell, C., Stevens, E., & Watters, M. (1999). Short-rotation forestry—Operations, productivity and costs based on experience gained in the UK. *Forest Ecology and Management, 121*, 123–136.

Monteiro, L. A., Lanças, K. P., & Guerra, S. P. S. (2011). Desempenho de um trator agrícola equipado com pneus radiais e diagonais com três níveis de lastros líquidos. *Engenharia Agrícola, 31*, 551–560 [Portuguese].

Nurminen, T., Korpunen, H., & Uusitalo, J. (2006). Time consumption analysis of the mechanized cut-to-length harvesting system. *Silva Fennica, 40*, 335.

Nyström, J., & Dahlquist, E. (2004). Methods for determination of moisture content in woodchips for power plants—A review. *Fuel, 83*, 773–779.

Pereira, I. M., & de Assis, I. S. (2013). Urban energy consumption mapping for energy management. *Energy Policy, 59*, 257–269.

Schweier, J., & Becker, G. (2012a). New Holland forage harvester's productivity in short rotation coppice: Evaluation of field studies from a German perspective. *International Journal of Forest Engineering, 23*, 82–88.

Schweier, J., & Becker, G. (2012b). Harvesting of short rotation coppice—Harvesting trials with a cut and storage system in Germany. *Silva Fennica, 46*, 287–299.

Sgroi, F., Di Trapani, A. M., Foderà, M., Testa, R., & Tudisca, S. (2015). Economic assessment of Eucalyptus (spp.) for biomass production as alternative crop in Southern Italy. *Renewable and Sustainable Energy Reviews, 44*, 614–619.

Soler, R. R., Lanças, K. P., Guerra, S. P. S., Sereghetti, G. S., Oguri, G., & Guimarães, R. T. S. (2013). Incremento médio anual pós colheita mecanizada em sistema de rebrota. In *8° Congresso Internacional de Bioenergia, São Paulo* [Portuguese].

Spinelli, R., & Magagnotti, N. (2010). Comparison of two harvesting systems for the production of forest biomass from the thinning of Picea abies plantations. *Scandinavian Journal of Forest Research, 25*, 69–77.

Spinelli, R., Hartsough, B. R., & Magagnotti, N. (2005). Testing mobile chippers for chip size distribution. *International Journal of Forest Engineering, 16*, 29–35.

Spinelli, R., Nati, C., & Magagnotti, N. (2009). Using modified foragers to harvest short-rotation poplar plantations. *Biomass and Bioenergy, 33*, 817–821.

Spinelli, R., Nati, C., Sozzi, L., Magagnotti, N., & Picchi, G. (2011). Physical characterization of commercial woodchips on the Italian energy market. *Fuel, 90*, 2198–2202.

Spinelli, R., Cavallo, E., Eliasson, L., & Facello, A. (2013). Comparing the efficiency of drum and disc chippers. *Silva Fennica, 47*(2), 1–11.

Stape, J. L., Binkley, D., & Ryan, M. G. (2004). Eucalyptus production and the supply, use and efficiency of use of water, light and nitrogen across a geographic gradient in Brazil. *Forest Ecology and Management, 193*, 17–31.

Tharakan, P., Volk, T., Abrahamson, L., & White, E. (2003). Energy feedstock characteristics of willow and hybrid poplar clones at harvest age. *Biomass and Bioenergy, 25*, 571–580.

Technology and Sustainability of Crop Fibre Uses in Bioproducts in Ontario, Canada: Corn Stalk and Cob Fibre Performance in Polypropylene Composites

Muhammad Arif, Muhammad Riaz, C. Joe Martin, Yarmilla Reinprecht, Leonardo Simon, Bill Dean, and K. Peter Pauls

Abstract Composites containing fibres from biological sources, such as residues from field crop production, are being increasingly used for manufacturing consumer products, automobile parts and construction materials because of their low costs, as well as their ecological and performance benefits. The chapter examines the sustainability of using plant fibres for bioproduct manufacturing in Ontario, Canada from annual and perennial crops. It also examines parameters that affect the performance of composites compounded with polypropylene and (*Zea mays*) corn fibres. In particular, the study identified relationships between specific performance characteristics of the corn fibres and their chemical compositions and confirmed that plant genetics and crop production environment play significant roles in both traits. Further, it identified cell wall traits, genomic regions and genes that might be used to select corn lines that have improved fibre characteristics for bioproduct manufacturing.

1 Introduction

Natural fibre composites are emerging as a viable alternatives to mineral, carbon and glass-reinforced composites for manufacturing automotive parts, construction materials, and for electrical and electronic uses because of their low costs, as well as their ecological and performance benefits. The global natural fibre composite market is anticipated to grow by 12% annually with a value of $5.8 billion (US) by

M. Arif • M. Riaz • C.J. Martin • Y. Reinprecht • B. Dean • K. Peter Pauls (✉)
Plant Agriculture, University of Guelph, 50 Stone Road E, Guelph, ON, Canada N1G 2W1
e-mail: marif@uoguelph.ca; mrsafi@gmail.com; cmarti07@uoguelph.ca; yreinpre@uoguelph.ca; bdeen@uoguelph.ca; ppauls@uoguelph.ca

L. Simon
Chemical Engineering, University of Waterloo, 200 University Ave W, Waterloo, ON, Canada N2L 3G1
e-mail: lsimon@uwaterloo.ca

S. Dabbert et al. (eds.), *Knowledge-Driven Developments in the Bioeconomy*, Economic Complexity and Evolution, DOI 10.1007/978-3-319-58374-7_13

2019.The growth of the global automotive and construction industries, promulgation of new environmental regulations, local availability of natural fibres, the suitability of natural fibres for various manufacturing processes (including compression and injection moulding) and consumer acceptance of bio-filled composite materials are some of the key drivers for the expansion of natural fibre-based composite market (http://www.researchandmarkets.com/reports/2881528/global-natural-fiber-composites-market-2014-2019; accessed on May 21, 2015).

The province of Ontario is one of the largest agricultural biomass producing regions in Canada. It and its neighbouring state of Michigan in the USA host more than 95% of the automotive industry in the North America. In 2014 Canada built 2.1 million vehicles—10th highest in the world (http://www.cbc.ca/news/business/loss-of-corolla-doesn-t-have-to-be-lethal-blow-for-canadian-auto-production-1.3033914). The use of A. Schulman's AgriPlas™ Wheat Straw Bio-filler for an injection-molded storage bin in the interior of the Ford Flex was an early adoption of this technology (http://www.prnewswire.com/news-releases/a-schulmans-agriplastm-wheat-straw-bio-filler-on-ford-flex-receives-innovation-recognition-from-spe-automotive-division-70197502.html). The advantages of using this material in automobiles include: weight savings of approximately 10%, increased dimensional stability, reduced energy use in manufacturing due to lower machine temperatures, and lower carbon footprint (1.30 kg less of carbon dioxide per kilogram of product, based on Ford's analysis).

The chapter examines the sustainability of using plant fibres for bioproduct manufacturing in Ontario and presents results from polypropylene reinforced with corn (*Zea mays*) fibres as a model for bio-filled thermoplastic composite manufacturing. Corn is examined in detail because large amounts of stalk and cob residues are readily available in Ontario and they are, currently, left in the field to decompose after the harvest of the grain. However, there is increasing interest in harvesting and utilizing the residue for energy or bioproduct manufacturing.

2 Sustainability of Biomass Production from Field Crops in Ontario

Approximately 5% of Ontario's land base is suitable for agriculture. In 2011, crop production occurred on 3.6 M hectares with corn, soybean, winter wheat, forage and pasture dominating (Statistics Canada, Census of Agriculture, 2011, http://www.statcan.gc.ca/eng/ca2011/index). In Canada, land capability is determined using the Canada Land Inventory (CLI) for agriculture. CLI is an interpretative system for assessing the effects of climate and soil characteristics on the limitations of land for growing common field crops. Class 1 soils have no significant limitations in use for crop production. The soils are deep, are well to imperfectly drained, hold moisture well, and in the virgin state were well supplied with plant nutrients. They can be managed and cropped without difficulty. Under good management they are

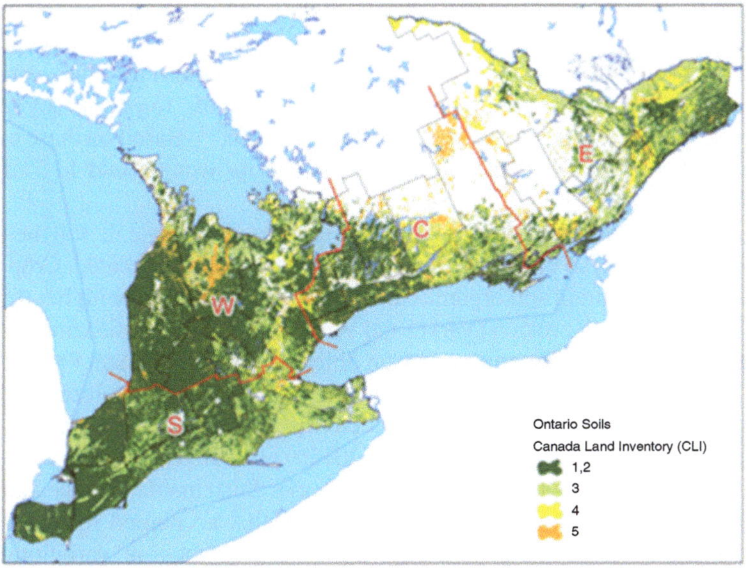

Fig. 1 Map of Class 1, 2, 3, 4 and 5 lands in Southern (S), Western (W), Central (C), and Eastern (E) Ontario (reproduced from Kludze et al. 2013a)

moderately high to high in productivity for a wide range of field crops. Class 2 soils have moderate limitations, and moderate conservation practices required to achieve moderately high to high in productivity. Class 3 soils have moderately severe limitations and the range of crops is restricted or special conservation practices required. Class 4 soils have severe limitations. Class 5 soils are restricted to forage crops and improvement practices are feasible. (http://sis.agr.gc.ca/cansis/nsdb/cli/class.html). Figure 1, shows the distribution of Class1–5 land in Ontario. It can be seen that Ontario has considerable acreage of Class 1–3 land. In Ontario, the majority of annual crop production occurs on Class 1–3 soils and corn-soybean are increasingly the dominant rotation (Gaudin et al. 2015). Pasture and forage production tends to occur on Class 4–5 soils which, as can be seen in Fig. 1, are more dispersed. In Ontario biomass for emerging markets can be derived from either (1) crop residues removed from crop rotations, or (2) dedicated biomass crops which would require displacement of existing crops and land use. In either case, the evaluation of sustainability of these systems must occur and must consider the context above.

2.1 Crop Residues

In Ontario, corn, soybean and winter wheat are the main candidate crops for residue removal with 0.8 M, 1.0 M and 0.44 M hectares grown in 2011, respectively

(Statistics Canada, Census of Agriculture, 2011). Like much of the U.S. Northern Corn Belt, corn-soybean rotations dominate the Ontario landscape (Gaudin et al. 2015), and corn-soybean-winter wheat rotation is less common. Five-year (2004–2008) average aboveground crop residue biomass yields for corn, soybean and winter wheat are 8.6, 2.6 and 4.6 Mg (=metric tonne) ha^{-1}. Based on these estimate, complete removal of all residue from these three crops would yield 11.5 M tonnes $year^{-1}$ of biomass.

Complete residue removal would have significant ecological, soil health and erosion implications, so complete removal is unlikely to be advocated. Crop residue requirements to maintain soil organic carbon and nutrient pools are higher than those required to control soil erosion (Varvel and Wilhelm 2008); thus, sustainable residue removal levels are often determined by an objective to maintain soil organic matter levels. This objective furthermore recognizes the importance of soil organic matter in maintaining soil health and crop productivity. Based on the assumption of soil organic matter maintenance, Kludze et al. (2013b) estimated how much crop residue can be removed from Ontario cropping systems. Sustainable residue removable rates were determined using a five-step approach that accounted for maintenance of soil organic matter in the presence of yield and rotation variations across Ontario counties. Under typical soil organic matter formation and decomposition conditions, and assuming typical corn-soybean and corn-soybean-winter wheat rotation scenarios, about 1.1 million Mg of residue could be sustainably removed each year, primarily from the major (Southern and Western, see Fig. 1) agricultural regions in the province. For a given region in Ontario, the amount of residue that could be sustainably removed was determined by rotation complexity and crop yield. Rotational complexity, through the addition of winter wheat, increased sustainable residue removable rates, while inclusion of soybean decreased available residue compared to maize and winter wheat. To increase sustainable residue removal level above 1.1 million Mg $year^{-1}$ will require greater rotation diversity, adoption of cover crops, and increased crop yields. Note that, in Ontario, no-till systems are assumed to not impact sustainable residue removal rates since no-till production systems are not associated with elevated soil organic matter (Powlson et al. 2014) and tend to be associated with yield reductions (Pittelkow et al. 2014).

Removal of corn and soybean residues can be challenging under the Ontario environment. Long season corn hybrids and soybean varieties are grown to maximize yield potential, consequently planting occurs as early as field conditions permit in the spring and harvest occurs late in the season. Spring collection of maize residue (Fig. 2) can interfere with timeliness of subsequent panting operations or may cause soil compaction if conducted when soil are still too wet. Fall collection of corn and soybean residues may interfere with fall operations, for example, soybean residue collection may delay timely planting of winter wheat which is associated with winter survival and spring vigour. Late season collection of corn and soybean also increases probability of high moisture levels in crop requiring drying or risk or storage losses. Winter wheat residue collection has fewer of these concerns however, currently a significant portion of winter wheat residue is already collected for competing markets such as livestock bedding, mushroom compost, mulch, or as a fibre source

Fig. 2 Baling corn stover (*left*) and handling bales (*right*) in a field in Ontario

for ruminant livestock. Corn cobs can be harvested in a one-pass system with a commercially available modified combine that collects cobs into a trailing wagon. Based on a cob to grain ratio of 0.16 and a 5-year (2004–2008) grain yield average, cob yield is 0.9–1.4 dry t ha^{-1} and total annual production potential in Ontario is 0.9 M tonne. Moisture content at harvest however, is 25% (% weight) or greater, so drying and storage conditions must be considered.

2.2 Dedicated Biomass Crops

Numerous crops have been evaluated in Ontario as candidates for dedicated biomass crops. These include: (1) perennial grass species such as switchgrass (*Panicum virgatum*), miscanthus (*Miscanthus* spp.), big bluestem (*Andropogon gerardii*) and reed canary grass (*Phalaris arundinacea*); (2) annual species such as hemp (*Cannabis sativa* L.) and energy corn; and (3) short rotation woody species such as poplar (*Populus trichocarpa*) and willow (*Salix viminalis*). In Ontario, the perennial grass species have received the most attention as potential biomass crops, particularly the C4 species, switchgrass and miscanthus. These species are adapted to Ontario conditions (Sage et al. 2015), have high yield potential (Heaton et al. 2004), low input requirements, utilize existing commercial equipment for production (e.g., Fig. 3), and have favourable greenhouse gas balances (Eichelmann et al. 2016). The use of only 5% of all arable land in Ontario (i.e., Classes 1–5 lands) can provide over 2 million t DM of either switchgrass or miscanthus biomass. Due to higher yield potential (approximately 2×), a miscanthus-based biomass system, compared with a switchgrass-based system, would require a significantly lower percentage of available land area to produce an equivalent amount of biomass.

Both miscanthus and switchgrass cultivation are associated with increases in soil organic matter and improvements in soil health; however the magnitude of this effect is dependent on previous land use (Sanscartier et al. 2014a, b). Figure 3 shows fall harvest of miscanthus (left) and switchgrass (right) in Ontario. Introduction of dedicated biomass crops in Ontario necessarily requires displacement of existing

Fig. 3 Fall harvest of miscanthus (*left*) and switchgrass (*right*) in Ontario

crop production, since there is little to no idle land in the area represented by Fig. 1. Displacement of pasture and forage production by C4 perennial grasses will have minimal impact on soil organic matter since these systems are already associated with high levels of soil organic matter. Also, since pasture and forage production is most common on Class 4–5 soils, and yield potential of miscanthus and switchgrass is lower on these soils, opportunity for increasing soil organic matter is reduced. In contrast, soybean dominated systems tend to be associated with reduced soil organic matter levels, and also occurs on Class 1–3 soils that have high yield potential for C4 perennial grasses increasing organic matter returns to soil. Displacement of soybean based rotations by C4 perennial grass can significantly increase soil organic matter and improve soil health. Although soil health and direct greenhouse gas benefits tend to be greatest on Class 1–3 soils, there is concern in Ontario regarding displacement of food crops by non-food crops; consequently there is considerable interest in producing dedicated biomass crops on Class 4–5 lands. However, there are significant challenges associated with producing biomass on Class 4–5 lands in Ontario, including:

1. Production of significant quantities would require the conversion of a relatively high percentage of the Class 4 and 5 lands in Ontario. If production is distributed across land classes, the impact on any given land class would be minimal.
2. Availability of Class 4 and 5 lands in Ontario is very region-dependent (Fig. 1) and restricting biomass production to marginal land would disadvantage the development of biomass in regions dominated by Class 1–3 land.
3. Production on marginal lands, due to lower yield potential and scattered distribution (Fig. 1), would significantly increase production costs and transportation costs.

In conclusion, there is significant opportunity for biomass production in Ontario. Feasibility from an economic, technical and sustainability perspective will depend, to a large part, on decisions regarding source of biomass and where it is produced in the province. It is expected that a combination of market forces and government policies will dictate these decisions.

3 Case Study of Corn Fibre Reinforced Polypropylene as a Model Material for Bio-filled Thermoplastic Composite Manufacturing

This study was initiated to investigate the performance characteristics of composite materials produced with polypropylene (PP) and fibres extracted from agricultural residues, in particular, fibres obtained from corn stalks and cobs. The study was structured to examine the influences of the genetic background (genotype) of the corn plant and the environment in which the crop was grown on the functional properties of the corn fibres after they are incorporated into composites (Fig. 4). We have examined the influences of genotype and environment on the functionalities of fibres obtained from soybean stems in PP composites (Reinprecht et al. 2015) and corn fibres in polylactic acid (PLA; Riaz 2012) but have not previously conducted tests with corn fibres in PP.

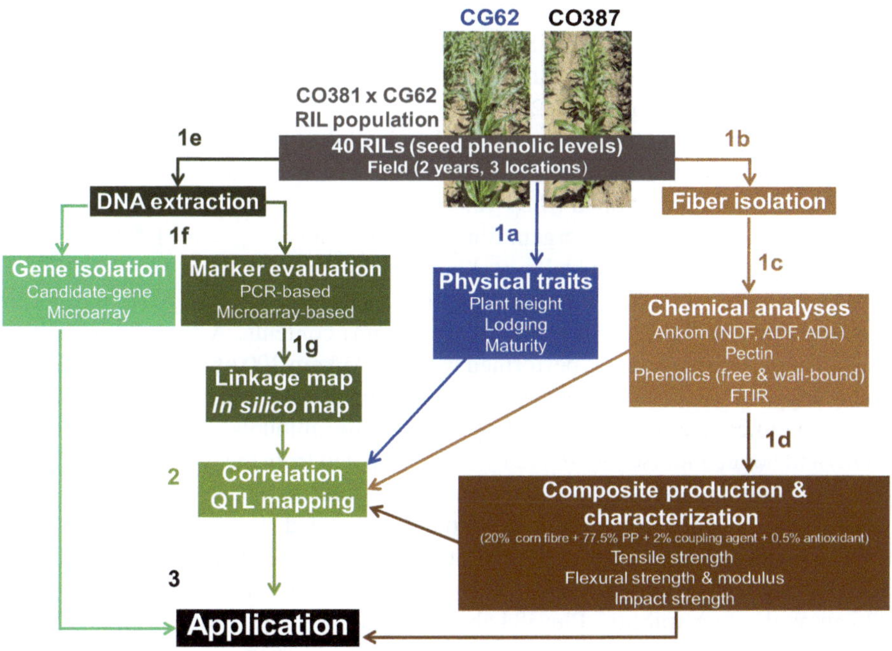

Fig. 4 An overview of the study to identify environmental and genetic effects on the functional properties of corn fibres in composites with a polypropylene (PP) matrix

3.1 *Stage I: Production of Corn Stalk and Cob Fibres PP Composites*

3.1.1 Materials and Experimental Design for Corn Fibre-based Composite Tests

In the current study 40 recombinant inbred lines (RILs) created by selfing F_2 s from a cross between parents (CG62 x CO387) that differed for a number of characteristics, including their disease resistance (Ali et al. 2005) and phenolic content (Bily et al. 2003), were used as sources of stalk and cob fibres. The selected RILs included the top 20 and the bottom 20 lines from a population of 144 RILs for the levels of phenolic compounds [esterified ferulic acid (EFA), dehydrodimers of ferulic acid (DFA) and *p*-coumaric acid (PCA)] in their kernels (Fig. 4).

The selected lines along with their parents were grown in two Ontario locations over 2 years and evaluated in the field for their physical traits (Step 1a). At maturity five plants were randomly selected for fibre analysis. Dried stalks and cobs were ground (Step 1b) with a Thomas Wiley Mill Model 4 (Thomas Scientific, Swedesboro, NJ, USA) to pass through a 2 mm sieve in stage I of the study. In stage II the materials were ground with a commercial fibre grinding machine (OMTEC Inc. Ridgetown, Ontario) with a 2 mm sieve.

The chemical compositions [cellulose, hemicellulose, lignin and phenolics (total free and cell wall-bound)], particle sizes, moisture contents and thermal stabilities of the fibres were determined (Step 1c) before compounding in the polypropylene (PP) matrix. A sequential detergent fibre analysis (Van Soest et al. 1991) was used to characterize ground corn stalks and cobs for neutral detergent fibre [NDF, cell wall (cellulose, lignin and hemicellulose)], acid detergent fibre (ADF, lignin and cellulose) and acid detergent lignin (ADL, lignin) contents. A three-step analysis using a filter bag method was performed with the Ankom 200 fibre analyzer (Ankom technology, Macedon, NY) according to the manufacturer's instructions (http://www.ankom.com/analytical-procedures.aspx). The amounts of cellulose (ADF–ADL) and hemicellulose (NDF–ADF) were estimated from the NDF, ADF and ADL measurements and expressed as a percent of dry fibre.

The free phenolic contents in the ground corn residues were determined with a 50% Folin-Ciocalteu's phenol reagent (Sigma Chemicals Company, St. Louis, USA) using a microwave-based protocol (Fletcher et al. 2005) and determined at 725 nm with a SpectraMax® Plus384 absorbance microplate reader using SOFTmax PRO 4.0 controller software (Molecular Devices Corporation, Sunnyvale, CA, USA) with gallic acid (Acros Chemical Company, NJ, USA) in 50% ethanol as a standard. The quantities of free phenolics were expressed as an average of three (10 mg) subsample measurements (from each replication/location/year) in μg phenolics mg^{-1} dry fibre. The concentration of cell wall-bound phenolics (*p*-coumaricacid and ferulic acid) was determined by high-performance liquid chromatography (HPLC) and expressed in mg phenolic acids g^{-1} dry fibre.

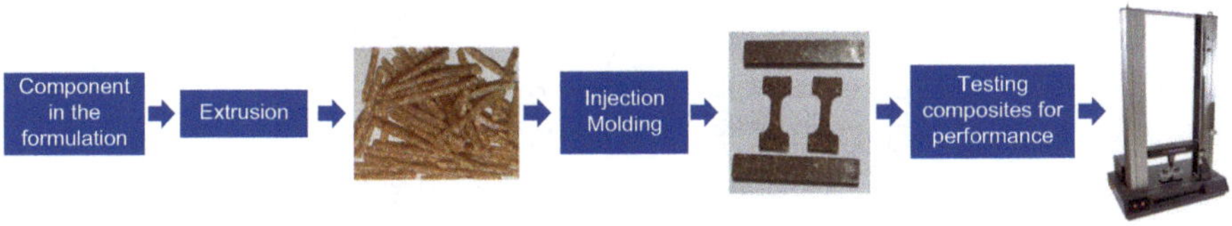

Fig. 5 Stages in corn fibre PP composite preparation

The thermal properties of the fibres, including thermal stability and onset degradation temperatures, were measured by heating 5–10 mg samples from 35 °C to 700 °C at a rate of 20 °C min^{-1} under nitrogen gas at a flow rate of 50 mL min^{-1} using TA Instrument Q500 TGA Model 19720 (TA Instruments, New Castle, DE, USA). TGA thermographs were used to measure fibre weight loss for RILs. Onset degradation temperatures were recorded as the temperatures (°C) after 150 °C at which the sample showed 1% weight loss.

Compounding was done to obtain homogeneous materials through the melt blend process (Step 1d). Corn stalk or cob residues [20 wt-% (stage I) and 30 wt-% (stage II)], were mixed with homopolymer polypropylene [PP, grade D180M (Sunoco Chemicals, Inc., Philadelphia, PA—Braskem America) with 18 melt flow index (MFI); 77.5 wt-% (stage I) and 67.5 wt-% (stage II)], coupling agent [Fusabond P-353, Maleic anhydride grafted (Dupont, Canada); 2 wt-%] and antioxidants [0.25 wt-% each Irganox 1010, Phenolic and Irgaphos 168, Phosphate (Ciba, Inc., Canada)]. Composite components were blended together using a conical twin-screw micro-extruder (HAAKE Mini Lab, Thermo Electron Corporation, Canada) with processing conditions of 190 °C and 40 rpm machine speed. Test specimens and bars were injection moulded according to the ASTM standards using a RR/TSMP machine (Ray-Ran, Warwickshire, UK) with a barrel temperature at 190 °C, mould tool temperature at 50 °C, and 15 s hold time at 100 psi. The specimens were annealed in an air circulating GC 5890A oven (Hewlett Packard, USA) at 151 °C for 11 min and then cooled down to room temperature at a rate of 10 °C min^{-1}. Five test specimens for each flexural (ASTM methods D790-10, 2010a), impact (ASTM methods D256-10, 2010b) measurement and five test bars for the tensile measurements (ASTM method D1708-10, 2010c) were used to measure the physical properties of the corn stalk and cob fibre PP composites (Fig. 5).

3.1.2 Effects of the Genotype and Growth Environments on the Chemical Composition of the Corn Fibres

Significant variations in the corn stalk and cob fibre chemical compositions (cellulose, hemicellulose, lignin, total free phenolics, *p*-coumaric acid and ferulic acid quantities) and onset degradation temperatures were identified among RILs (Table 1). The environments in which the corn crops were grown significantly affected the chemical compositions of the fibres (data not shown).

Table 1 Chemical compositions and onset degradation temperatures of corn stalk and cob fibres (average of four environments)

Fibre	Genotypes		NDF	ADF	ADL	Ce	Hc	Fp	Pca	Fa	ODT
Stalk	Population (40RILs)	Avg.	68.2	38.5	3.8	34.7	29.7	**6.0↑**	**8.0↑**	3.6	173.4
		Min.	61.8	34.7	3.3	31.3	27.1	4.6	6.2	2.8	165.0
		Max.	75.2	43.5	6.0	39.3	32.1	8.7	9.9	4.3	188.2
	Parents	CG62	63.4	35.3	3.5	31.8	28.1	6.1	6.9	3.2	167.6
		CO387	68.0	37.8	3.1	34.8	30.2	4.9	7.5	3.5	177.5
Cob	Population (40 RILs)	Avg.	**83.5↑**[a]	**44.4↑**	**5.5↑**	**38.9↑**	**39.0↑**	3.1	7.3	**4.1↑**	**206.7↑**
		Min.	78.4	41.4	4.3	36.9	36.3	1.5	5.3	3.0	189.9
		Max.	87.5	46.6	6.8	41.2	41.3	4.2	9.0	5.1	216.5
	Parents	CG62	80.4	42.0	4.3	37.7	38.4	3.0	8.3	4.0	202.9
		CO387	83.2	44.0	5.4	38.6	39.2	3.9	6.1	5.1	207.9

NDF neutral detergent fibres = crude fibre = cellulose + hemicellulose + lignin, *ADF* acid detergent fibres = cellulose + lignin, *ADL* acid detergent lignin, *Hc* hemicellulose = NDF–ADF, *Ce* cellulose = ADF–ADL (%), *Fp* free phenolics (μg phenolics mg^{-1} dry fibre), *Pca* *p*-coumaric acid, *Fa* ferulic acid (mg phenolic acids g^{-1} dry fibre), *ODT* onset degradation temperature (°C)

[a]underlined values significantly higher in the cob fibres compared to the stalk fibres, P = 0.05

Fibres from stalks and cobs had different chemical compositions. In general, cob fibres had higher onset degradation temperatures and contained larger amounts of cellulose, hemicellulose lignin, and ferulic acid than the stalk fibres. In contrast, stalk fibres had higher quantities of *p*-coumaric acid and twice the levels of free phenolics, compared to the cob fibres.

3.1.3 Characteristics of Corn Fibre PP Composites

In stage I of the study, a total of 336 corn stalk and cob fibres PP composites (20% wt/wt from 40 RILs and two parents grown in four environments) were tested for flexural, tensile and impact properties. The corn fibre PP composites were significantly different from pure PP with respect to their physical properties. In particular, the addition of 20% of corn fibres (wt/wt) from all 40 RILs improved the flexural properties (strength and modulus) of composites compared to the pure PP (Fig. 6). However, tensile and impact strengths were all reduced, compared to pure PP.

The environments in which the corn plants were grown (Elora 2008 [E8]; Elora 2009 [E9]; Ottawa 2008 [O8]; Ottawa 2009 [O9]) from which the stalk and cob fibres were extracted, had significant effects on the composite properties (Table 2). Composites produced with stalk fibres had significantly higher values for flexural strength, flexural modulus, and tensile strength than cob fibre composites, across the environments. Impact strengths of the corn stalk fibre and cob fibre composites (23.0 and 24.0 Jm^{-1}) were not significantly different.

3.1.4 Relationships Among Corn Fibre Chemical Compositions and Composite Mechanical Properties

A principal component analysis (PCA) was conducted to determine the associations among fibre chemical compositional traits and the performance characteristics of the fibres in composites.

The first and second principal components comparisons of composites containing corn stalk and cob fibres with composites prepared with wheat and soybean residues show that the corn containing materials perform as well as the latter (Fig. 7). In fact, for the impact strength the corn fibre composites had better values than the wheat straw composites (Fig. 8). The comparison to the wheat straw composites is particularly interesting because wheat straw PP/composites have commercial uses by the automotive industry. Cob fibres have an additional benefit that is associated with their high onset degradation temperatures (200 °C compared to 187 °C for wheat) which allows the composite materials to be processed at higher temperatures and consequently faster speeds. The use of abundant corn fibres for the production of composite materials would significantly expand the opportunities for using field crops to manufacture bioproducts in Ontario for the reasons described above.

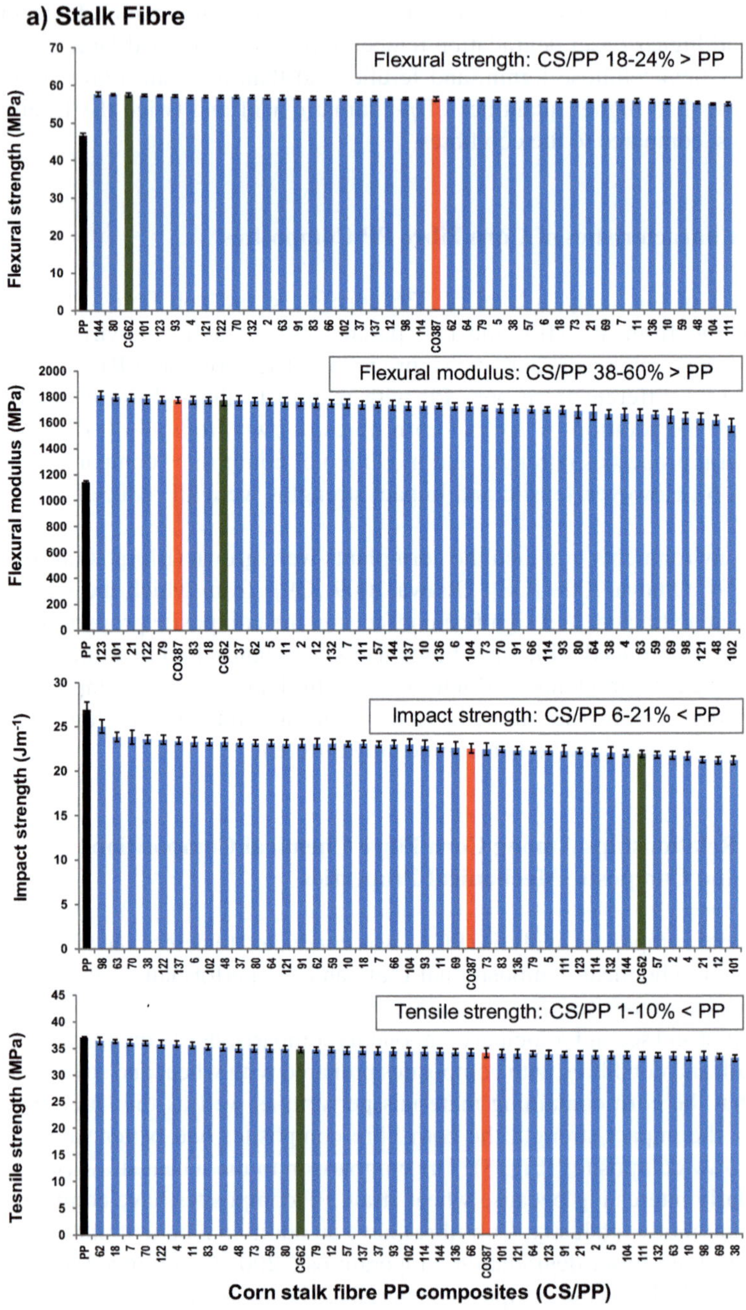

Fig. 6 Mechanical properties of corn stalk and cob fibre PP composites from 40 CG62xCO387 RILs and two parents compared to pure PP. Values are averages of composites prepared from (**a**) stalk fibres (CS/PP) and (**b**) cob fibres (CC/PP) from corn samples obtained from four environments

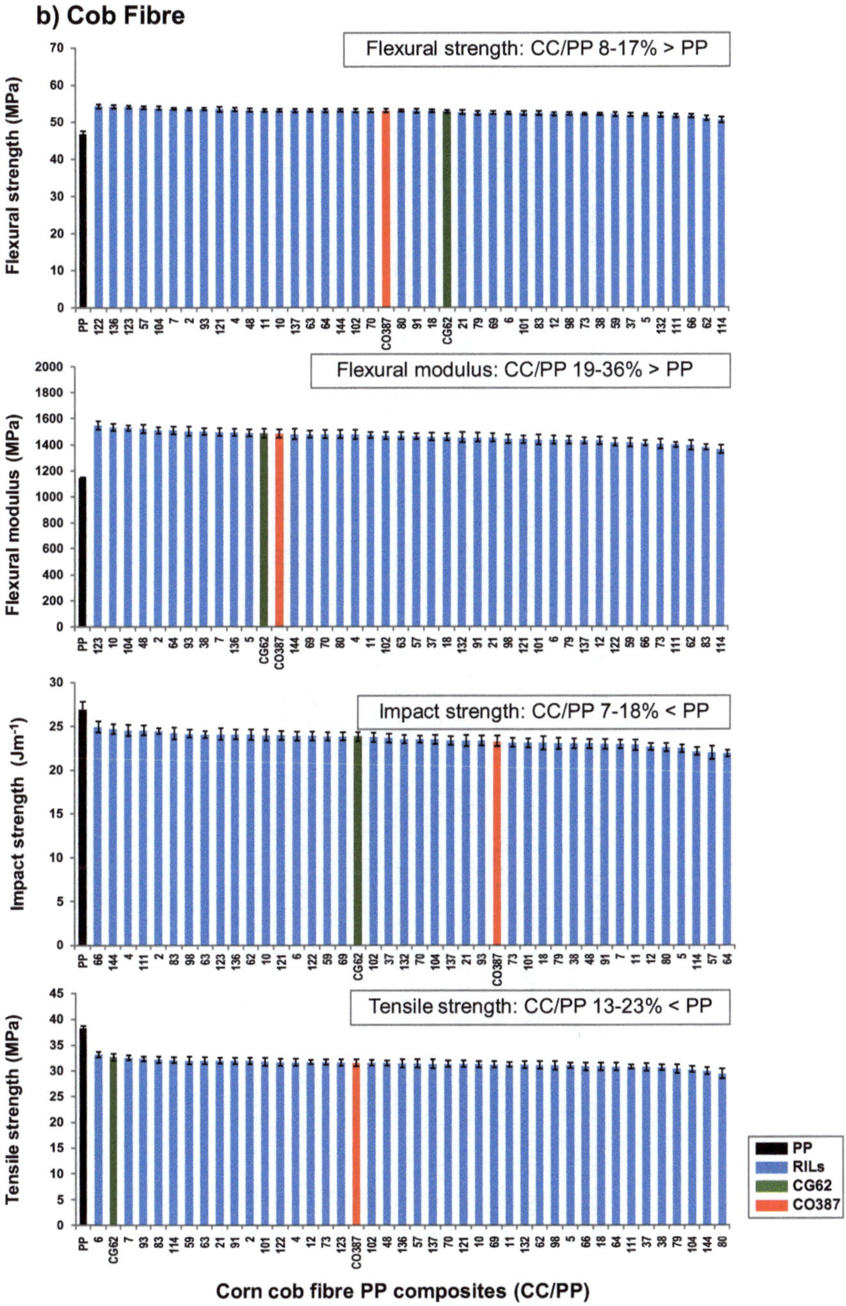

Fig. 6 (continued)

Table 2 Mean mechanical properties (±SE[a]) of corn RILs stalk (CS/PP) and cob (CC/PP) fibre polypropylene composites evaluated in two Ontario locations in 2008 and 2009 and pure polypropylene (PP) ($N = 5$)

Env[b]	Flexural strength (MPa)			Flexural modulus (MPa)			Impact strength (Jm^{-1})			Tensile strength (MPa)		
	PP[c]	CS/PP	CC/PP	PP	CS/PP	CC/PP	PP	CS/PP	CC/PP	PP	CS/PP	CC/PP
E8	51±0.4	54±0.5	53±0.6	1196±12.2	1598±45.6	1401±27.6	31±0.5	23±0.6	24±0.6	37±0.5	33±0.6	32±0.7
O8		57±0.64	53±0.8		1636±57.2	1392±57.2		23±0.6	24±0.6		35±0.6	31±0.8
E9	43±0.5	58±0.5	53±0.7	1087±16.9	1829±32.6	1540±32.6	23±0.4	23±0.5	23±0.5	37±0.6	35±0.8	31±.08
O9		57±0.5	53±0.7		1813±37.0	1521±37.0		23±0.7	23±0.7		35±0.9	31±0.9

[a] *SE* standard error

[b] *Env* environment, *E8* Elora 2008; *E9* Elora 2009, *O8* Ottawa 2008, *O9* Ottawa 2009

[c] Polypropylene used in this study was from two batches with the similar melt flow index of 18. Therefore, differences noted for flexural strength, flexural modulus and impact strength could be attributed to the batch differences

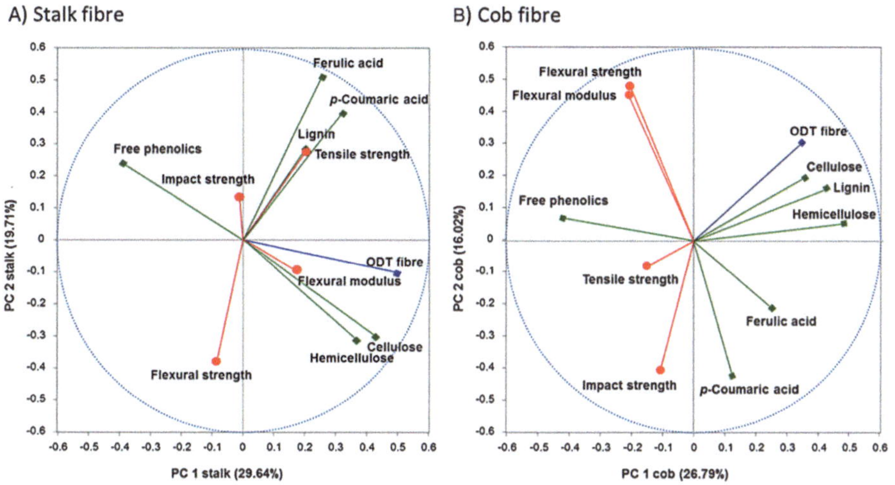

Fig. 7 Principal component analyses of corn fibre chemical traits and the physical traits of composites produced with corn fibres. Relationships among corn stalk (**a**) and corn cob (**b**) fibre samples

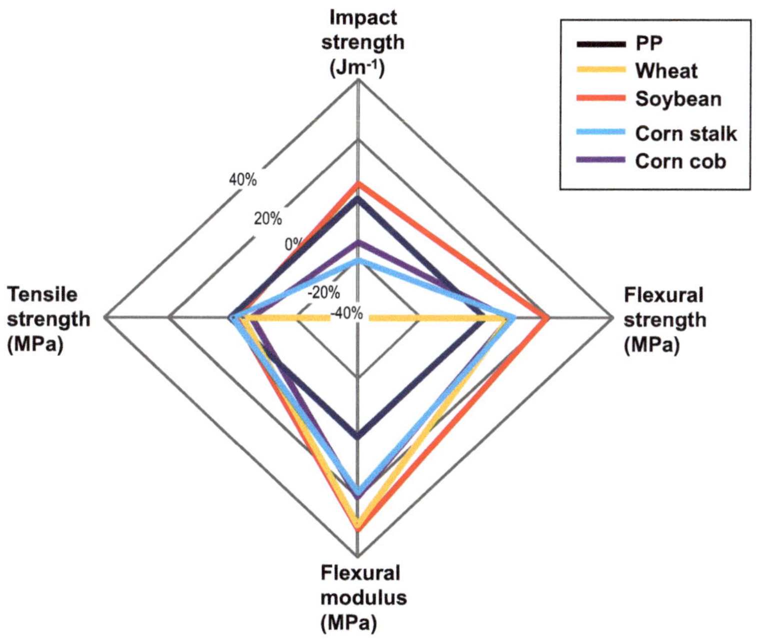

Fig. 8 Mechanical properties of polypropylene (PP)-based composites reinforced (20% fibre) with corn fibre (stalk and cob), soybean stem fibre, and wheat straw fibre compared to pure PP. The composites were manufactured from fibres obtained from single corn and soybean lines and a commercial source of wheat straw. Properties are shown as a percent of pure polypropylene (PP)

3.2 *Stage II: Scale-Up Tests with Corn Stalk Fibre PP Composites*

Based on the stage I results obtained from measurements of 336 corn fibre PP composite samples (stalk and cob fibres from 40 RILs and two parents grown in four environments), corn stalk samples from two corn lines (CC37 and CC122) were selected for scale-up composite production. The corn stalk, soybean and wheat straw fibres were produced in a commercial facility (OMTEC Inc., a commercial natural fibre processor) located in Ridgetown, Ontario. The PP composites were compounded with 30% (wt/wt) stalk fibre for stage II of the project. The increase in the amount of corn stalk fibre used in composites from (20% wt/wt to 30%wt/wt) for both corn lines significantly increased their flexural modulus and strength values, but decreased the tensile strengths of the composites (Fig. 9).

The corn stalk fibre PP composite (from line CC122) was also compared to soybean and wheat straw composites produced in similar ways (Fig. 10). In general, the corn stalk fibre composite material had similar physical properties to the wheat straw PP composite, which is used commercially in passenger bins inside the Ford Flex cabins. Overall, the this study demonstrated that PP composites, reinforced

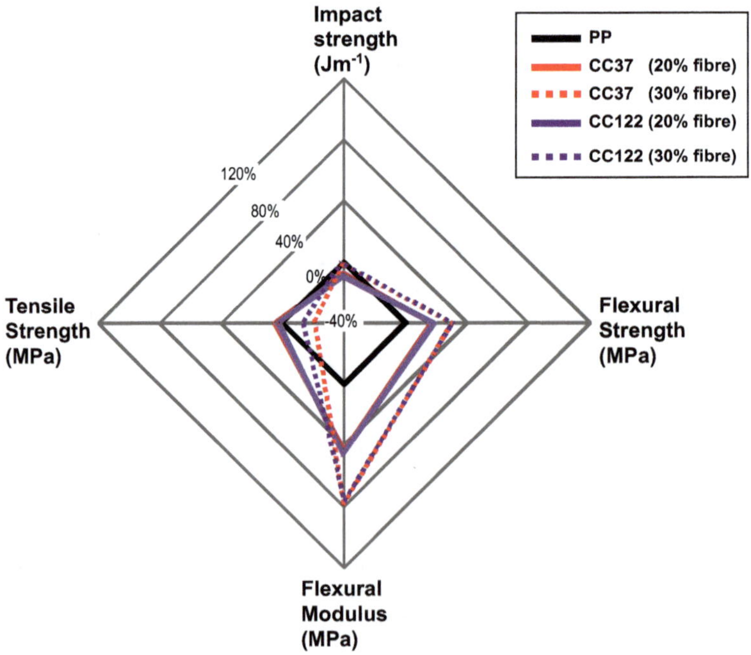

Fig. 9 Mechanical properties of polypropylene (PP)-based composites reinforced with different amounts corn stalk fibres from two RILs (CC37 and CC122) compared to pure PP. Properties are shown as a percent of pure polypropylene (PP)

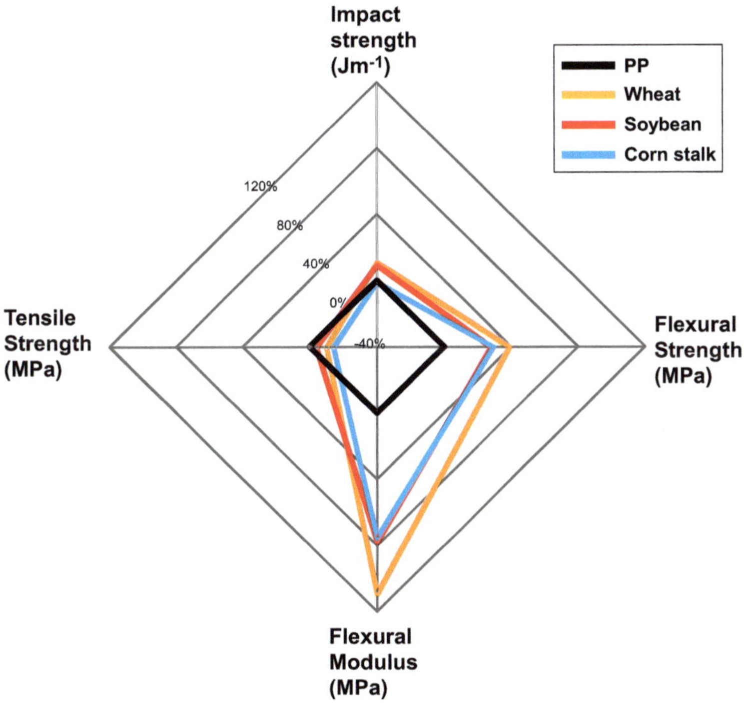

Fig. 10 Mechanical properties of polypropylene (PP)-based composites reinforced (30% fibre) with corn stalk fibre, soybean stem fibre, and wheat straw fibre compared to pure PP. Properties are shown as a percent of pure polypropylene (PP)

with as much as 30% corn stalk fibres, have properties that are very similar to wheat straw composites, with flexural properties that are significantly better than pure PP.

4 Genome Locations Related to Corn Fibre Properties

The relationships between the compositions and the performance properties of the corn fibres and the differences among the composites prepared from different RILs raised the possibility that the composite performance characteristics are under genetic control. In order to identify regions of the corn genome that contain regions and genes (Martin 2011) that determine composite performance the set of 40 RILs from the CG62 x CO387 mapping population (Ali et al. 2005) was genotyped with 162 molecular markers and analyzed for QTL (with MapQTL6 software, Van Ooijen 2009). In total, 149 putative QTL were identified for nine fibre traits [onset degradation temperature and contents of NDF, ADF, ADL (lignin) cellulose, hemicellulose, free phenolics, *p*-coumaric acid and ferulic acid] and four composite

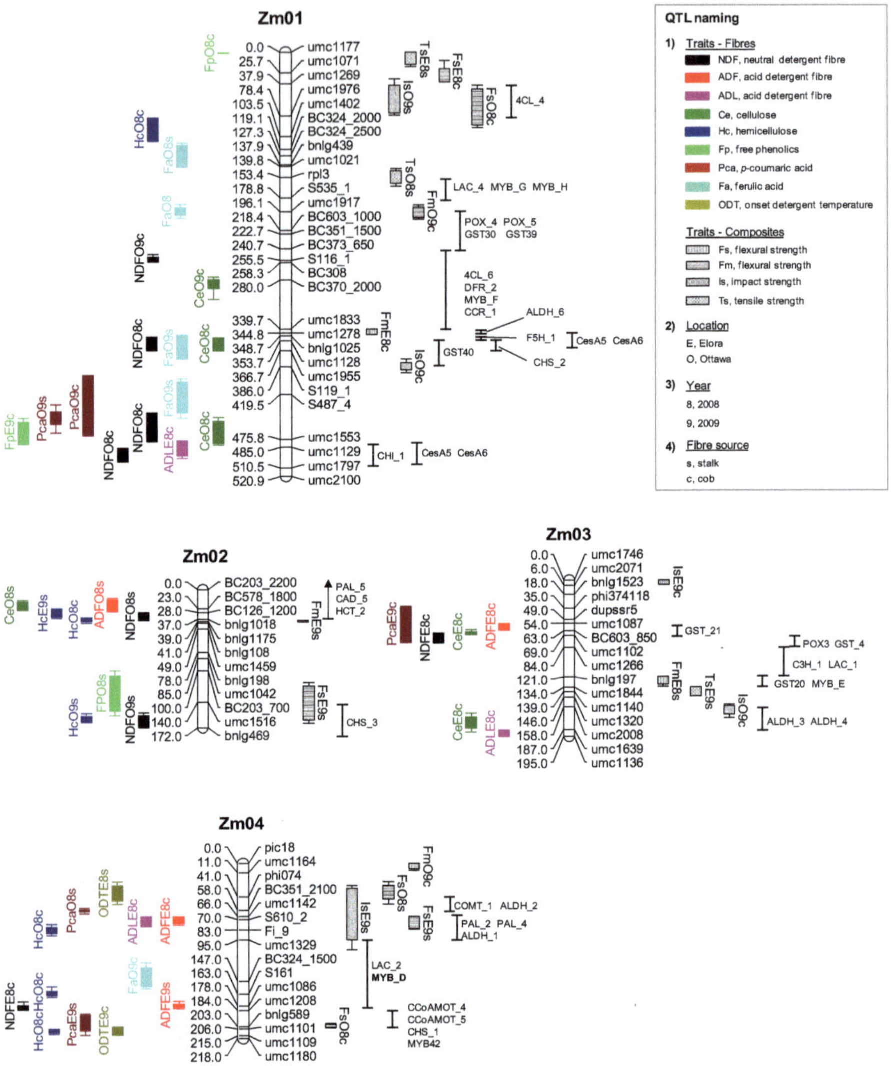

Fig. 11 Corn CG62 x CO387 stalk and cob fibre PP composite QTL map. QTL were detected using MapQTL6. Because of the novelty of some of the mapping traits, QTL at LOD threshold values ≥2.0 were considered as putative QTL

properties (flexural strength, flexural modulus, impact strength and tensile strength) in four environments. The QTL were distributed throughout corn genome and explained significant portions of phenotypic variability for individual traits (Fig. 11).

Stalk QTL Thirty nine QTL associated with chemical composition and onset degradation temperature of the corn stalk fibres were identified on nine linkage

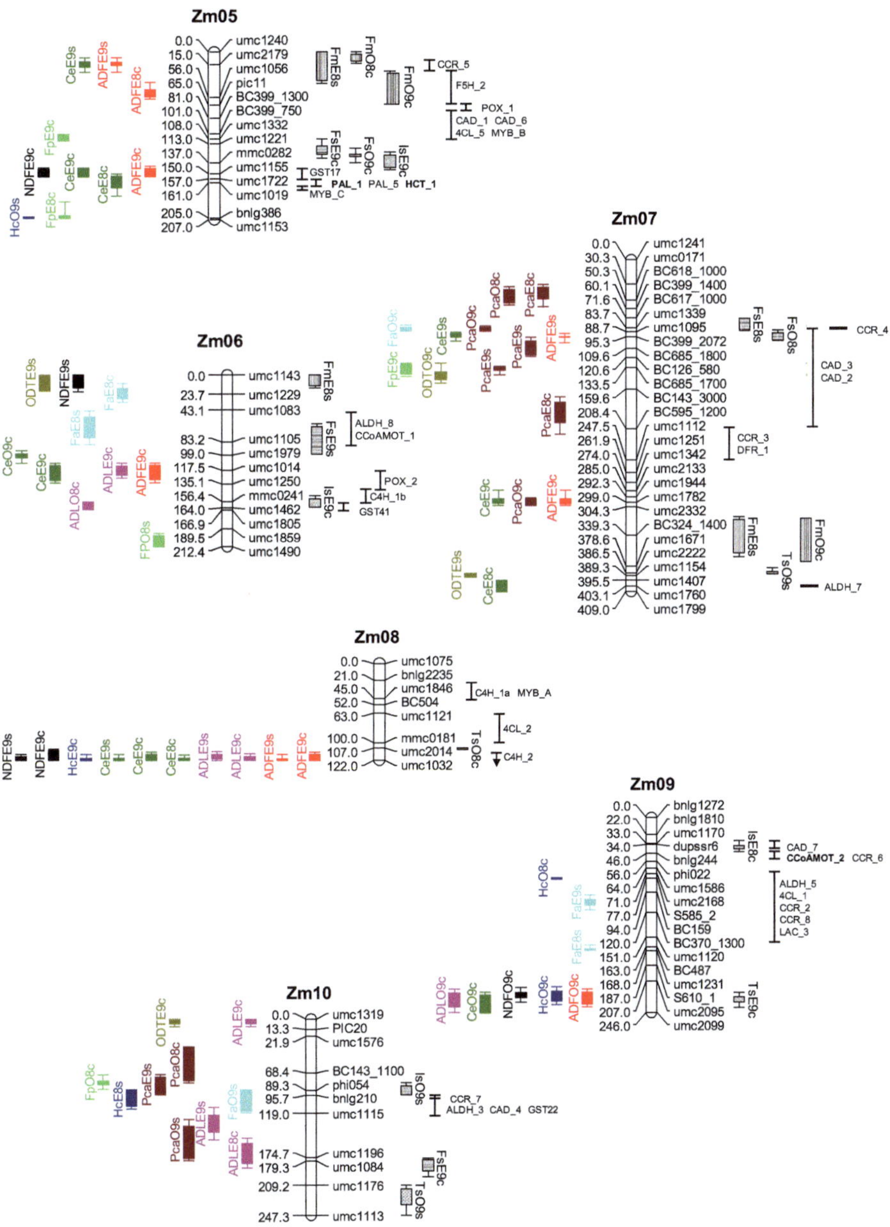

Fig. 11 (continued)

groups (LG, Fig. 11), shown left on each bar). They ranged from two on LG Zm09 to seven on LG Zm02 and explained 11–27% of the total phenotypic variability for traits. No QTL associated with chemical composition of the stalk fibres were identified on LG Zm03. Twenty-two QTL were determined for the mechanical properties of the corn stalk PP composite traits (seven QTL for flexural strength, seven for flexural modulus, three for impact strength and five for tensile modulus) on nine linkage groups (Fig. 10, shown right on each bar). They ranged from one on LG Zm08 to five on LG Zm07 and explained 9–29% of the variability for individual traits. No QTL associated with the mechanical properties of the corn stalk fibre composites were identified on LG Zm09.

Cob QTL Sixty seven QTL were identified for the cob fibre chemical composition and onset degradation temperature on all ten linkage group. They ranged from one QTL on LG Zm02 to 12 QTL mapped on LG Zm01 and explained 5–28% of the total phenotypic variability for the traits. Twenty one QTL were identified for the cob fibre PP composites mechanical traits (six QTL for flexural strength, six for flexural modulus, seven for impact strength and two for tensile modulus) on the nine linkage groups. They ranged from a single QTL mapped on several LG (Zm06, Zm07 and Zm10) to five QTL mapped on LG Zm01 and Zm05 and explained from 12 to 28% of the variability for the traits. No QTL associated with the mechanical properties of the corn cob fibre composites were identified on LG Zm02 (Fig. 11).

QTL Consistency Because of the significant GxE interactions QTL analysis was performed for each environment, separately. The variability in the expression of these QTL was consistent with the low correlations observed among the fibre compositional and composite mechanical performance properties in different locations and years. The enzymes and regulatory proteins involved in the biosynthesis and modification of corn cell wall are encoded by hundreds of genes, which often belong to large gene families, whose expression is affected by environmental conditions (Penning et al. 2009; Xu et al. 1996; Shen et al. 2009; Yokoyama and Nishitani 2004). The inconsistencies of these QTL over environments could limitat their use in marker assisted selection (MAS) for a specific trait. To be useful for breeding, QTLs need to be consistent in different environments and/or genetic backgrounds.

A number of fibre compositional QTL were consistently identified in several environments, including: *p*-coumaric acid QTL (PcaO8c and PcaE8c) on LG Zm07 and cellulose QTL (CE8c and CE9c) on LG Zm05 and Zm08 (CE8c AND CE9c). Similarly, some composite performance QTL were identified in several environment, including: flexural modulus QTL (FmO8c and FmO9c) and flexural strength QTL (FsE9c and FsO9c) on LG Zm05. These QTL might be important in breeding for cell wall compositional traits.

QTL Clustering and Co-Localization Clustering of QTL associated with the compositional traits and composite mechanical properties QTL was identified on several LG for both stalk and cob fibres. In addition, our ability to connect the linkage map with the annotated genome sequence for corn through the RFLP/SSR markers identified several examples where these QTL were associated with the genome regions containing cell wall biosynthetic genes (Fig. 11). For example, five compositional QTL (NDFO8s, ADFO8s, CeO8s, HcO8c and HcE9s) on LG Zm02 co-localized with the flexural modulus QTL (FmE9s) and lignin biosynthetic genes PAL_5, CAD_5 and HCT_2. Some genomic regions contained clusters of linked and/or peiotropic loci that affected a number of traits (Fig. 11). For example, marker interval umc1278-umc1128 on the chromosome Zm01 was associated with QTL for three fibre compositional traits (NDFO8c, CeO8c, FaO8s) and two composite QTL (FmE8s and FmE8c). Similarly, marker interval on LG Zm08 was associated with QTL for several traits (ten fibre compositional and three composite). Pleoitropy is usually associated with major gene effects. A common genetic basis might explain some correlations between fibre compositional and composite traits. Alternatively, clustering of genes for different traits might be the basis for overlapping QTL. However, a more saturated map would help to determine if the regions significant for more than one trait were the result of pleiotropy or gene linkage.

5 Conclusion

In conclusion, our work confirmed the importance of corn stem and cob fibres for the production of bio-filled thermoplastic composites. Thermal stability analysis revealed that some corn stalks fibre can be processed on elevated temperatures (200 °C). High thermal stability of a genotype can be manipulated in plant breeding programs for high thermal stable corn lines/hybrids. Normally, corn stalks fibre improved flexural properties without negative effects on the impact and tensile properties of the composites. It indicated that abundantly available corn stalk and cob fibres are good alternative to wood and other natural fibres. In some cases, the composites had equal or better performance characteristics than the pure PP matrices. However, standardization of the fibre quality is necessary prior to commercial scale composite manufacturing.

This study also confirmed that plant genetics and environment play significant roles in fibre composition and their performance when incorporated into PP-based composites. Results identified cell wall traits, genome regions and potential candidate genes that are important for the use of plant fibres in composites. This can lead to selection and/or development of corn lines and hybrids valuable for composite production.

Acknowledgements The work was supported by the Ontario Ministry for Agriculture, Food and Rural Affairs.

References

Ali, M. L., Taylor, J. H., Lie, J., Sun, G., William, M., Kasha, K. J., Reid, L. M., & Pauls, K. P. (2005). Molecular mapping of QTLs for resistance to Gibberella ear rot, in corn, caused by *Fusarium graminearum*. *Genome, 48*, 521–533.

ASTM Standard D790-10. (2010a, April 1). Standard test methods for flexural properties of unreinforced and reinforced plastics and electrical insulating materials. West Conshohocken, PA: ASTM International. doi: 10.1520/D0790-10.; www.astm.org

ASTM Standard D256-10. (2010b, May 1). Standard test methods for determining the izod pendulum impact resistance of plastics. West Conshohocken, PA: ASTM International. doi: 10.1520/D0256-10.; www.astm.org

ASTM Standard D1708-10. (2010c, April 1). Standard test method for tensile properties of plastics by use of microtensile specimens. West Conshohocken, PA: ASTM International, doi: 10.1520/D1708-10.; www.astm.org

Bily, A. C., Reid, L. M., Taylor, J. H., Johnston, D., Malouin, C., Burt, A. J., Bakan, B., Regnault-Roger, C., Pauls, K. P., Arnason, J. T., & Philogene, B. J. R. (2003). Dehydrodimers of ferulic acid in maize grain pericarp: Resistance factors to *Fusarium graminearum. Phytopathology, 93*, 712–719.

Eichelmann, E., Wagner-Riddle, C., Warland, J., Deen, B., & Voroney, P. (2016). Carbon dioxide exchange dynamics over a mature switchgrass stand. *Global Change Biology, 8*, 428–442. doi:10.1111/gcbb.12259.

Fletcher, R. S., Slimmon, T., McAuley, C. Y., & Kott, L. S. (2005). Heat stress reduces the accumulation of rosmarinic acid and the antioxidant activity of Spearmint (*Mentha spicata* L.) *Journal of the Science of Food and Agriculture, 85*, 2429–2436.

Gaudin, A. C. M., Janovicek, K., Deen, B., & Hooker, D. C. (2015). Wheat improves nitrogen use efficiency of maize and soybean-based cropping systems. *Agriculture, Ecosystems & Environment, 210*, 1–10. doi:10.1016/j.agee.2015.04.034.

Heaton, E. A., Voigt, T. B., & Long, S. P. (2004). A quantitative review comparing the yields of two candidate C4 perennial biomass crops in relation to nitrogen, temperature and water. *Biomass and Bioenergy, 27*, 21–30.

Kludze, H., Deen, B., Weersink, A., van Acker, R., Janovicek, K., & De Laporte, A. (2013a). Impact of land classification on potential warm season grass biomass production in Ontario, Canada. *Canadian Journal of Plant Science, 93*, 249–260.

Kludze, H., Deen, B., Weersink, A., van Acker, R., Janovicek, K., De Laporte, A., & McDonald, I. (2013b). Estimating sustainable crop residue removal rates and costs based on soil organic matter dynamics and rotational complexity. *Biomass and Bioenergy, 56*, 607–618. doi:10.1016/j.biombioe.2013.05.036.

Martin, C. J. (2011). *Development of sequence-specific molecular markers based on phenylpropanoid pathway genes for resistance to Fusarium graminearum (Schwabe) in Zea mays (L.)*. MSc Thesis: University of Guelph.

Penning, B. W., Hunter III, C. T., Tayengwa, R., Eveland, A. L., Dugard, C. K., Olek, A. T., Vermerris, W., Koch, K. E., McCarty, D. R., Davis, M. F., Thomas, S. R., McCann, M. C., & Carpita, N. C. (2009). Genetic resources for maize cell wall biology. *Plant Physiology, 151*, 1703–1728.

Pittelkow, C. M., Liang, X., Linquist, B. A., van Groenigen, K. J., Lee, J., Lundy, M. E., & van Kessel, C. (2014). Productivity limits and potentials of the principles of conservation agriculture. *Nature, 517*, 365–368. doi:10.1038/nature13809.

Powlson, D. S., Stirling, C. M., Jat, M. L., Gerard, B. G., Palm, C. A., Sanchez, P. A., & Cassman, K. G. (2014). Limited potential of no-till agriculture for climate change mitigation. *Nature Climate Change, 4*, 678–683. doi:10.1038/nclimate2292.

Reinprecht, Y., Arif, M., Simon, L. C., & Pauls, K. P. (2015). Genome regions associated with functional performance of soybean stem fibers in polypropylene thermoplastic composites. *PLoS ONE, 10*(7), e0130371. doi:10.1371/journal.pone.0130371.

Riaz, M. (2012). *Characterization of corn fibres for manufacturing automotive plastic parts*. MSc Thesis, University of Guelph.

Sage, R. F., Peixoto, M., Friesen, P., & Deen, B. (2015). C4 bioenergy crops for cool climates, with special emphasis on perennial C4 grasses. *Journal of Experimental Botany*. doi:10.1093/jxb/erv123.

Sanscartier, D., Deen, B., Dias, G., Maclean, H. L., Dadfar, H., Mcdonald, I., & Kludze, H. (2014a). Implications of land class and environmental factors on life cycle GHG emissions of *Miscanthus* as a bioenergy feedstock. *GCB Bioenergy, 6*, 401–413. doi:10.1111/gcbb.12062.

Sanscartier, D., Dias, G., Deen, W., Dadfar, H., McDonald, I., & MacLean, H. L. (2014b). Life cycle greenhouse gas emissions of electricity generation from corn cobs in Ontario, Canada. *Biofuels, Bioproducts and Biorefining, 8*, 568–578. doi:10.1002/bbb.1485.

Shen, H., Yin, Y., Chen, F., Xu, Y., & Dixon, R. A. (2009). A bioinformatic analysis of NAC genes for plant cell wall development in relation to lignocellulosic bioenergy production. *Bioenergy Research, 2*, 217–232.

Van Ooijen, J. W. (2009). *MapQTL® 6, Software for the mapping of quantitative trait loci in experimental populations of diploid species*. Wageningen, Netherlands: Kyazma B.V.

Van Soest, P. J., Robertson, J. B., & Lewis, B. A. (1991). Methods for dietary neutral detergent fiber and non starch polysaccharides in relation to animal nutrition. *Journal of Dairy Science, 74*, 3583–3597.

Varvel, G. E., & Wilhelm, W. W. (2008). Corn cob biomass production in the western corn belt. *Bioenergy Research, 1*(3e4), 223e8.

Xu, W., Campbell, P., Vargheese, A. K., & Braam, J. (1996). The *Arabidopsis XET*-related gene family: Environmental and hormonal regulation of expression. *The Plant Journal, 9*, 879–889.

Yokoyama, R., & Nishitani, K. (2004). Genomic basis for cell-wall diversity in plants. A comparative approach to gene families in rice and *Arabidopsis*. *Plant and Cell Physiology, 45*, 1111–1121.

Related Web Resources

http://www.researchandmarkets.com/reports/2881528/global-natural-fiber-composites-market-2014-2019; accessed on May 21, 2015

http://www.omafra.gov.on.ca/english/stats/crops/estimate_new.htm; accessed on May 21, 2015

Part IV
Bioeconomy Applications: Optimizing Processes and Management of the Bioeconomy

Strategic Supply Chain Planning in Biomass-Based Industries: A Literature Review of Quantitative Models

Stephan Fichtner and Herbert Meyr

Abstract Fossil resources are limited and will run short. Moreover, the extensive usage of fossil resources is discussed as a key driver for climate change which means that a changeover in basic economic and ecological thinking is necessary. Especially for the energy production, there has to be a movement away from the usage of fossil resources and towards renewable resources like wind, water, sun or biomass. In this chapter we present a structured review of recent literature on the long-term, strategic planning of biomass-based supply chains. Firstly, we structure the overall research field "bioeconomy" by means of the various utilization pathways of biomass and bring together the demand-oriented view of supply chain management models and the supply-oriented view of bioeconomy. Secondly, we provide a literature review of Operations Research models and methods for strategic supply chain planning in biomass-based industries. Thirdly, we analyze trends and draw conclusions about research gaps.

1 Introduction

In recent years global economy has continuously improved (World Bank, 2015). The leading industrial nations have achieved an enormous wealth. Furthermore, there is also an increasing wealth in threshold countries. However, this global wealth is largely based on the usage of finite fossil resources like crude oil, coal and natural gas. Fossil resources are limited and will run short. In addition, the extensive usage of fossil resources is recognized as a key driver for climate change. As a consequence, a changeover in basic economic and ecological thinking is necessary. Especially for energy production, there has to be a movement away from the usage of fossil resources towards renewable resources like wind, water, sun or biomass. For the remainder of this chapter, we define all not yet fossil materials with organic origin as types of biomass. That means that plants and animals as well as

S. Fichtner (✉) • H. Meyr
Department of Supply Chain Management, University of Hohenheim, 70593 Stuttgart, Germany
e-mail: Stephan.Fichtner@uni-hohenheim.de; H.Meyr@uni-hohenheim.de

S. Dabbert et al. (eds.), *Knowledge-Driven Developments in the Bioeconomy*,
Economic Complexity and Evolution, DOI 10.1007/978-3-319-58374-7_14

their residues are biomass, but also dead phytomass, as for example straw, can be biomass, as long as it is not yet fossil.[1]

Note that using biomass for various production processes is not a new idea. Decades ago, before the industrial revolution, the global economy was significantly more biomass-based than today. Nevertheless, in the future a more efficient use of biomass will be required not only to tackle the above challenges, but also to mitigate the increasing world food problem. The industrial use of biomass can provide a building block of a more sustainable economy, which in the following will be called "bioeconomy". A crucial barrier for leveraging bioeconomy are costs. In current times of extremely low crude oil prices, producing energy and other products like textiles from biomass is significantly more expensive than exploiting fossil resources. Hence, the costs of the bioeconomy have to be decreased. This could be achieved by technological innovation as well as by optimized organization. Since many parties are involved in converting biomass in more valuable final items like energy and (or) in transporting it to the final consumers (for usage as food), a large part of the total costs of biomass usage is caused by logistical processes. Thus, a clever management of biomass-based supply chains (SCs) is necessary.

In this chapter we present a structured review of recent literature on the long-term, strategic planning of biomass-based SCs. We try to embrace the whole research field of bioeconomy and not only a single, specific branch like the fuel area. However, we focus on publications applying quantitative simulation and optimization models, as commonly done in operations research (OR). All in all we consider several dozen publications. Thus it is not possible to discuss each model in detail. Instead, we identify the—in our opinion—most important characteristics of these models to introduce a classification scheme which allows an easy comparison of the different modeling approaches. This helps to reveal current trends and gaps as well as opportunities for future research. To the best of our knowledge, there is no prior review with a focus on all of these aspects.

The remainder of this chapter is organized as follows. In Sect. 2 we structure the overall research field "bioeconomy" by means of the various utilization pathways of biomass. Note that supply chain management (SCM) is—despite of its name—rather demand- than supply-oriented. The customer and her/his requested final items are in the center of the thoughts. All parties cooperating to fulfill the final customer's demand should try to integrate as good as possible in order to offer highest customer service for lowest costs. Thus, the members of a single SC should act as partners in a team; but different SCs have to compete with each others for this final customer's demand. Section 2 additionally takes a different view. Here, the scarce resource biomass, i.e., the ultimate supply, is the starting point. If one likes to judge about the best possible usage of this scarce resource, all different utilization pathways originating from and competing for the same biomass need to be identified. This may comprise hundreds and thousands of different final items and customers. Thus, Sect. 2 lays the ground for bringing together the demand-oriented view of

[1]A more detailed definition will follow in Sect. 2.

SCM models and the supply-oriented view of bioeconomy. Section 3 provides the literature review of OR models and methods for strategic SC planning in biomass-based industries. Section 4 draws conclusions by analyzing trends and research gaps. Finally, Sect. 5 summarizes the results and identifies opportunities for future research.

2 Biomass-Based Supply Chains

In Sect. 2.1 biomass is used as the starting material to identify different utilization pathways and their resulting final products. These final products are then the starting point to determine the corresponding types and members of supply chains in Sect. 2.2 and to structure the literature review of Sect. 3.

2.1 Utilization Pathways of Bioeconomy

A biomass-based utilization pathway is a specific sequence of process steps or processes (e.g., harvesting/collection, pre-processing, conversion) to generate a biomass-based final product. These pathways are investigated in the research area "bioeconomy", which is a composite out of biology and economy (Kaltschmitt, 2009). It is the knowledge-based production and utilization of renewable resources for products, processes and services in all industrial sectors and thus a pre-requisite to form a sustainable economy (BMBF and BMEL, 2014). As already mentioned, the basis of such a bioeconomy are renewable feedstocks in form of biomass. All living materials with organic origin are types of biomass. That means that plants and animals as well as their residues are part of biomass. Furthermore, dead phytomass, as for example straw, is biomass, too, if it is not yet fossil. Moreover, everything which can be put into a biowaste container could be used as biomass feedstock. The process of rotting constitutes the differentiating characteristic between biomass and fossil resources. For example, peat is already rotten and thus not regarded as biomass anymore (Kaltschmitt, 2009).

Figure 1 offers a simplified overview of utilization pathways that are investigated in bioeconomy. Traditional illustrations, e.g. by Kaltschmitt (2009, Fig. 1.2) for energetic usage of biomass, are more detailed and show a stronger focus on the conversion technologies that are currently technologically feasible. However, for our purposes a more simplistic view will be sufficient. Thus, four different types of biomass are identified: plants, wood, residuals and living beings. These four groups contain all the materials which have in the above definition been characterized as biomass. Those different types of biomass have to be converted after their cultivation, harvesting and collection. Mainly three different conversion technology groups need to be distinguished, which are the thermo-chemical, the bio-chemical and the phyiscal-chemical conversion. Intermediate products are created through

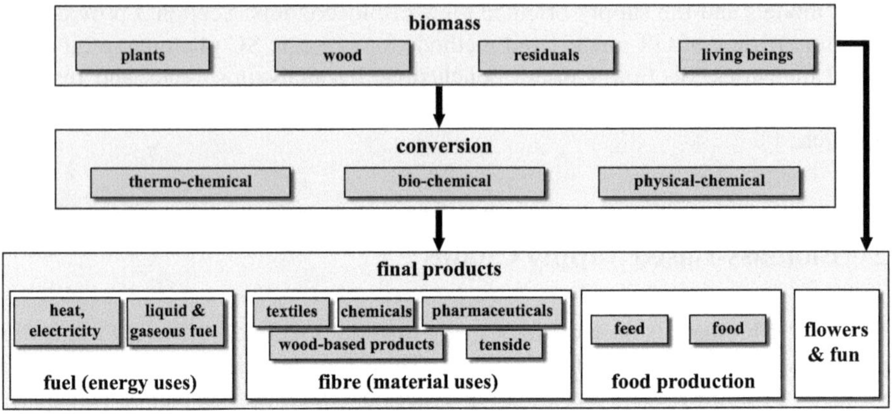

Fig. 1 Simplified utilization pathways of bioeconomy

those technologies and finally transformed into a large variety of final products. The German "Bioökonomierat" groups the final products into the five different types food, feed, fibre, fuel and "flowers and fun" (Bioökonomierat, 2015). Because feed is usually indirectly used to produce food, we pool both in a more comprehensive group "food production".

Many different ways are possible through this network. With respect to the identified groups of final products, Sects. 2.1.1–2.1.4 give some further information on these possibilities. Furthermore, Sect. 2.1.5 uses the example of municipal waste to illustrate how the different utilization pathways of a specific type of biomass can look like.

2.1.1 Fuel

A large variety of utilization pathways ends up in the final product "fuel". All four groups of biomass can be the starting material. For example, many different types of plants can be used to produce fuel in general. Possible plants are lignocellulosic plants like miscanthus, reed or millet, oleiferous plants like rape, sunflower or soy, starchy plants like potatoes, maize or grain and sugar-containing plants like sugar beet and cane. Additionally, wood out of short-rotation coppice, the traditional forest wood and residuals could be used. The residuals can further be classified into waste timber, agricultural residuals like straw or liquid manure and municipal waste. It is also possible to generate fuel from micro- or macro-algae which would be categorized as "living beings". These types of algae cannot yet be produced and exploited on an industrial scale. However, they could be an opportunity for the future (Kaltschmitt, 2009).

Biomass utilization pathways are always feedstock-oriented. As a consequence, for every type of biomass like for example rape or municipal waste, a specific utilization pathway has to be distinguished. Exemplary utilization pathways for municipal waste will be shown in Sect. 2.1.5 below. Nevertheless, the steps of such a path can roughly be divided into the biomass collection, a specific chemical conversion technology and the production of the final product out of intermediate products of the conversion. Thermo-chemical conversion technologies used for fuel production are, for instance, combustion, gasification and pyrolysis, whereas fermentation and aerobic decomposition are examples for bio-chemical conversion technologies. Additionally, also physical-chemical conversion technologies can be used to produce fuel, e.g., by extracting biodiesel from rape. All in all, a large variety of biomass can be processed using these three different types of conversion technologies, which again can be subdivided into various specific treatments (Kaltschmitt, 2009).

Thus many different utilization pathways for the final product "fuel" do exist. However, as Fig. 1 shows, "fuel" is again just used as a generic term for a whole class of final products transforming biomass into energy. Such final products are, for instance, liquid and gaseous fuels like bioethanol, biodiesel and hydrogen. Apart from that there are also heat and electricity generation subsumed under this type. Heat and electricity are often generated simultaneously by using so called "combined heat and power plants" (e.g., burning of biogas). However, it is also possible to produce heat apart from electricity, e.g., by combustion.

2.1.2 Fibre

A similarly large variety of biomass types as for fuels can be used to produce final products of the fibre type. Fibre denotes tangible products that are neither used for energetic purposes nor for food production. As Fig. 1 shows, these "material uses" of biomass are manifold and can further be classified into textiles (and textile fibres, respectively), wood-based products (like wood fibres, paper, cartoon, but also furniture, floorboards or timbers), chemicals, pharmaceuticals and tensides.

The already described conversion technologies can also be applied to gain fibres. However, because of the many and very different final products subsumed here, with pulping, cutting and chemical synthesis some additional conversion technologies of the physical-chemical type can be used. Thus, a greater number and a greater heterogeneity of utilization pathways can be found. To give an example: it is obvious that the production process of timber (mainly sawing, drying and moulding, maybe jointing) is very different to the production of bioplastics, which is rather a sort of chemical product. Thus the different pathways can span from traditional and comparably simple conversion technologies like cutting and extraction to more advanced ones like fermentation. All in all, this great diversity of pathways and final products is characteristic for the group "fibre" (Türk, 2014).

2.1.3 Food Production

The area "food production" concerns both the intermediate product "feed" for the breeding of farm animals as well as the various types of food for human beings. Thus, all types of groceries (ranging from fruit, vegetables and cereals, via fish, meat or other products of animal origin like eggs and cheese, to combinations thereof as, for example, convenience foods), but also feed for animals and even fertilizers are included (Bioökonomierat, 2015).

In contrast to the utilization pathways so far presented, the ones ending up in feed or food do not use all types of biomass. For example, woody biomass does not matter in food production.

The relevant biomass is not only processed by means of the earlier described conversion technologies, but also directly used. This "unconverted utilization" is the main distinguishing feature between the pathways described so far and the ones considered here. Although the variety of potential feed and food conversion technologies is still high, it appears lower than for fuel and fibre because both source materials and final products are less heterogeneous. For instance, thermo-chemical conversion techniques are mainly used for cooking or baking convenience food. Here, source materials and final products are homogeneous enough to allow a cost-efficient conversion on an industrial scale. Although there are some exceptions, in general the structure of the biomass feedstock is less modified than in the prior pathways.

2.1.4 Flowers and Fun

The final products' group "flowers" comprises uneatable horticultural products like ornamental plants and turf rolls. The importance of this industry heavily depends on regional aspects. For instance, the flower industry in Belgium or the Netherlands is much more important than the one in Germany. "Fun" focuses on the usage of biomass for extraordinary leisure activities. Examples are turf-rolls for football stadiums or golf courses that are built on former farm land. Obviously, even less and more specific types of biomass are relevant for this area. Utilization pathways are also fewer and simpler because conversion is of minor importance or no importance at all. Instead, efficiency and speed of transportation might become more crucial. All in all, as compared to the other groups of final products discussed before, this area only plays an insignificant role. It will thus not further be discussed in the remainder of this chapter.

2.1.5 Example: Municipal Biowaste as Starting Material

To illustrate how different utilization pathways can origin from the source material "biomass", municipal waste is taken as an example. Municipal biowaste consists

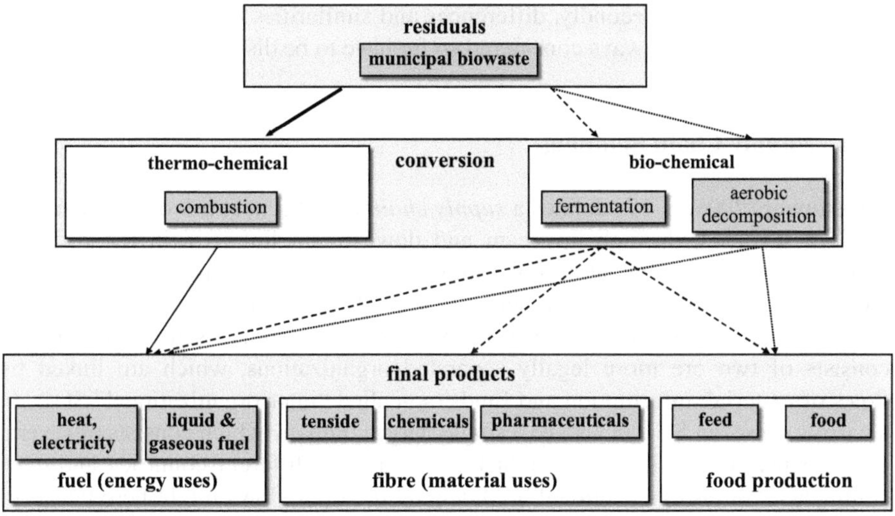

Fig. 2 Utilization pathways of municipal waste

of biodegradable garden waste and compostable food waste like fruit and vegetable peelings, i.e., it belongs to the "residuals" class of biomass presented in Fig. 1.

Exemplary utilization pathways of municipal waste are shown in Fig. 2. Municipal waste can either be processed using the thermo-chemical conversion technology "combustion" (solid arrow) or using the bio-chemical conversion technologies "fermentation" (dashed arrow) or "aerobic decomposition" (dotted arrow).The classical way of converting municipal waste into cascading products is combustion. In this case, it is possible to produce heat apart from electricity. Note that physical-chemical conversion is not applied to municipal waste (Thrän et al., 2009).

Besides electricity and heat also liquid and gaseous fuels could be produced from municipal waste. Material uses do at most occur for tensides, chemicals and pharmaceuticals, but not for textiles and wood-based products. Since fertilizer results as a by-product of bio-chemical conversion (e.g., in biogas refineries), municipal waste might indirectly contribute to feed and food production when cultivating the farm land (Diepenbrock, 2014). However, such utilization pathways do not play an important role.

2.2 Supply Chains

In Sect. 2.1, the various utilization pathways have been considered that start from the source material "biomass". However, the aim of this chapter is to review the literature about the strategic planning of biomass-based *supply chains*. Thus it is first necessary to define the terms "supply chain", "supply chain planning" and "strategic planning". Secondly, differences and similarities between supply chains and the utilization pathways considered so far have to be discussed.

2.2.1 Supply Chain Planning

Christopher (2005, p. 17) defines a *supply chain* as a "... network of organizations that are involved, through upstream and downstream linkages, in the different processes and activities that produce value in the form of products and services in the hands of the ultimate consumer." Stadtler (2015, p. 3f.) further differentiates between SCs in a broad and in a narrow sense. In a broad sense, such an SC consists of two ore more legally separated organizations, which are linked by flows of material, information and funds (so-called "inter-organizational" SCs). In a narrow sense, an SC can also be a single, large company which consists of several departments and/or sites that might be spread over different countries and even continents ("intra-organizational"). Managing the same flows might here be easier since all of the different parties belong to the same company. Nevertheless, because of the size of such companies, this is usually still very complex. Thus decision making needs to be supported.

SC planning offers this decision support for the various planning tasks arising in supply chains by building simplified models of the real SCs, deriving solutions for these models, and interpreting these solutions in order to solve the original, real-world problem. Such models might be forecasting models, which try to predict the future, optimization models, which try to find the best solution out of a huge number of alternative, feasible solutions, or simulation models, which cost-efficiently try to mimic the behavior of complex multi-stage SCs in sufficiently detailed computer models (Fleischmann and Meyr, 2003; Fleischmann et al., 2015).

Strategic SC planning aims at offering this decision support for all planning problems that concern the design and long-term structure of supply chains. Such decisions typically involve substantial investments (e.g., for establishing a cooperation, building a new factory or introducing new products in unexploited markets) and show long-lasting effects over several years or even decades because they cannot easily be revised again. Because of their high importance, these decisions are usually made by the top management(s) of the company (or the companies) involved. Nevertheless, they can be pre-selected and evaluated in terms of their advantages and disadvantages by computer systems and the staff departments of the companies. These decisions should be comprehensive and thus consider all relevant material and financial flows (e.g., fixed costs for investments, variable sales revenues and operational costs) of the SC as a whole—from raw material supply, through the various conversion processes in the different production sites up to the sales to the ultimate consumers. This includes the necessary transportation and storage processes to bridge space- and time-related discrepancies.

2.2.2 Differences Between Supply Chains and Utilization Pathways

When looking at these definitions some differences between supply chains and the utilization pathways discussed so far become obvious.

As already mentioned, utilization pathways are feedstock-oriented, starting with (in the future probably scarce) biomass as the source material. In contrast supply chains are customer- and product-focused. The final customer and her/his desired product should be in the center of all thoughts.

Furthermore, utilization pathways are mainly focusing on the material flow, i.e., the materials involved and the technologies to convert these materials. Often life-cycle-analyses (LCAs) are executed to evaluate and compare the ecological impacts of different pathways. If economic effects are considered at all, they are usually estimated for whole (sectors of) economies. In contrast, supply chains are only interested in the economic benefits of their own members. This does usually either comprise a single company (intra-organizational SC) or only a few collaborating companies (inter-organizational SC). However, this economic benefit is of very high importance because it justifies the existence and ensures the survival of the SC. Thus it is necessary to stress monetary aspects that are not at all considered in utilization pathways. Such aspects are, for example, the allocation of profits to the various participants of the SC or the sharing of costs for joint appliances or services.

Besides these monetary effects, other benefits (e.g., increased visibility) and risks (e.g., loss of autonomy, cheating) may depend on cooperation and coordination aspects of SC planning.

Figure 3 gives an example of the different parties that may be involved in biomass-based SCs. By using Fig. 1 as a basis, we link the customer-oriented view of SCs with the biomass-based view of utilization pathways. As can be seen, many different parties have to work together if a supply chain wants to become and stay successful. For sake of clarity, we did not even include service providers

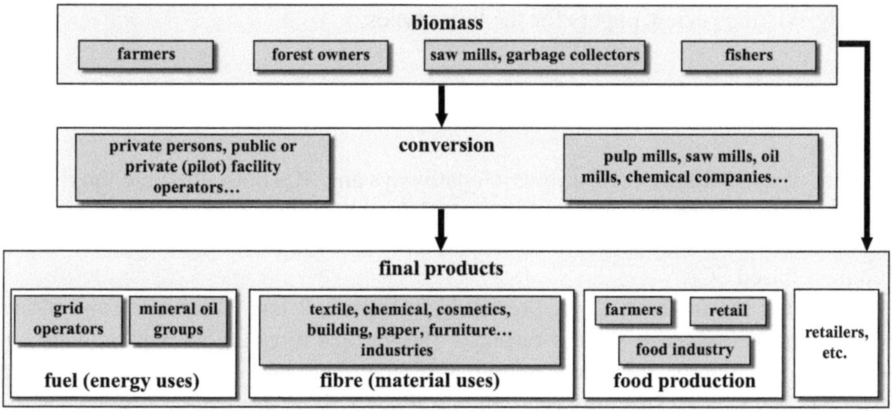

Fig. 3 Potential members of biomass-based supply chains

of support processes like transportation, storage, cutting/compacting and drying. Remember that SCs producing similar—or, to be more precise, in the customers' perception substitutable—final products compete with each other. Thus innovative biomass-based products are in competition with traditional fossil-based products. For example, energy from biogas refineries competes with energy from power stations burning natural gas. Although governmental subsidies can support the development and market entry of ecologically preferable, biomass-based products, they are usually only granted for a limited time span. Afterward, the corresponding biomass-based supply chains have to be profitable on their own and, furthermore, stand the competition with other biomass-based and the traditional fossil-based SCs. Thus, a mere concentration on utilization pathways would be too short-sighted. SC planning aspects, as briefly addressed here and reviewed in Sect. 3, should be taken into consideration from the very first beginning.

Figures 1 and 3 can give some clues on the competition between SCs offering substitutable products. For example, in the case of electricity, biomass-based electricity has to compete with other renewable energies like solar power or wind, with nuclear power and with fossil-based energies stemming from coal, mineral oil and natural gas. The grid operators and energy companies transport and sell the electricity to the customers. Thus, they have to ensure that a customer can buy the specific mix of energy (s)he wants to get. As another example, liquid automotive fuels are mainly sold by mineral oil companies via their network of petrol stations. There is a competition within a single petrol station between products containing different shares of biomass-based ethanol, but also between the fuels of different mineral oil companies. When considering the SCs providing these final products we can see that innovative biomass-based SCs for fuel and fibre mainly compete with traditional SCs of the energy, mineral oil, chemical, pharmaceutical and textile industries. SCs offering wood-based fibres and food, however, have always relied on biomass as their primary source material. A lot of scientific research has already been done for these latter types of SCs. Our review in Sect. 3 will thus rather concentrate on the former "innovative" types biomass-based SCs and just refer to already existing review papers for the latter types.

2.2.3 Similarities

Despite of these differences, utilization pathways and SCs both illustrate the flow of materials from the supply of the raw materials, through a network of transforming facilities to the final products. Thus utilization pathways help to identify and model biomass-based SCs.

Note that the material flows presented in Figs. 1, 2 and 3 had been simplified to allow and emphasize the clustering of the various biomass sources, conversion technologies, final products and potential SC members into catchy classes. For the review of Sect. 3, the structure presented in Fig. 4 will be more appropriate.

Up to three stages of an SC will be distinguished. Stage 1 includes the different biomass collection processes, like harvesting or the collection of residuals, and the

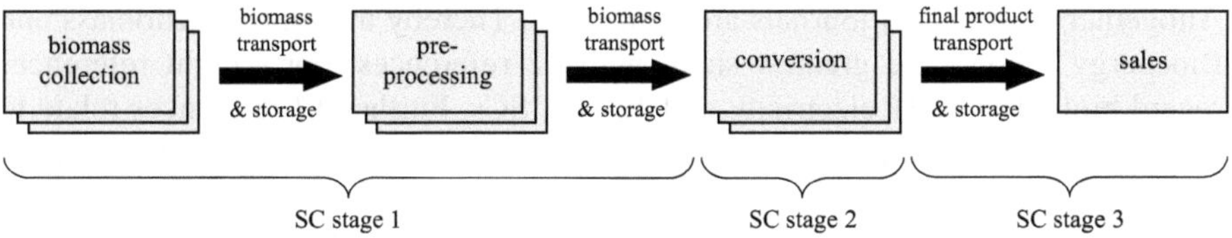

Fig. 4 Supply chain stages

transport to and storage at certain collection points. Seasonal storage of biomass feedstock is necessary because of the seasonal availability of biomass and the great uncertainty concerning quantity and quality. Additionally, but not necessarily, pre-processing activities and their respective transport and storage processes are also included in stage 1. Pre-processing is often necessary because the original biomass shows a high content of water and a low energy density. By processes like cutting, compacting and drying, both energy density and transportation efficiency are increased.

Stage 2 comprises the whole (in itself maybe again multi-stage) production network of the SC, i.e., all conversion, transportation and storage processes that are necessary to transform (pre-processed) biomass into the final product. This may regard only a single, but also several alternative conversion technologies if a single SC simultaneously comprises several utilization pathways. The specific conversion technologies are depending on the type of the final products. Chemical products are mainly produced in biorefineries which are similar to classical petroleum refineries.

Stage 3 finally includes the transport and storage of the final products and the corresponding sales activities. Note that supply and demand usually do not occur at the same point in time. Supply is, for example, bound to the harvesting times of biomass, which may be seasonal. Demand is determined by the customers' wishes and expectations. Thus storage processes are necessary to bridge the time lag. In general, these storage processes may occur at any stage of the SC. However, it has to be taken into account that biomass and its resulting intermediate products often are perishable and either can only be stored for a limited amount of time or have to be made more durable somehow, e.g., by drying. Nevertheless, as a rule of thumb, it is preferable that storage processes occur early in the material flow, i.e., upstream in the SC. At early stages of the SC the value of the respective intermediate products is still low so that storage costs are not yet crucial.

3 Literature Review

In the following we survey the recent research on quantitative models for the long-term, strategic planning of biomass-based SCs. Seventy journal publications, regarding biomass-based fuel and fibre SCs, have been analyzed. They have been published since the year 1997. However, their majority stems from the year 2011.

Altogether 33 different journals are concerned. Thereby the journal "Biomass and Bioenergy" shows the greatest share with 12 references. Forty eight references regard biofuel SCs, 21 electricity and 12 heat SCs. Further 12 references relate to fibre SCs. Thus some journal publications refer to several types of biomass-based SCs.

The references have been analyzed using a common scheme that is expressed by the columns of Table 2. First of all, we are interested in the type of quantitative model that has been used (column "O./S."). Here optimization ("O") and simulation ("S") models are distinguished. The optimization models are further classified into deterministic ("det") and stochastic ("sto") models. Deterministic means that all input parameters of the model are assumed to be deterministically known, whereas stochastic models assume that at least one uncertain random variable exists. Furthermore, we are interested in the optimization models' objective functions (column "obj"). The models pursue monetary ("mon"), ecological ("eco") or social ("soc") objectives, either separately or simultaneously as part of a multi-objective optimization (indicated by a "/"). The column "SC stages" shows which of the three stages introduced in Fig. 4 is/are actually considered by the model. The column "biomass type" finally denotes the type of biomass concerned. All four types of biomass introduced in Sect. 2.1 and Fig. 1 are possible, i.e., plants, wood, residuals ("resi.") and living beings ("beings").

Sections 3.1 and 3.2 discuss strategic models of the fuel and the fibre areas in sufficient detail. Readers who would additionally be interested in quantitative models on the tactical and operational planning of biomass-based SCs are referred to the reviews of An et al. (2011b), Awudu and Zhang (2012) and Ba et al. (2015). As mentioned before, the area "food production" will not be discussed in detail. Recent reviews concerning quantitative models to plan feed and food SCs can be found in Table 1.

Table 1 Literature food/feed

Ahumada and Villalobos (2009)	Application of planning models in the agri-food supply chain: A review
Amorim et al. (2013)	Managing perishability in production-distribution planning: A discussion and review
Akkerman et al. (2010)	Quality, safety and sustainability in food distribution: A review of quantitative operations management approaches and challenges
Dabbene et al. (2014)	Traceability issues in food supply chain management: A review
Soysal et al. (2012)	A review on quantitative models for sustainable food logistics management
Tsolakis et al. (2014)	Agrifood supply chain management: A comprehensive hierarchical decision making framework and a critical taxonomy
Zhang and Wilhelm (2011)	OR/MS decision support models for the specialty crops industry: A literature review

3.1 Fuel

Due to the substantial research effort in biofuel SCs, in the following we further distinguish between biofuel (Sect. 3.1.1) and electricity and heat (3.1.2) supply chains.

3.1.1 Biofuel

Ahn et al. (2015) consider a three-stage SC producing biodiesel from microalgae, i.e., living beings are the biomass feedstock. A multiperiod, deterministic optimization model takes decisions about transportation quantities and biorefinery locations. The objective is to minimize the total costs. Akgul et al. (2012b) assess multi-objective performance aspects in hybrid first and second generation bioethanol SCs. They decide about local biomass and import quantities, conversion quantities, biorefinery locations and capacities. A deterministic, multi-objective optimization model is used. The two objectives are to minimize the total costs as well as the carbon emissions of the SC. Again, the model covers the "whole" three-stage SC. With plants, wood and residuals, three different types of biomass can serve as input. Akgul et al. (2012a) take a hybrid first and second generation biofuel SC of UK's biofuel industry into account. In contrast to the model before, only a single objective is pursued. This deterministic model also covers the whole SC with plants, wood and residuals as biomass feedstock. In another work of Akgul et al. (2011), several models to optimally design a three-stage bioethanol SC are presented. The deterministic, single-objective models try to optimize the locations and capacities of bioethanol production facilities as well as the biomass and bioethanol transport flows by minimizing the total SC costs. Only plants are the possible biomass feedstock. Within these three models a development is noticeable from single- to multiple-feedstock and from single- to multi-objective modelling.

Aksoy et al. (2011) present a model configuring an SC with four different conversion technologies. All these technologies use woody biomass and mill wastes as feedstocks. The objective of the deterministic optimization model is to minimize the total costs. The model considers only the second stage of the SC, where decisions about the conversion technology are made. An et al. (2011a) also present a model to design a lignocellulosic biofuel SC. The deterministic optimization model is multi-period and multi-commodity. This means that, in contrast to other analyzed models, several kinds of biofuels are considered. The objective is to maximize the discounted profit of the SC. The whole SC from the biomass feedstock supplier up to the biofuel customer is taken into consideration. Again, plants, wood and residuals are the possible feedstocks.

Andersen et al. (2012) design and plan a three-stage biodiesel SC. The characteristic feature of their model is to consider land competition. The multi-period, deterministic optimization model maximizes the net present value. Only plants are possible feedstocks. Bai et al. (2011) propose a deterministic optimization model for biorefinery location planning by minimizing the total system costs. Special about their three-stage SC is that traffic congestion can be taken into account. Only plants are considered as possible feedstocks. Bernardi et al. (2013) propose a multi-objective model to design and plan a three-stage bioethanol SC, which includes first and second generation biorefineries. The deterministic, multi-period optimization model maximizes the net present value, minimizes carbon emissions and minimizes water consumption. Possible feedstocks are plants and residuals.

Bowling et al. (2011) place a biorefinery into an SC consisting of only the first two stages. Specific final products are not distinguished, but, for example, biofuel could be produced. Their deterministic model maximizes total profits with respect to nonlinear economies of scale. Information about possible feedstocks is missing. Cambero et al. (2015) deterministically optimize the mix of bioenergy and biofuel production within a three-stage SC using forest residuals as input. The objective is to maximize the net present value of investments in the conversion technologies. One of the few stochastic optimization models stems from Chen and Fan (2012). Bioethanol is produced out of waste while supply and demand of the three-stage SC are assumed to be uncertain. The two-stage stochastic model minimizes the expected total system costs of investments, production and transport.

Cobuloglu et al. (2014) focus on the farmers' point of view of switchgrass production. They consider both economic and environmental aspects. The deterministic, multi-objective optimization model maximizes the revenues of harvested switchgrass and positive ecological impacts. Their model covers just the first stage of an SC with plants as the only biomass feedstock. Correll et al. (2014) present a combined simulation and optimization approach to design the first stage of an SC for bioenergy and biobased products. Therefore, the model can be considered as a subproblem of a biofuel production network. The model compares diversified feedstocks with monocultures. The optimization part of the model is deterministic with the objective of minimizing capital investment and purchasing costs. The necessary input data are generated by simulation. Only plants are possible feedstocks. Corsano et al. (2011) design a three-stage sugar-to-ethanol SC where plants are the biomass input. The deterministic model is able to consider recycling processes while maximizing net profits, which are defined as the difference of the total revenues and the costs for sugar cane supply, production and transportation and for investments in conversion facilities and warehouses, respectively.

Dal-Mas et al. (2011) regard the design and planning of capacity investments for an ethanol SC. Uncertainties concerning both biomass production costs and the final products' selling prices are considered. Hence, the optimization model is stochastic. The objectives are to maximize the expected net present value and to minimize the financial risks. The model covers the whole SC. Only plants are possible feedstocks. The early work of De Mol et al. (1997) compares a simulation and an optimization approach concerning biomass collection for biofuel production. The former one assumes the network structure for the biomass collection as given. Whereas the latter one optimizes this structure by deterministically minimizing the total collection costs. Both models do only consider the first stage of the SC. Wood is the biomass feedstock.

Dunnett et al. (2008) simultaneously optimize production and logistics of three-stage, lignocellulosic bioethanol SCs by deterministically minimizing the respective costs. They decide whether the processing structure is rather decentral or central, i.e., whether the biomass is either pre-processed in decentral hubs and afterward converted in a centralized plant or whether it is completely processed in central facilities. Plants, wood and residuals are used as biomass feedstocks. Ekşioğlu et al. (2010) investigate the impact of intermodal facilities on the design of three-stage corn-to-bioethanol SCs. Their deterministic optimization model minimizes the total delivery costs of bioethanol. In earlier work, Ekşioğlu et al. (2009) had already analyzed and designed biomass-to-biorefinery SCs. The links between the biomass harvesting sites and the conversion plants were modeled as part of a deterministic network design problem. The objective was to minimize the total SC costs. However, only the first two stages of the SC were covered by this model. With plants, wood and residuals several biomass types were possible feedstocks.

Frombo et al. (2009a) plan the logistics of energy production from woody biomass. They compare several conversion technologies to produce different final products. Their deterministic optimization model minimizes the total costs. Wood and woody residuals are inputs to a three-stage SC. Giarola et al. (2013) design bioethanol SCs under risk management aspects. First and second generation production technologies are considered in a multi-period, stochastic optimization model with multiple objectives, maximizing the net present value and minimizing greenhouse gas emissions. The model covers only the first two stages of the SC. Plants and residuals can be biomass feedstocks. This work builds on earlier deterministic models published by Giarola et al. (2012, 2011), who considered a three-stage SC, however. Apart from that, the models share the same characteristics.

Huang et al. (2010) optimize three-stage, waste-based bioethanol SCs. Their deterministic model assesses economic potentials and infrastructure requirements by minimizing the total SC costs. Only residuals are considered as possible feedstocks. Ivanov and Stoyanov (2016) design integrated biodiesel and fossil-based fuel SCs. The deterministic optimization model considers all three stages of the SC using only plants as possible feedstocks. In addition to minimizing the total SC costs, the total life cycle greenhouse gas emissions are also minimized. Kanzian (2009) plans the logistics of wood to produce solid fuel. Using a deterministic optimization model, minimizing the total transportation costs, different demand scenarios for this fuel are evaluated and different network structures, with and without terminals, are compared. Only the first stage of the SC and only wood and woody residuals are taken into consideration.

Kim et al. (2011a,b) tackle similar problems. In both models, the whole three-stage biomass processing network to produce biofuel is designed. The first one is a deterministic optimization model. Its objective is to maximize the overall profits. It is assumed that the network can process plants, wood and residuals. The second, stochastic model considers uncertainty concerning supply quantities, market demand and price, as well as technology. All other modeling characteristics remain unchanged. Leão et al. (2011) optimize the logistic structure of two-stage biodiesel

SCs by deterministically minimizing their costs. Small farmers providing plants are the only feedstock suppliers.

Leduc et al. (2010) plan the location of methanol production facilities converting lignocellulosic plants, wood and residuals. A deterministic optimization model minimizes the costs of the three-stage network. An earlier work of Leduc et al. (2008) tackled a similar problem. However, there only the gasification of wood and wood residuals was considered. Lin et al. (2014) integrate the strategic and tactical planning of large-scale bioethanol SCs. Apart from typical strategic decisions about the number, capacities and locations of facilities also operating schedules and inventory planning are considered. The authors deterministically minimize the annual costs of biomass-to-bioethanol conversion. All three stages, from the farmers (providing plants as the only biomass type) to the distribution of the bioethanol, are represented. The multi-objective, two-stage stochastic model of Marufuzzaman et al. (2014) concerns the production of biodiesel through wastewater treatment. It respects the impacts of different carbon regulation policies. Objectives are to minimize the annual costs of three-stage SCs and the resulting emissions.

Marvin et al. (2013) plan the locations of biomass conversion facilities and the selection of the appropriate conversion technology. Their deterministic optimization model maximizes the net present value of a three-stage SC using plants, wood and residuals as feedstocks. Marvin et al. (2012) tackle bioethanol production from lignocellulosic biomass. There, a bio-chemical conversion technology is applied. Five different types of agricultural residues are considered as lignocellulosic feedstocks. A single-objective, deterministic optimization model maximizes the net present value of the SC. Only the first two stages of the SC are planned. Biomass feedstock types are plants, wood and residuals. Mele et al. (2011) try to increase the sustainability of three-stage sugarcane-to-bioethanol SCs. The combined production of sugar and ethanol is considered by a multi-objective deterministic optimization model. The objectives are to maximize the net present value and to minimize the environmental damage, which is calculated using LCA. Plants are the only possible feedstocks.

Mohseni and Pishvaee (2016) and Mohseni et al. (2016) tackle similar problems. In both models, the whole three-stage SC network to produce microalgae-based biofuel is designed. Both approaches are using robust optimization with sensitivity analysis to minimize the total costs of the SC. As microalgae is used, living beings are the biomass feedstock. Osmani and Zhang (2013) consider the production of bioethanol from lignocellulosic plants, wood and residuals. Their stochastic model takes uncertain biomass prices, uncertain bioethanol demand and uncertain sales prices for bioethanol into account while maximizing the expected profit of a three-stage SC.

Santibañez-Aguilar et al. (2011) compare different, alternative utilization pathways. Their deterministic, multi-objective model maximizes their corresponding profits and minimizes their environmental impacts in order to grasp economic and environmental aspects, simultaneously. Only the upstream part of the SC (i.e., the first two stages) is modeled. Plants, wood and residuals are considered as possible feedstocks. Schwaderer (2012) integrates location, capacity and technology planning for SCs that use residuals as biomass. His deterministic optimization model minimizes the costs of the first two stages of such SCs. It can deal with final products of both the fuel and fibre type. The model of Tittmann et al. (2010) tackles the techno-economic planning of biofuel production. Their deterministic optimization model decides about locations and technologies of conversion facilities. Total profits are maximized—also for electricity, which comes up as a by-product of biofuel generation. Again only the first two SC stages are considered with plants, wood and residuals being the feedstocks.

Walther et al. (2012) design regional SCs for the production of second generation synthetic biodiesel. Their deterministic, multi-period optimization model maximizes the net present value of the three-stage network. Possible feedstocks are plants and their residuals. The—to our knowledge—first and only model, which also covers social objectives, has been published by You et al. (2012). The model tries to establish a sustainable, three-stage SC for producing biofuel from cellulosic biomass. Their multi-objective, deterministic model minimizes the annual total costs and greenhouse gas emissions, respectively, and maximizes the number of new jobs generated. Plants, wood and residuals can serve as cellulosic feedstocks. You and Wang (2011) plan three-stage biomass-to-liquid SCs with respect to economic and environmental aspects. They also apply a multi-objective deterministic model. However, theirs only minimizes the annual costs and the life cycle greenhouse gas emissions. Plants, wood and residuals are possible feedstocks. Zamboni et al. (2009) design three-stage bioethanol production SCs. As in the previous model, economic and environmental aspects are considered by deterministically minimizing both total costs as well as greenhouse gas emissions. Here, only plants are the feedstock. Finally, Zhang and Hu (2013) combine the strategic and operational planning of second generation drop-in-fuel production. Apart from the usual long-term aspects, further decisions on production patterns and inventories are made. Their deterministic optimization model minimizes the total annual costs. Only plants and residuals are possible feedstocks of a three-stage SC.

All references described above and their characteristics are summarized in Table 2. As can be seen, all authors formulate optimization models. Sometimes an additional simulation model is proposed. Only six out of 45 references use a stochastic approach in order to take uncertainty into account. Approximately a quarter of the references pursue an ecological objective additionally to the monetary one. Only a single paper furthermore considers a social objective. Most models cover the whole three-stage SC. Otherwise, at least the biomass supply or biomass conversion is represented. Many models allow several types of biomass. However, then often these types share a common characteristic, for example, all of them are lignocellulosic. This is convenient from a technological point of view because they show similar conversion properties. However, it might be less convenient from a logistical point of view because, for example, logistical processes to collect wood residuals in and from saw mills are very different from harvesting in agricultural fields or forests.

Table 2 Literature concerning biofuel production

	O./S.	det./sto.	obj.	SC stages			Biomass type			
				1	2	3	Plants	Wood	Residuals	Beings
Ahn et al. (2015)	O	det	mon	x	x	x				x
Akgul et al. (2012a)	O	det	mon	x	x	x	x	x	x	
Akgul et al. (2012b)	O	det	mon/eco	x	x	x	x	x	x	
Akgul et al. (2011)	O	det	mon	x	x	x	x			
Aksoy et al. (2011)	O	det	mon		x			x	x	
An et al. (2011a)	O	det	mon	x	x	x	x	x	x	
Andersen et al. (2012)	O	det	mon	x	x	x	x			
Bai et al. (2011)	O	det	mon	x	x	x	x			
Bernardi et al. (2013)	O	det	mon/eco	x	x	x	x		x	
Bowling et al. (2011)	O	det	mon	x	x					
Cambero et al. (2015)	O	det	mon	x	x	x			x	
Chen and Fan (2012)	O	sto	mon	x	x	x			x	
Cobuloglu et al. (2014)	O	det	mon/eco	x			x			
Correll et al. (2014)	O/S	det	mon	x			x			
Corsano et al. (2011)	O	det	mon	x	x	x	x			
Dal-Mas et al. (2011)	O	sto	mon/mon	x	x	x	x			
De Mol et al. (1997)	O/S	det	mon	x				x		
Dunnett et al. (2008)	O	det	mon	x	x	x	x	x	x	
Ekşioğlu et al. (2009)	O	det	mon	x	x		x	x	x	
Ekşioğlu et al. (2010)	O	det	mon	x	x	x	x			
Frombo et al. (2009a)	O	det	mon	x	x	x		x	x	
Giarola et al. (2011)	O	det	mon/eco	x	x	x	x		x	

Giarola et al. (2012)	O	det	mon/eco	x	x	x	x		x	
Giarola et al. (2013)	O	sto	mon/eco	x	x		x		x	
Huang et al. (2010)	O	det	mon	x	x	x			x	
Ivanov and Stoyanov (2016)	O	det	mon/eco	x	x	x	x			
Kanzian (2009)	O	det	mon	x				x	x	
Kim et al. (2011a)	O	sto	mon	x	x	x	x	x	x	
Kim et al. (2011b)	O	det	mon	x	x	x	x	x	x	
Leão et al. (2011)	O	det	mon	x	x		x			
Leduc et al. (2008)	O	det	mon	x	x	x		x	x	
Leduc et al. (2010)	O	det	mon	x	x	x	x	x	x	
Lin et al. (2014)	O	det	mon	x	x	x	x			
Marufuzzaman et al. (2014)	O	sto	mon/eco	x	x	x			x	
Marvin et al. (2012)	O	det	mon	x	x		x	x	x	
Marvin et al. (2013)	O	det	mon	x	x	x	x	x	x	
Mele et al. (2011)	O	det	mon/eco	x	x	x	x			
Mohseni and Pishvaee (2016)	O	det	mon	x	x	x				x
Mohseni et al. (2016)	O	det	mon	x	x	x				x
Osmani and Zhang (2013)	O	sto	mon	x	x	x	x	x	x	
Santibañez-Aguilar et al. (2011)	O	det	mon/eco	x	x		x	x	x	
Schwaderer (2012)	O	det	mon	x	x				x	
Tittmann et al. (2010)	O	det	mon	x	x		x	x	x	
Walther et al. (2012)	O	det	mon	x	x	x	x		x	
You et al. (2012)	O	det	mon/eco/soc	x	x	x	x	x	x	
You and Wang (2011)	O	det	mon/eco	x	x	x	x	x	x	
Zamboni et al. (2009)	O	det	mon/eco	x	x	x	x			
Zhang and Hu (2013)	O	det	mon	x	x	x	x		x	

3.1.2 Electricity and Heat

In the following we concentrate on biomass-based SCs producing electricity or heat as final products. Table 3 contains an overview of the respective literature. Since biofuel can be used as both a final product and an intermediate product for generating electricity, some references do appear in Tables 2 and 3 as well. Of course, those will not be discussed in detail a second time.

The production of heat is always a joint product in power generation. The ambition of so-called combined heat and power (CHP) plants is to produce electricity. Heat automatically emerges during this process. Since a few years, this waste product "heat" is more and more systematically used, e.g., to heat private houses or to dry (or cool) materials of nearby industrial parks. Because of these various possible uses, various stakeholders are increasingly interested in the final product heat. Table 3 illustrates this intended usage. Since none of its references uses living beings as biomass we have replaced the column "beings" with a new column "final products". This column contains the entry "e" if electricity is the intended final product and "h" if heat is an intended final product. To ease readability, references allowing to also produce biofuel are marked with an additional "b". An "f" indicates whether—besides electricity or heat—also fibre can be produced. However, the production of fibre will be discussed in more detail in Sect. 3.2.

Akgul et al. (2014) represent the co-firing of biomass with fossil fuels and a capturing and storage of CO_2 in a multi-objective, deterministic optimization model. The objectives are to minimize the total annual costs and the total annual emissions. All three stages of the SC are covered with plants and residuals being possible feedstocks. Ayoub et al. (2007) offer decision support for a general, three-stage bioenergy SC. A geographical information system (GIS) and a simulation model help to estimate the potential biomass supply of wood and wood residuals and to identify promising locations of conversion facilities. Another simulation model is proposed by Caputo et al. (2005), which considers biomass-based energy generation through combustion and gasification facilities. The authors want to evaluate the effects of different logistical alternatives on the costs of conversion. The model does only comprise the first two stages of the SC. Information about possible biomass feedstocks is missing. Feng et al. (2010) investigate bio-refinery design within a three-stage forest product SC producing (woody) fibre, electricity and heat. A deterministic, multi-period optimization model maximizes the net present value of the SC, which is fed by wood and wood residuals.

Frombo et al. (2009b) use a deterministic optimization model to produce energy and heat from woody biomass (plants, wood and residuals) in a three-stage SC. Its objective is to minimize the difference of the total costs (purchasing, transportation and plant costs) and the benefits deriving from energy sales. Judd et al. (2012) design a logistics system for bioenergy production, using satellite storage locations (SSLs). These SSLs are temporary and uncovered feedstock depots which are decentrally located around a biomass conversion facility. The autors' deterministic optimization model minimizes storage costs of the SSLs and transportation costs for only the first stage of an SC. Biomass is used as feedstock, but the type of biomass is not further

Table 3 Literature concerning electricity and heat production

				SC stages			Biomass type			Final
	O./S.	det./sto.	obj.	1	2	3	Plants	Wood	Residuals	prod.
Akgul et al. (2014)	O	det	mon/eco	x	x	x	x		x	–/e/–/–
Aksoy et al. (2011)	O	det	mon		x			x	x	b/e/h/–
Ayoub et al. (2007)	S			x	x	x		x	x	–/e/–/–
Cambero et al. (2015)	O	det	mon	x	x	x			x	b/e/h/–
Caputo et al. (2005)	S			x	x					–/e/–/–
Feng et al. (2010)	O	det	mon	x	x	x		x	x	–/e/h/f
Frombo et al. (2009a)	O	det	mon	x	x	x		x	x	b/e/h/–
Frombo et al. (2009b)	O	det	mon	x	x	x	x	x	x	–/e/h/–
Judd et al. (2012)	O	det	mon	x						–/e/–/–
Lam et al. (2013)	O	det	mon	x					x	–/e/–/–
Meyer et al. (2015)	O	det	mon/eco	x	x		x	x		–/e/h/–
Meyer et al. (2016)	O	det	mon/eco	x	x		x		x	–/e/h/–
Paulo et al. (2015)	O	det	mon	x	x	x			x	–/e/–/–
Rauch and Gronalt (2011)	O	det	mon	x				x		–/e/h/–
Reche López et al. (2008)	O	det	mon	x	x			x	x	–/e/–/–
Rentizelas et al. (2009)	O	det	mon	x	x	x			x	–/e/h/–
Rentizelas and Tatsiopoulos (2010)	O	det	mon	x	x	x			x	–/e/h/–
Roni et al. (2014)	O	det	mon	x	x		x	x	x	–/e/–/–
Santibañez-Aguilar et al. (2011)	O	det	mon/eco	x	x		x	x	x	b/e/h/f
Tittmann et al. (2010)	O	det	mon	x	x		x	x	x	b/e/–/–
Wang et al. (2012)	O	det	mon	x	x		x			–/e/h/–

specified. Lam et al. (2013) design the first stage of a green bioenergy SC basing on waste as feedstock. They propose a deterministic, two-stage optimization model, which maximizes the profit on a micro decision level and minimizes the costs on a macro decision level. First, the conversion processes of each conversion facility are optimized by choosing the best feedstock-to-product allocation (micro level). Then the whole SC is optimized by balancing supply and demand at minimal costs (macro level).

Meyer et al. (2015, 2016) combine the strategic and tactical planning of bioenergy and heat production in a two-stage SC. They introduce a basic, multi-objective, deterministic optimization model called OPTIMASS. Its objectives are to maximize the profits and energy outputs as well as to minimize the global warming potential. The model of the previous work from 2015 considers plants and wood as possible biomass feedstocks. The latter one considers plants and residuals. Paulo et al. (2015) use a deterministic optimization model to design a bioelectricity SC based on forestry residuals. Within the model the production capacities and locations are defined. They cover all three stages of the SC and consider several uncertainties by using sensitivity analysis. Minimizing the total SC costs is the single objective.

Rauch and Gronalt (2011) examine the relation between increasing energy costs and the transport mode choice in a forest fuel SC network, i.e., only woody biomass is considered as possible feedstock. Different modes of transport are analyzed to ensure the supply for combined heat and power plants. Therefore, only the first SC stage is covered. The objective of the presented deterministic optimization model is to minimize the total costs. Reche López et al. (2008) present a deterministic optimization model to determine locations and sizes of power facilities within a two-stage SC. They only focus on the supply side of the conversion facilities, which use wood and wood residuals as input. They apply particle swarm optimization to maximize a profitability index taking costs and benefits into consideration.

Rentizelas et al. (2009) support strategic decision making for residual-to-bioenergy conversion, more specifically for so-called "tri-generation applications" comprising electricity, heating and cooling. Their deterministic optimization model maximizes the net present value of a three-stage SC by choosing the optimal location for the biomass conversion facility, its size and the optimal mix of specific biomass residuals. Rentizelas and Tatsiopoulos (2010) optimize the locations of biomass-to-bioenergy conversion facilities producing electricity and heat for district energy applications. Their deterministic optimization model maximizes the net present value of a three-stage SC that is only fed by residuals. Roni et al. (2014) consider co-firing of biomass (plants, wood and residuals) in coal-fired power facilities. They propose a deterministic optimization model to design the first two stages of an SC as a hub-and-spoke structure. The model minimizes the costs of transport and investments in locations for a given energy demand. Finally, Wang et al. (2012) determine the supply of energy crops as well as the locations and capacities of conversion facilities generating heat and power. They propose a deterministic optimization model maximizing the profits of the two-stage SC.

As Table 3 shows, again optimization models are preferred to simulation models. Only a single "pure" simulation model has been proposed. Similarly to the last section, there are only a few multi-objective models. Three references do only consider a single-stage SC, two of them concentrating on the biomass supply, one of them on the biomass conversion. Most references comprise either the first two or all three stages of the SC. Not surprisingly, the variety of biomass used is similar to biofuel production. As mentioned before, heat could be produced without generating electricity. However, none of the references found intends to do this.

3.2 Fibre

Table 4 summarizes literature on the quantitative, strategic planning of biomass-based SCs that aims at producing final products of (at least) the fibre group. Again we will only discuss references in detail which have not yet been introduced in the preceding sections.

Gunn (2009) describes an optimization model to produce forest products. Just the first SC stage is considered with wood being the only feedstock. As only the first stage of the SC is considered, no information about specific forest products is given. The developed optimization model is deterministic with the objective to maximize the profits. Gunnarsson et al. (2005) integrate the search for terminal locations of various pulp products and for their outbound shipping routes in a deterministic optimization model minimizing total distribution costs. Thus only

Table 4 Literature concerning fibre production

				SC stages			Biomass type			Final
	O./S.	det./sto.	obj.	1	2	3	Plants	Wood	Resi.	prod.
Bowling et al. (2011)	O	det	mon	x	x					b/–/–/f
Chen and Fan (2012)	O	sto	mon	x	x	x			x	b/–/–/f
Correll et al. (2014)	O/S	det	mon	x			x			b/–/–/f
Ekşioğlu et al. (2009)	O	det	mon	x	x		x	x	x	b/–/–/f
Feng et al. (2010)	O	det	mon	x	x	x		x	x	–/e/h/f
Gunn (2009)	O	det	mon	x				x		–/–/–/f
Gunnarsson et al. (2005)	O	det	mon			x				–/–/–/f
Kelley et al. (2013)	O	det	mon	x						–/–/–/f
Philpott and Everett (2001)	O	det	mon	x	x	x		x		–/–/–/f
Santibañez-Aguilar et al. (2011)	O	det	mon/eco	x	x		x	x	x	b/e/h/f
Schwaderer (2012)	O	det	mon	x	x				x	b/–/–/f
Troncoso and Garrido (2005)	O	det	mon	x	x			x		–/–/–/f

the third SC stage, downstream of some pulp mills in Scandinavia, is considered. Specific information on biomass feedstocks is missing, but pulp mills are usually fed by forest wood.

Kelley et al. (2013) design a transportation network in a mainly roadless region of Amazonian Ecuador in order to transport indigenous goods to the markets. A deterministic optimization model minimizes the total costs of storage and of the various transportation vehicles. Only the first stage of the SC is considered. The types of biomass feedstocks are not mentioned. Philpott and Everett (2001) optimize an SC of the paper industry. They propose a deterministic optimization model to allocate suppliers to paper mills and customers and their requested products to paper machines, respectively. The objective is to maximize the overall profits of the three-stage, wood-based SC. Troncoso and Garrido (2005) deterministically minimize the costs of the production and logistics processes of a forest SC by choosing the optimal location and size of conversion facilities. Additionally, they consider production quantities and freight flows. Wood is the only feedstock. Specific information on the final items is missing because the authors do only consider the first two stages of such SCs.

Note that there is a whole stream of literature on quantitative (and also strategic) SC planning in the pulp and paper industry. The respective references discussed above are only a few typical examples. As mentioned in Sect. 2.2.2, including all relevant work would have led to a loss of focus on more innovative types of biomass-based SCs.

According to Table 4 again optimization models are dominating. Moreover, only a single stochastic and a single multi-objective model can be found. The fibre research rather concentrates on the upstream instead of downstream part of the SC. Despite of that, information on the type of biomass used is more often missing. No research has been identified, which intends to exclusively produce non-wood based fibres like chemicals, pharmaceuticals or tensides.

4 Conclusions

All in all, when comparing Tables 2, 3 and 4, we recognize that most research has been done on biofuel production, whereas electricity, heat and fibre production have less frequently been considered. However, the research effort in biofuel SCs seems to be decreasing. The peak in the number of published articles was in 2011. In contrast, the effort in biomass-based electricity and heat SCs as well as in fibre SCs is relatively stable since 2009, yet on a much lower level. Thus, the overall research effort on the strategic planning of biomass-based SCs appears rather decreasing. Overall, deterministic optimization models, deciding about the structure and facilities of a two- to three-stage supply chain starting from the biomass supply, are dominating. Usually they pursue only a single monetary objective, which is to either minimize costs or maximize profits. In multi-period models, which do not only determine the type of investment, but also the timing of investments, net present values are taken into account. Except for living beings (which rather play a role in food SCs), all types of biomass are considered—often even simultaneously as substitutable or supplementary feedstocks. However, it seems that research during the years 2011 and 2012 focused stronger on plants-based fuel and fibre SCs, whereas earlier on and later on wood- and residual-feedstocks appealed higher interest.

Figures 3 and 4 of Sect. 2 have revealed that many different participants may be involved in three-stage, biomass-based supply chains. However, none of the models of Sect. 3 takes an inter-organizational perspective, caring about problems concerning the cooperation between legally separate companies (e.g., trust building, aligning incentives or sharing of information, risks, joint profits or joint costs). The vast majority of the models is characterized by a centralized point of view, meaning that the decision maker is a centralized SC planner in an intra-organizational SC, having all necessary information about the SC (i.e., deterministic model) and being able to decide for the supply chain as a whole. Sometimes this planner is characterized by a rather macroeconomic point of view, considering influences on the economy as a whole like environmental or social aspects (for instance, see the multi-objective models and simulation models). Apart from those, there are some models with a conversion facility owner's point of view or with a farmer's or third-party logistics service provider's point of view. The SC stages depicted in Tables 2, 3 and 4 give a hint on the specific decision maker: If all three stages are considered, this indicates the planning of the whole SC from a central SC planner's point of view. Exceptions are models on the planning of a single biorefinery with both upstream and downstream stages. A partial consideration indicates either again the biorefinery operator's point of view (if only the second stage is considered) or the farmer's and third-party logistics service provider's point of view, respectively.

Comparing biomass-based SCs with their traditional fossil-based counterparts helps to stress further specific characteristics of the models of Sect. 3.

Fossil-based fuel is produced in a few, large refineries of mineral oil companies (Roitsch and Meyr, 2015). The only input material "crude oil" may show a different chemical composition if it stems from different oil fields in different regions of the world. Nevertheless, as compared to biomass, it is a very homogeneous material. By using pipelines or tankers, this raw material can cost-efficiently be transported to the refineries. The refineries are of industrial size, what also allows cost-efficient conversion processes. If used as automotive fuel, merely the distribution from the refineries to the multitude of petrol stations requires small-sized transports by trucks. Storage is possible at any stage of the supply chain. It is necessary to save costs (e.g., varying market prices for crude oil, lotsizing) in and hedge against risks (e.g., varying lead times) of transportation and production.

As opposite, biomass feedstocks are typically more heterogeneous. They are decentrally located within a specific region and have there to be collected or harvested. The biomass shows a high content of water and a low energy density. If it should also be brought to a few, centrally located conversion facilities, either high transportation costs would result or a further, decentral pre-processing step would be necessary, which increases the energy density and thus decreases transportation costs. However, this pre-processing may also incur fix costs for investments and variable costs for transformation. Some types of biomass are perishable (e.g., plants). That means, either decentral pre-processing additionally conserves the biomass so that it can be stored. Or it has—more or less immediately—to be transported to the central conversion facilities. For other types of biomass (e.g., wood), decentral storage may even save a costly pre-processing step (like drying). Anyway, decisions concerning the number, locations and capacities of (pre-) processing facilities for *biofuel* have to be made by managing the tradeoff between, at least, investment costs for these facilities and transportation costs of unprocessed biomass. Potential solutions, provided in the analyzed literature, are to use only one central conversion facility, several centralized conversion facilities, or central conversion facilities and several upstream, decentral pre-processing facilities. The decision about these potential network structures is crucially dependent on the scalability of the conversion facilities, which are also denoted as "biorefineries". From a transformation point of view, biorefineries for biofuel production play a similar role as petroleum refineries. However, the characteristics of the used supply are totally different.

Examples for many decentral conversion facilities can also be found, but rather for the production of electricity and heat. Whereas typical *fossil-based power plants* either also profit from cost-efficient transportation means (pipelines) for their supplying material (natural gas) or are located in a region of highly concentrated supply (coal), *biogas refineries* usually are decentrally located and small-sized. To save transportation costs, they can only cost-efficiently be fed by biomass from their immediate vicinity. Their small-sized conversion technology is currently only profitable if subsidies are guaranteed by law. The main advantage of this type of biomass-based supply chain is that its primary final product "electricity" can easily and cheaply be brought to the final consumer by feeding it into the already existing power network. Unfortunately, its co-product "heat" cannot as easily be transported. Thus, a clever usage has to be found, for example, through cooperations with neighboring industrial parks, close-by housing areas etc. Intra-organizational decisions about locations of (pre-)processing facilities are less important here. At most, investments in alternative conversion technologies and facility designs could be optimized. Overall, however, it is rather necessary to establish successful regional, inter-organizational cooperations between suppliers of biomass, operators of the biogas refinery and adjacent consumers of heat. As Sect. 3.1.2 has shown, SCM research does not yet offer much support for this.

Supply chains producing fibres like chemicals, tensides and pharmaceuticals from biomass compete with traditional fossil-based (natural gas, coal, crude oil) SCs of the chemical and pharmaceutical industries. *Fossil-based (organic) chemicals* usually are produced on an industrial scale in integrated production sites consisting of several production facilities which are interconnected by a pipeline system. They are supplied by the mineral oil industry with large quantities of intermediate materials like Naphtha that results as a by-product of typical refinery processes (see above). These fossil intermediate materials are split into basic chemicals, which are re-composed into intermediate chemicals so that both can finally react to final chemicals (Kirschstein, 2015, Chap. 2). In order to save transportation costs, chemical production sites are often located in close vicinity to oil refineries. Thus they also profit from economies of scale in transportation and production. In contrast, *biomass-based fibre SCs* struggle with the same problems as biomass-based fuel SCs. Bioethanol could play a similar role for biomass-based fibre SCs as Naphtha does for fossil-based SCs. It can already be produced on an industrial scale. For example, the PlantBottle™, a beverage bottle developed and used by the Coca-Cola Company, is partly made out of bioethanol, which is produced on large scale from sugarcane of Brazil (Coca-Cola, 2016). However, for the reasons mentioned above, this is currently more costly than using fossils as an input. Small-scaled biorefineries, which could—similarly to biogas refineries—process the biomass decentrally into bioethanol or even into final products of the fibre type are still too expensive for an operational usage. Since it is not yet clear what technological research will bring, research on the strategic planning of such types of SCs appears premature.

As already mentioned, SCs for pulp and paper and other wood-based fibres traditionally already base on biomass as source material. The same is, of course, true for food production. Therefore, much research has already been done to find out how to place conversion facilities into these types of SCs. Examples are given by Carlsson et al. (2009), Cambero and Sowlati (2014) or Ahumada and Villalobos (2009). It can be learned that especially the upstream processes in biomass-based SCs are characterized by manifold uncertainties. For instance, the quality and quantity of biomass supply is depending on the weather and thus uncertain. Moreover, the harvesting or collection time may be seasonal and uncertain, too. Hence, the upstream processes, meaning the supply side of the SC, are characteristic and crucial for many of the downstream processes. Due to the seasonal and uncertain supply, storage would be desired. However, if the biomass is perishable, this may hardly be possible (e.g., for fresh food) or (pre-)processing steps are necessary to enlarge durability. As Sect. 3 has shown, some of the models for the more innovative fuel and fibre SCs tackle the same problems. They deal with uncertainty by using stochastic modeling techniques. Here, typically the biomass supply is modeled as being uncertain. Additionally, the demand is uncertain, too, in some of the models. Perishability is less in the focus than it is in food SCs (see, e.g., Amorim et al. 2013). It is rather indirectly considered by potentially introducing pre-processing facilities.

5 Summary and Outlook

This chapter provided an overview of the latest literature on the long-term, strategic planning of biomass-based supply chains using quantitative models of operations research. We structured the overall research field "bioeconomy" by means of various utilization pathways of biomass. In such utilization pathways, the scarce resource "biomass", i.e., the ultimate supply, is the starting point. In contrast, supply chains are rather demand- than supply-oriented. All participants in an SC cooperate to fulfill the final customer's demand. Section 2 bridged the gap between the demand-oriented view of SC management models and the supply-oriented view of bioeconomy. Subsequently, several dozen publications have been analyzed with respect to the modeling characteristics used, the sections of the SC covered and the types of biomass considered.

As results of and conclusions from the analysis, some characteristics of biomass-based SCs, some trends of current research on the strategic planning of biomass-based SCs and some research gaps have been identified. On the one hand, it is noticeable that the research effort on the strategic planning of SCs producing fuels and rather "innovative" fibres from biomass seems to be decreasing. This is caused by a decrease of research on biofuel production. Overall, peak efforts were recorded in the years 2011 and 2012, research thereby mainly focusing on plants-based biomass. On the other hand, the following characteristics of biomass-based SCs have been identified: Biomass-based SCs are inter-organizational and characterized by a great heterogeneity of parties involved. This heterogeneity should be tackled by means of inter-organizational cooperation and intra-organizational coordination. However, most of the analyzed models assume an intra-organizational view with a central, omniscient and omnipotent planner. High uncertainty concerning the biomass supply is another important characteristic of biomass-based SCs, which is considered by some models. High transportation costs of unprocessed biomass, caused by its high water content and low energy density, are a further characteristic. Because of them, decisions on locations for pre-processing and conversion facilities are crucial and thus considered by most of the analyzed models.

Further research should also take inter-organizational aspects of SC management into account. Biomass-based SCs have to become profitable on their own, i.e., without governmental subsidies, and have to compete with their fossil-based counterparts. Clever cooperation between the partners of biomass-based SCs would help to save costs and to become more competitive. Nevertheless, the current intra-organizational models with a central view are not useless. They can serve as a benchmark of what could be achieved if an SC would be truly integrated. Thus these models need to be brought as close to reality as possible, for example, by increasingly incorporating the risks arising through supply and demand uncertainties. And they permanently need to be adapted to new surrounding constraints, which, for instance, emerge from new laws or changed governmental support programs.

References

Ahn, Y.-C., Lee, I.-B., Lee, K.-H., & Han, J.-H. (2015). Strategic planning design of microalgae biomass-to-biodiesel supply chain network: Multi-period deterministic model. *Applied Energy, 154*, 528–542.

Ahumada, O., & Villalobos, J. R. (2009). Application of planning models in the agri-food supply chain: A review. *European Journal of Operational Research, 196*(1), 1–20.

Akgul, O., Mac Dowell, N., Papageorgiou, L. G., & Shah, N. (2014). A mixed integer nonlinear programming (MINLP) supply chain optimisation framework for carbon negative electricity generation using biomass to energy with CCS (BECCS) in the UK. *International Journal of Greenhouse Gas Control, 28*, 189–202.

Akgul, O., Shah, N., & Papageorgiou, L. G. (2012a). Economic optimisation of a UK advanced biofuel supply chain. *Biomass and Bioenergy, 41*, 57–72.

Akgul, O., Shah, N., & Papageorgiou, L. G. (2012b). An optimisation framework for a hybrid first/second generation bioethanol supply chain. *Computers and Chemical Engineering, 42*, 101–114.

Akgul, O., Zamboni, A., Bezzo, F., Shah, N., & Papageorgiou, L. G. (2011). Optimization-based approaches for bioethanol supply chains. *Industrial and Engineering Chemistry Research, 50*(9), 4927–4938.

Akkerman, R., Farahani, P., & Grunow, M. (2010). Quality, safety and sustainability in food distribution: A review of quantitative operations management approaches and challenges. *OR Spectrum, 32*(4), 863–904.

Aksoy, B., Cullinan, H., Webster, D., Gue, K., Sukumaran, S., Eden, M., & Sammons, N. (2011). Woody biomass and mill waste utilization opportunities in Alabama: Transportation cost minimization, optimum facility location, economic feasibility, and impact. *Environmental Progress and Sustainable Energy, 30*(4), 720–732.

Amorim, P., Meyr, H., Almeder, C., & Almada-Lobo, B. (2013). Managing perishability in production-distribution planning: A discussion and review. *Flexible Services and Manufacturing Journal, 25*, 389–413.

An, H., Wilhelm, W. E., & Searcy, S. W. (2011a). A mathematical model to design a lignocellulosic biofuel supply chain system with a case study based on a region in Central Texas. *Bioresource Technology, 102*(17), 7860–7870.

An, H., Wilhelm, W. E., & Searcy, S. W. (2011b). Biofuel and petroleum-based fuel supply chain research: A literature review. *Biomass and Bioenergy, 35*(9), 3763–3774.

Andersen, F., Iturmendi, F., Espinosa, S., & Diaz, M. S. (2012). Optimal design and planning of biodiesel supply chain with land competition. *Computers and Chemical Engineering, 47*, 170–182.

Awudu, I., & Zhang, J. (2012). Uncertainties and sustainability concepts in biofuel supply chain management: A review. *Renewable and Sustainable Energy Reviews, 16*(2), 1359–1368.

Ayoub, N., Martins, R., Wang, K., Seki, H., & Naka, Y. (2007). Two levels decision system for efficient planning and implementation of bioenergy production. *Energy Conversion and Management, 48*(3), 709–723.

Ba, B. H., Prins, C., & Prodhon, C. (2015). Models for optimization and performance evaluation of biomass supply chains: An Operations Research perspective. *Renewable Energy*, doi: 10.1016/j.renene.2015.07.045.

Bai, Y., Hwang, T., Kang, S., & Ouyang, Y. (2011). Biofuel refinery location and supply chain planning under traffic congestion. *Transportation Research Part B: Methodological, 45*(1), 162–175.

Bernardi, A., Giarola, S., & Bezzo, F. (2013). Spatially explicit multiobjective optimization for the strategic design of first and second generation biorefineries including carbon and water footprints. *Industrial and Engineering Chemistry Research, 52*(22), 7170–7180.

Bioökonomierat (Ed.) (2015). *Nachhaltige Bereitstellung von biobasierten agrarischen Rohstoffen*. Office of the Bioeconomy Council.

BMBF, BMEL. (2014). *Bioökonomie in Deutschland: Chancen für eine biobasierte nachhaltige Zukunft*.

Bowling, I. M., Ponce-Ortega, J. M., & El-Halwagi, M. M. (2011). Facility location and supply chain optimization for a biorefinery. *Industrial and Engineering Chemistry Research, 50*(10), 6276–6286.

Cambero, C., & Sowlati, T. (2014). Assessment and optimization of forest biomass supply chains from economic, social and environmental perspectives – A review of literature. *Renewable and Sustainable Energy Reviews, 36*, 62–73.

Cambero, C., Sowlati, T., Marinescu, M., & Röser, D. (2015). Strategic optimization of forest residues to bioenergy and biofuel supply chain. *International Journal of Energy Research, 39*(4), 439–452.

Caputo, A. C., Palumbo, M., Pelagagge, P. M., & Scacchia, F. (2005). Economics of biomass energy utilization in combustion and gasification plants: Effects of logistic variables. *Biomass and Bioenergy, 28*(1), 35–51.

Carlsson, D., D'Amours, S., Martel, A., & Rönnqvist, M. (2009). Supply chain planning models in the pulp and paper industry. *INFOR: Information Systems and Operational Research, 47*(3), 167–183.

Chen, C.-W., & Fan, Y. (2012). Bioethanol supply chain system planning under supply and demand uncertainties. *Transportation Research Part E: Logistics and Transportation Review, 48*(1), 150–164.

Christopher, M. (2005). *Logistics and supply chain management, creating value-adding networks* (3rd ed.). Harlow: Financial Times Prentice Hall.

Cobuloglu, H. I., Büyüktahtakın, & Esra, İ. (2014). A mixed-integer optimization model for the economic and environmental analysis of biomass production. *Biomass and Bioenergy, 67*, 8–23.

Coca-Cola. (2016). PlantBottle frequently asked questions: The Coca-Cola company. http://www.coca-colacompany.com/stories/plantbottle-frequently-asked-questions/. Accessed 13 Jun 2017.

Correll, D., Suzuki, Y., & Martens, B. J. (2014). Logistical supply chain design for bioeconomy applications. *Biomass and Bioenergy, 66*, 60–69.

Corsano, G., Vecchietti, A. R., & Montagna, J. M. (2011). Optimal design for sustainable bioethanol supply chain considering detailed plant performance model. *Computers and Chemical Engineering, 35*(8), 1384–1398.

Dabbene, F., Gay, P., & Tortia, C. (2014). Traceability issues in food supply chain management: A review. *Biosystems Engineering, 120*, 65–80.

Dal-Mas, M., Giarola, S., Zamboni, A., & Bezzo, F. (2011). Strategic design and investment capacity planning of the ethanol supply chain under price uncertainty. *Biomass and Bioenergy, 35*(5), 2059–2071.

De Mol, R. M., Jogems, M. A. H., Beek, P. V., & Gigler, J. K. (1997). Simulation and optimization of the logistics of biomass fuel collection. *Netherlands Journal of Agricultural Science, 45*, 219–228.

Diepenbrock, W. (2014). *Nachwachsende rohstoffe* (1st ed., Vol. 4189). Stuttgart: Eugen Ulmer KG.

Dunnett, A. J., Adjiman, C. S., & Shah, N. (2008). A spatially explicit whole-system model of the lignocellulosic bioethanol supply chain: An assessment of decentralised processing potential. *Biotechnology for Biofuels, 1*(1), 13–30.

Ekşioğlu, S., Li, S., Zhang, S., Sokhansanj, S., & Petrolia, D. (2010). Analyzing impact of intermodal facilities on design and management of biofuel supply chain. *Transportation Research Record: Journal of the Transportation Research Board, 2191*, 144–151.

Ekşioğlu, S. D., Acharya, A., Leightley, L. E., & Arora, S. (2009). Analyzing the design and management of biomass-to-biorefinery supply chain. *Computers and Industrial Engineering, 57*(4), 1342–1352.

Feng, Y., D'Amours, S., LeBel, L., & Nourelfath, M. (2010). *Integrated bio-refinery and forest products supply chain network design using mathematical programming approach*. Montreal: CIRRELT.

Fleischmann, B., & Meyr, H. (2003). Planning hierarchy, modeling and advanced planning systems. In A. De Kok & S. C. Graves (Eds.), *Supply Chain management: Design, coordination, operation*. Handbooks in Operations Research and Management Science (Chap. 9, Vol. 11, pp. 457–523). Amsterdam: Elsevier.

Fleischmann, B., Meyr, H., & Wagner, M. (2015). Advanced planning. In H. Stadtler, C. Kilger, & H. Meyr (Eds.), *Supply Chain management and advanced planning* (Chap. 4, 5th ed., pp. 71–95). Berlin: Springer.

Frombo, F., Minciardi, R., Robba, M., Rosso, F., & Sacile, R. (2009a). Planning woody biomass logistics for energy production: A strategic decision model. *Biomass and Bioenergy, 33*(3), 372–383.

Frombo, F., Minciardi, R., Robba, M., & Sacile, R. (2009b). A decision support system for planning biomass-based energy production. *Energy, 34*(3), 362–369.

Giarola, S., Bezzo, F., & Shah, N. (2013). A risk management approach to the economic and environmental strategic design of ethanol supply chains. *Biomass and Bioenergy, 58*, 31–51.

Giarola, S., Zamboni, A., & Bezzo, F. (2011). Spatially explicit multi-objective optimisation for design and planning of hybrid first and second generation biorefineries. *Computers and Chemical Engineering, 35*(9), 1782–1797.

Giarola, S., Zamboni, A., & Bezzo, F. (2012). Environmentally conscious capacity planning and technology selection for bioethanol supply chains. *Renewable Energy, 43*, 61–72.

Gunn, E. (2009). Some perspectives on strategic forest management models and the forest products supply chain. *INFOR: Information Systems and Operational Research, 47*(3), 261–272.

Gunnarsson, H., Rönnqvist, M., & Carlsson, D. (2005). A combined terminal location and ship routing problem. *Journal of the Operational Research Society, 57*(8), 928–938.

Huang, Y., Chen, C. W., & Fan, Y. (2010). Multistage optimization of the supply chains of biofuels. *Transportation Research Part E: Logistics and Transportation Review, 46*(6), 820–830.

Ivanov, B., & Stoyanov, S. (2016). A mathematical model formulation for the design of an integrated biodiesel-petroleum diesel blends system. *Energy, 99*, 221–236.

Judd, J. D., Sarin, S. C., & Cundiff, J. S. (2012). Design, modeling, and analysis of a feedstock logistics system. *Bioresource Technology, 103*(1), 209–218.

Kaltschmitt, M. (2009). Biomasse als nachwachsender Energieträger. In M. Kaltschmitt, H. Hartmann, & H. Hofbauer (Eds.), *Energie aus biomasse: Grundlagen, techniken und verfahren* (Chap. 1.1, 2nd ed., pp. 1–7). Berlin/Heidelberg: Springer.

Kanzian, C. (2009). Regional energy wood logistics – Optimizing local fuel supply. *Silva Fennica, 43*(1), 113–128.

Kelley, J., Kuby, M., & Sierra, R. (2013). Transportation network optimization for the movement of indigenous goods in Amazonian Ecuador. *Journal of Transport Geography, 28*, 89–100.

Kim, J., Realff, M. J., & Lee, J. H. (2011a). Optimal design and global sensitivity analysis of biomass supply chain networks for biofuels under uncertainty. *Computers and Chemical Engineering, 35*(9), 1738–1751.

Kim, J., Realff, M. J., Lee, J. H., Whittaker, C., & Furtner, L. (2011b). Design of biomass processing network for biofuel production using an MILP model. *Biomass and Bioenergy, 35*(2), 853–871.

Kirschstein, T. (2015). *Integrated planning of chemical supply chains*. Wiesbaden: Springer Fachmedien.

Lam, H. L., Ng, W. P., Ng, R. T., Ng, E. H., Aziz, M.K.A., & Ng, D. K. (2013). Green strategy for sustainable waste-to-energy supply chain. *Energy, 57*, 4–16.

Leão, R., Hamacher, S., & Oliveira, F. (2011). Optimization of biodiesel supply chains based on small farmers: A case study in Brazil. *Bioresource Technology, 102*(19), 8958–8963.

Leduc, S., Schwab, D., Dotzauer, E., Schmid, E., & Obersteiner, M. (2008). Optimal location of wood gasification plants for methanol production with heat recovery. *International Journal of Energy Research, 32*(12), 1080–1091.

Leduc, S., Lundgren, J., Franklin, O., & Dotzauer, E. (2010). Location of a biomass based methanol production plant: A dynamic problem in northern Sweden. *Applied Energy, 87*(1), 68–75.

Lin, T., Rodríguez, L. F., Shastri, Y. N., Hansen, A. C., & Ting, K. C. (2014). Integrated strategic and tactical biomass-biofuel supply chain optimization. *Bioresource Technology, 156*, 256–266.

Marufuzzaman, M., Eksioglu, S. D., & Huang, Y. (2014). Two-stage stochastic programming supply chain model for biodiesel production via wastewater treatment. *Computers and Operations Research, 49*, 1–17.

Marvin, W., Schmidt, L. D., Benjaafar, S., Tiffany, D. G., & Daoutidis, P. (2012). Economic optimization of a lignocellulosic biomass-to-ethanol supply chain. *Chemical Engineering Science, 67*(1), 68–79.

Marvin, W., Schmidt, L. D., & Daoutidis, P. (2013). Biorefinery location and technology selection through supply chain optimization. *Industrial and Engineering Chemistry Research, 52*(9), 3192–3208.
Mele, F. D., Kostin, A. M., Guillén-Gosálbez, G., & Jiménez, L. (2011). Multiobjective model for more sustainable fuel supply chains. A case study of the sugar cane industry in Argentina. *Industrial and Engineering Chemistry Research, 50*(9), 4939–4958.
Meyer, A. D., Cattrysse, D., & van Orshoven, J. (2015). A generic mathematical model to optimise strategic and tactical decisions in biomass-based supply chains (OPTIMASS). *European Journal of Operational Research, 245*(1), 247–264.
Meyer, A. D., Cattrysse, D., & van Orshoven, J. (2016). Considering biomass growth and regeneration in the optimisation of biomass supply chains. *Renewable Energy, 87*, 990–1002.
Mohseni, S., & Pishvaee, M. S. (2016). A robust programming approach towards design and optimization of microalgae-based biofuel supply chain. *Computers and Industrial Engineering, 100*, 58–71.
Mohseni, S., Pishvaee, M. S., & Sahebi, H. (2016). Robust design and planning of microalgae biomass-to-biodiesel supply chain: A case study in Iran. *Energy, 111*, 736–755.
Osmani, A., & Zhang, J. (2013). Stochastic optimization of a multi-feedstock lignocellulosic-based bioethanol supply chain under multiple uncertainties. *Energy, 59*, 157–172.
Paulo, H., Azcue, X., Barbosa-Póvoa, A. P., & Relvas, S. (2015). Supply chain optimization of residual forestry biomass for bioenergy production: The case study of Portugal. *Biomass and Bioenergy, 83*, 245–256.
Philpott, A., & Everett, G. (2001). Supply chain optimisation in the paper industry. *Annals of Operations Research, 108*(1/4), 225–237.
Rauch, P., & Gronalt, M. (2011). The effects of rising energy costs and transportation mode mix on forest fuel procurement costs. *Biomass and Bioenergy, 35*(1), 690–699.
Reche López, P., Jurado, F., Ruiz Reyes, N., García Galán, S., & Gómez, M. (2008). Particle swarm optimization for biomass-fuelled systems with technical constraints. *Engineering Applications of Artificial Intelligence, 21*(8), 1389–1396.
Rentizelas, A. A., & Tatsiopoulos, I. P. (2010). Locating a bioenergy facility using a hybrid optimization method. *International Journal of Production Economics, 123*(1), 196–209.
Rentizelas, A. A., Tatsiopoulos, I. P., & Tolis, A. (2009). An optimization model for multi-biomass tri-generation energy supply. *Biomass and Bioenergy, 33*(2), 223–233.
Roitsch, M., & Meyr, H. (2015). Oil industry. In H. Stadtler, C. Kilger, & H. Meyr (Eds.), *Supply chain management and advanced planning* (Chap. 24, 5th ed., pp. 443–458). Berlin: Springer.
Roni, M., Eksioglu, S. D., Searcy, E., & Jha, K. (2014). A supply chain network design model for biomass co-firing in coal-fired power plants. *Transportation Research Part E: Logistics and Transportation Review, 61*, 115–134.
Santibañez-Aguilar, J. E., González-Campos, J. B., Ponce-Ortega, J. M., Serna-González, M., & El-Halwagi, M. M. (2011). Optimal planning of a biomass conversion system considering economic and environmental aspects. *Industrial and Engineering Chemistry Research, 50*(14), 8558–8570.
Schwaderer, F. (2012). *Integrierte Standort-, Kapazitäts- und Technologieplanung von Wertschöpfungsnetzwerken zur stofflichen und energetischen Biomassenutzung*. Dissertation, Karlsruher Institut für Technologie, Karlsruhe.
Soysal, M., Bloemhof-Ruwaard, J. M., Meuwissen, M., & van der Vorst, J. (2012). A review on quantitative models for sustainable food logistics management. *International Journal of Foord System Dynamics, 3*(2), 136–155.
Stadtler, H. (2015). Supply chain management - An overview. In H. Stadtler, C. Kilger, H. Meyr (Eds.), *Supply chain management and advanced planning* (Chap. 1, 5th ed., pp. 3–28). Berlin: Springer.
Thrän, D., Scholwin, F., & Körner, I. (2009). Sonstige biomasse. In M. Kaltschmitt, H. Hartmann, & H. Hofbauer (Eds.), *Energie aus biomasse: Grundlagen, techniken und verfahren* (Chap. 4.3, 2nd ed., pp. 157–171). Berlin/Heidelberg: Springer.

Tittmann, P. W., Parker, N. C., Hart, Q. J., & Jenkins, B. M. (2010). A spatially explicit techno-economic model of bioenergy and biofuels production in California. *Journal of Transport Geography, 18*(6), 715–728.

Troncoso, J. J., & Garrido, R. A. (2005). Forestry production and logistics planning: An analysis using mixed-integer programming. *Forest Policy and Economics, 7*(4), 625–633.

Tsolakis, N. K., Keramydas, C. A., Toka, A. K., Aidonis, D. A., & Iakovou, E. T. (2014). Agrifood supply chain management: A comprehensive hierarchical decision-making framework and a critical taxonomy. *Biosystems Engineering, 120*, 47–64.

Türk, O. (2014). *Stoffliche Nutzung nachwachsender Rohstoffe: Grundlagen - Werkstoffe - Anwendungen*. Wiesbaden: Springer Vieweg

Walther, G., Schatka, A., & Spengler, T. S. (2012). Design of regional production networks for second generation synthetic bio-fuel – A case study in Northern Germany. *European Journal of Operational Research, 218*(1), 280–292.

Wang, S., Hastings, A., & Smith, P. (2012). An optimization model for energy crop supply. *GCB Bioenergy, 4*(1), 88–95.

World Bank. (2015). World development indicators: Growth of output. http://wdi.worldbank.org/table/4.1. Accessed 13 Jun 2017.

You, F., Tao, L., Graziano, D. J., & Snyder, S. W. (2012). Optimal design of sustainable cellulosic biofuel supply chains: Multiobjective optimization coupled with life cycle assessment and input-output analysis. *AIChE Journal, 58*(4), 1157–1180.

You, F., & Wang, B. (2011). Life cycle optimization of biomass-to-liquid supply chains with distributed–centralized processing networks. *Industrial and Engineering Chemistry Research, 50*(17), 10102–10127.

Zamboni, A., Shah, N., & Bezzo, F. (2009). Spatially explicit static model for the strategic design of future bioethanol production systems. 1. Cost minimization. *Energy and Fuels, 23*(10), 5121–5133.

Zhang, L., & Hu, G. (2013). Supply chain design and operational planning models for biomass to drop-in fuel production. *Biomass and Bioenergy, 58*, 238–250.

Zhang, W., & Wilhelm, W. E. (2011). OR/MS decision support models for the specialty crops industry: A literature review. *Annals of Operations Research, 190*(1), 131–148.

Structuring the Planning Tasks in Biomass-Based Supply Chains

Hendrik Butemann and Katja Schimmelpfeng

Abstract Due to the dependency on crude oil, mitigating, greenhouse gas emissions, energy and food security, support for rural economic development and the effort for environmental sustainability, renewable energy sources become more and more important. This led to an increasing research on biomass-based supply chains in recent years. Conventional supply chains, i.e., the commodity flow including all the stakeholders from the supplier to the end customer, have been studied intensively in the past. Biomass-based supply chains, however, feature different characteristics and uncertainties that have to be considered. In this paper, we identify the differences between the two types of supply chains and elaborate the stakeholders involved in the supply chain process and the different planning tasks structured according to the functional areas. As several possible pathways from feedstocks to different end-products exist, we focus on bio-fuels as the final product. We conclude by reviewing the literature that deals with supply chain optimization using operations research (OR) models to present the relevant planning tasks in biomass-based supply chains.

1 Introduction

In recent years, the research on biomass-based supply chains has risen substantially (Baños et al., 2011). Due to the dependency on crude oil, mitigating, greenhouse gas emissions, energy and food security, support for rural economic development and the effort for environmental sustainability, renewable energy sources become more and more important (Scott-Kerr et al., 2009). Conventional supply chains, i.e., the commodity flow including all the stakeholders from the supplier to the end customer, have been studied intensively in the past. Biomass-based supply chains, however, feature different characteristics and uncertainties that have to be considered. In this paper, we identify the differences between the two types of supply chains and present the relevant planning tasks in biomass-based supply

H. Butemann (✉) • K. Schimmelpfeng
Chair of Procurement and Production (580B), University of Hohenheim, 70593 Stuttgart, Germany
e-mail: Hendrik.Butemann@uni-hohenheim.de; Katja.Schimmelpfeng@uni-hohenheim.de

S. Dabbert et al. (eds.), *Knowledge-Driven Developments in the Bioeconomy*, Economic Complexity and Evolution, DOI 10.1007/978-3-319-58374-7_15

chains. As several possible pathways from feedstocks to different end-products exist, we focus on bio-fuels as the final product. We review the literature that deals with supply chain optimization using operations research (OR) models to determine the different planning tasks.

The remainder of this paper is structured as follows: in Sect. 2 we introduce the fundamentals about biomass, biofuels and supply chain management. In Sect. 3 we elaborate the specific properties of biomass-based supply chains. Section 4 illustrates the stakeholders involved in the supply chain process and the different planning tasks structured according to the functional areas. Finally, in Sect. 5 we conclude the paper's most important findings and give some directions for future work.

2 Fundamentals, Terms and Definitions

Before analyzing the planning tasks in biomass-based supply chains, further information on terms and definitions have to be provided. In this section, we introduce different types of biomass, conversion technologies and potential products. Furthermore, we address the fundamentals of conventional supply chain management.

2.1 *Biomass and Biofuels*

Biomass is basically anything organic that is or has once been a living organism (Saidur et al., 2011). It can be classified by its nature, namely into terrestrial and aquatic biomass. *Terrestrial biomass* contains feedstock rich in sugar or lipids including corn grain, sugarcane, oil seed, soy been etc. and cellulosic biomass like agricultural residues (e.g., corn stover), forest residues (e.g., leftovers from logging operations) and energy crops (e.g., switchgrass) (Yue et al., 2014). Also the organic fraction of industrial waste, municipal waste, human and animal excrements belong to this kind of biomass (Allen et al., 1998). Photosynthetic algae and cyanobacteria like microalgae and seeweeds represent *aquatic biomass* (DOE/EERE, 2010b).

While **growing**, biomass gains carbon by carbon dioxide fixation. Usually, it can easily be cultivated, collected and utilized (Yue et al., 2014). In addition, it is available worldwide, can be converted efficiently and allows a CO_2-neutral consumption (OECD/FAO, 2007).

After the **harvest**, in many cases biomass requires some kind of **pre-treatment**. Subject to the type of feedstock this may be a mechanical or chemical process like ensiling, drying, pelletization, torrefaction and pyrolisis (Gold and Seuring, 2011). These processes help to simplify the succeeding operations (e.g., handling, storage and transportation) and to reduce the corresponding cost (Kumar and Sokhansanj, 2007). They also lead to a lower moisture content, remove contaminants and thus ensure a better feedstock quality, stability and conversion performance (Bals et al., 2010).

After the biomass underwent the relevant pre-treatment processes and transportation, usually, the **energy conversion** is the next step. In biorefineries the biomass is converted into heat, power, biofuels or a combination of the mentioned outputs (Papapostolou et al., 2011). With starch-based, sugar-based, oil-based and lignocellulosic-based biorefineries four different types exist (Sadaka, 2009). Subject to the raw material and the desired final product three types of processes can be applied to obtain biofuel. Solid, liquid or gaseous fuel is made by thermo-chemical conversion (e.g., gasification, pyrolysis and charcoal production), whereas bio-chemical conversion can produce alcohol and biogas. Pressing and transesterification are examples for physicochemical processes and provide liquid fuels like biodiesel from energy crops (Iakovou et al., 2010). The method's selection also depends on several criteria like the quantity of biomass, environmental standards and financial resources (Saidur et al., 2011). In Fig. 1 different possible pathways from biomass to bioenergy are illustrated.

The biofuels bioethanol and biodiesel can be subdivided into different generations. The *first generation of bioethanol* is based on starch and sugar-based feedstock like corn, sugarcane and maize which compete with the food supply. Urban woody waste and land fills, forest biomass, herbaceous energy crops and short rotation woody crops lead to the *second generation bioethanol*. While palm, soybeans and rapeseeds are raw materials for the *first generation biodiesel*, the *second generation* is produced by jastropa and cassava. Finally, the *third generation of biodiesel* is made of microalgae, seaweeds and cynobacteria (EUBIA, 2007).

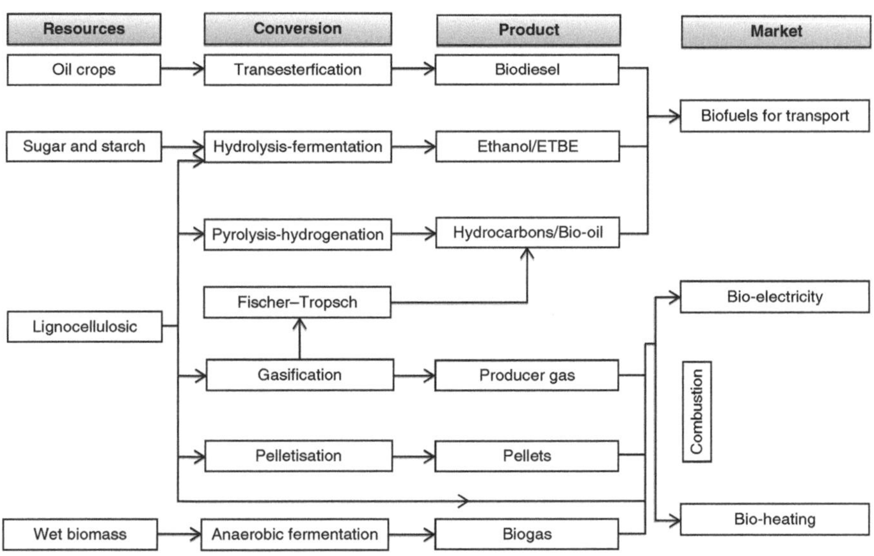

Fig. 1 Possible pathways from biomass to bioenergy (Sharma et al., 2013)

2.2 Supply Chain Management

Supply chain management is defined as "the systemic, strategic coordination of the traditional business functions and the tactics across these business functions within a particular company and across businesses within the supply chain, for the purposes of improving the long-term performance of the individual companies and the supply chain as a whole" (Mentzer and DeWitt, 2001). The business functions in this case are procurement, production, distribution and sales. Products, services, finances, and information are modeled in an upstream and a downstream flow through these business functions from a supplier to a customer (Mentzer and DeWitt, 2001) with the objective to provide the customer with the end-product of the desired quality and service level at the right time at minimal cost (Simchi-Levi et al., 2008). Usually, the suppliers, manufacturers, distribution centers and customers build the supply chain (Beamon, 1998). The performance measured by the degree of coordination and integration is highly influenced by uncertainty (Peidro et al., 2009).

Figure 2 shows the Supply Chain Planning Matrix (SCP Matrix) by Stadtler et al. (2015). It provides an overview of modules for the different planning tasks one company in a supply chain has to handle. The SCP Matrix uses the dimensions "planning horizon" and "supply chain process". The latter represents the business functions whereas the "planning horizon" represents the partition of strategic (long-term), tactical (medium-term) and operational (short-term) decisions (Mula et al., 2010).

Long-term decisions usually hold for up to 5 years and deal with, e.g., investments. Aggregated data is taken as a basis since no precise data is available. Tactical decisions are valid for some months concerning, e.g., aggregated production quantities. Decomposed tactical decisions result in operational decisions such as exact production plans for just a few days (Ba et al., 2016).

The module *"Strategic Network Design"* incorporates all the functional areas procurement, production, distribution, and sales. It contains decisions regarding storage and production capacities, new locations, and the strategic selection of suppliers and customers. *"Demand Planning"* belongs to the sales segment and

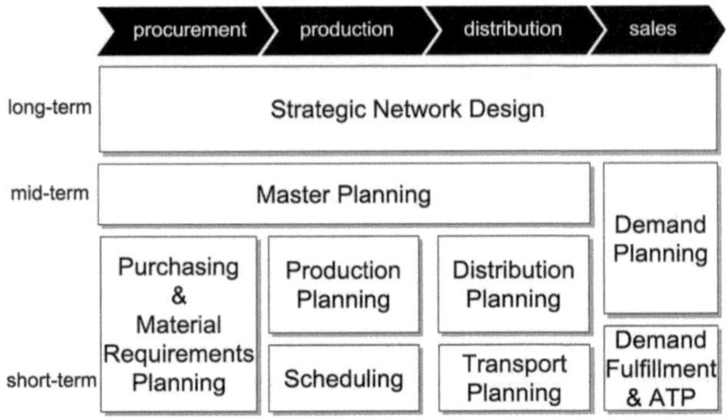

Fig. 2 The supply chain planning matrix (Stadtler et al., 2015)

is responsible for the demand forecasts applying various statistical methods. The module *"Available to Promise"* (ATP) applies this information to accept customer orders in the short-term. *"Master Planning"* uses these forecasts to synchronize the flows of goods, money and information. Results of the module *"Master Planning"* are procurement, production, transport and storage quantities as well as labor utilization on the tactical level. Detailed production schedules for the different plants are generated within *"Production Planning & Scheduling"*. Besides the order releases, also the processing sequence is determined. *"Distribution & Transport Planning"* ensures the cost-minimal supply of the customers by means of vehicle routing and vehicle loading. Finally, the module *"Material Requirements Planning"* generates the explicit purchase order of materials and provides information about the availability of materials for *"Production Planning & Scheduling"* (Rohde et al., 2000).

Please note, that the planning tasks of these modules are valid for just one company in conventional supply chains. Therefore, the SCP Matrix provides an overview, how planning tasks can be arranged. For biomass-based supply chains we cannot simply adapt the matrix because of the special features shown in Sect. 3. Especially the challenges and risks associated with handling and processing biomass have to be identified and analyzed.

3 Characteristics of Biomass-Based Supply Chains

3.1 Challenges

To model biomass-based supply chains we have to address certain challenges. Not only technological and financial issues, but also social and environmental aspects have to be considered. Besides the already mentioned factors, governmental decisions resp. institutional and organizational topics directly and indirectly affect the decisions to be made. Therefore, they play a crucial role in structuring the planning tasks.

First, we examine the **technological challenges**. Evolving an efficient feedstock supply system including harvesting/collection, preprocessing, storage and transport stages is important to continuously provide high-quality biomass for the conversion at low cost (Yue et al., 2014). Since in most cases biomass can only be harvested during a short period of time, but the conversion plant needs a stable supply, logistics play an important role. This is stressed by the fact that biomass is comparatively cheap, so logistic cost preponderate (Ba et al., 2016). Harvesting or collection at the supply chain's beginning has to be performed in a sustainable and efficient way (McKendry, 2002a). A lack in biomass supply due to shortsighted decisions in the past that did not consider the biomass' seasonality affects the whole supply chain (Hoogwijk et al., 2003).

Against this background, supply-chain-efficiency and inventory planning are substantial elements of biomass-based supply chain planning. Choosing an appropriate harvesting and collection schedule and adopting the proper supply-chain-configuration are the first steps to assure efficiency (Yılmaz and Selim, 2013). If different types of biomass are considered for the conversion, a complex decision concerning storage arises with a high degree of uncertainty about the quality and quantity of biomass feedstock (Gold and Seuring, 2011). Multiple types of biomass tend to reduce total cost, but require a sophisticated way of the conversion due to the different physical and chemical characteristics (Rentizelas et al., 2009). Concerns about the conversion facilities are ineffective techniques, insufficient maintenance of the equipment (Saidur et al., 2011) and inconsistencies in quality and moisture content of the feedstock (Kudakasseril Kurian et al., 2013). Altogether, the technical and technological limitations cause a complex problem concerning locations, technologies to apply, capacities and routes with multiple conflicting objectives (Čuček et al., 2012).

The analysis of **financial issues** is important as biomass energy production is capital intensive (Rentizelas et al., 2009). Thus, the operational cost mainly consisting of harvesting, collection, storage and transportation cost need to be reduced (Diamantopoulou et al., 2011). Since the scale of delivery, the conversion plants' size and location, their availability and the type of biomass have a significant impact, the life cycle cost have to be taken into account (Mafakheri and Nasiri, 2014). Spreading the risk by diversifying the revenue sources, e.g., electricity, heating and cooling, increases the flexibility of the biomass-based supply chain but also enlarges the planning efforts (Rentizelas et al., 2009).

Biomass-based supply chains effect more **social aspects** than the conventional counterparts. The most discussed concern is the competition with food supply. Biomass planted for energy conversion cuts the available amount of land for agriculture which is intensified by government decisions in favor of bioenergy in recent times (Ignaciuk et al., 2006). Especially developing countries are affected, since cultivated energy crops are popular commodities to export to the industrialized world (Kerckow, 2007). Furthermore, an inadequately designed biomass-based supply chain may increase the prices of other goods and materials due to the interdependencies with other supply chains. Particularly the transport sector could suffer from a higher traffic density in rural areas, an increasing number of accidents, problems with staff retention in the transport sector, road damages, spoilage of the landscape and even higher rates of migration from rural areas (Pimentel et al., 2009). Traffic congestions by excessive truck transports can also lead to resistance of affected communities (Kumar et al., 2006), but pipeline transportation can not serve as an alternative as the carrier fluid reduces the heating value of biomass significantly (Searcy et al., 2007). The lack of participation in decision making processes concerning bioenergy projects and insufficient information about the land use are additional causes for conflicts with communities (Upreti, 2004). Furthermore, a study has shown that injuries and diseases occur more often in connection with biomass-based supply chains than with conventional ones (Saidur et al., 2011). A lot of social issues can be prevented by thoroughly selecting the used land, the used type of biomass and the production quantities (Koh and Ghazoul, 2008). Beneath the risks, there are social benefits of bioenergy like regional development, higher net labor incomes and self-sufficiency in energy production (Krajnc and Domac, 2007). Also new job opportunities are established along the whole bioenergy supply chain (Thornley et al., 2008).

One fundamental idea in developing and applying biofuels is the considerate land exploitation. Therefore, **environmental challenges** play an important role dealing with biomass-based supply chains. In this context, we have to take ecological services and land-use, emissions, resource efficiency and waste management into account (Baños et al., 2011). All activities related to biomass-based supply chains, i.e., harvesting, storage, transportation and production, must be assessed with respect to carbon emissions to ensure environmental benefits by switching to biofuels (Charles et al., 2007). E.g., the import of biomass from overseas usually causes high carbon emissions thus influencing the ecological performance in a negative way (Grant and Clark, 2010). But long-distance shipping releases substantially lower CO_2 per unit of length than truck transport (Perry and Rosillo-Calle, 2008). A study showed that the transport of biomass from Scandinavia to conversion plants in the Netherlands still can be beneficial (Forsberg, 2000). Hence, the life cycle impacts of the supply chain have to be taken into consideration (Mafakheri and Nasiri, 2014). Further environmental challenges are the loss of natural habitats and wildlife, the loss of biodiversity and soil degradation (Awudu and Zhang, 2012).

Decisions of governments and other institutions highly impact the energy sector. A lot of policy measures and regulations aim to support the bioenergy production (Mafakheri and Nasiri, 2014). E.g., in 2003 the European Union fixed the biofuels' share of the total transportation fuel to 10% (European Union, 2003). Opposite to the desired effect, this led to a massive import of soybeans and sugar cane. Thus, less sustainable and less economical biomass feedstocks may be promoted by misled incentives (Charles et al., 2007). Therefore, a competition in the bioenergy sector and sustainable supply-chain-solutions should be supported (Roos et al., 1999).

Institutional and organizational issues arise due to the different stakeholders involved in biomass-based supply chains. The parties' institutional values, interests and rules differ thus leading to diverse ownership arrangements, standards and organizational norms (Costello and Finnell, 1998). Also supply priorities, attitudes towards risks and decision making processes may vary and have to be incorporated in the planning of the supply chain (Altman and Johnson, 2008). Change management practices between the stakeholders are necessary because of the uncertainties associated with the biomass-supply. They represent an important factor to achive operational agility (Christopher and Towill, 2001).

3.2 Uncertainties

Uncertainties are located on every stage of a biomass-based supply chain. They are related to all activities and occur in different types resp. varying forms (Awudu and Zhang, 2012). Conventional supply chains mostly cope with unknown demand whereas in biomass-based supply chains the supply-uncertainty and biomass availability are crucial (Sharma et al., 2013).

The following sources of variability have to be considered as well (Iakovou et al., 2010; Cundiff et al., 1997; Gold and Seuring, 2011):

- weather uncertainty,
- seasonality,
- physical and chemical characteristics,
- geographical distribution,
- low bulk density of biomass feedstocks,
- structure of biomass suppliers and their willingness to grow biomass crops,
- local transportation and distribution infrastructure,
- supplier contracts, and
- government policies.

Also biomass cost, technologies, expansion plans, biofuel price and natural or human disasters can confront the biomass-based supply chain planning with uncertainties (Sharma et al., 2013). Approaches to deal with the different kinds of variability in operations management include amongst others scenario analyses, stochastic programming, robust optimization, fuzzy programming, chance-constraint optimization and stochastic inventory theory (Sahinidis, 2004).

3.3 *Supply Chain Structure*

In the literature, different components of the "typical" biomass-based supply chain are identified. Usually, harvesting and collection, pre-treatment, storage, transport and energy conversion are included in all definitions (e.g., Mafakheri and Nasiri 2014). Although they seem to be similar to those in a petroleum-based supply chain, we have to perform the analysis in a different way. This is due to the differences in the feedstocks and the products leading to other strategies for production, transportation and storage (An et al., 2011). The "distribution of the end-products to the customers" is a stage that some authors ignore. Altogether, three main segments can be distinguished:

- the upstream segment,
- the midstream segment and
- the downstream segment.

The upstream segment contains all operations from harvesting and collecting to the conversion facility (An et al., 2011). The midstream segment represents the energy conversion itself and is often referred to as a black box where biomass is put in and bioenergy comes out (Iakovou et al., 2010). Each decision made in this segment, e.g., concerning the type of the conversion technology or the size and location of the facility, influences the upstream processes (Allen et al., 1998). The storage of bioenergy and the distribution to the customers serve as downstream activities

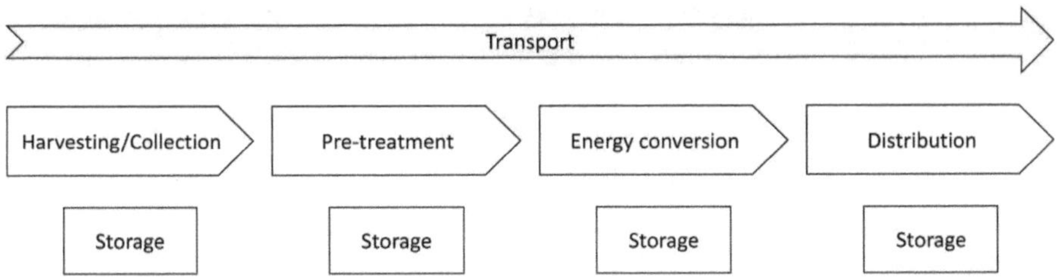

Fig. 3 Biomass-based supply chain structure

(An et al., 2011). Making the supply chain robust and flexible against changes of weather, biomass perishability, competition and market conditions is the objective of a proper supply chain structure (Iakovou et al., 2010). Figure 3 displays a typical biomass-based supply chain including the following considerations:

- Transports are necessary between all different stages.
- Storage/inventory is possible at the harvesting and collection site, the pre-treatment facility, the energy conversion plant and the customer's location.

4 Planning Tasks in Biomass-Based Supply Chains

Unlike conventional supply chains, biomass-based supply chains may contain entities resp. stakeholders responsible for several functional duties along the commodity flow. Therefore, we identify the various potential stakeholders analyzing the existing literature concerning biomass-based supply chains. Subsequently, we determine and classify the detected planning tasks.

4.1 Stakeholders

As already pointed out in Sect. 3, the functional parts of a biofuel supply chain can be classified into harvesting/collection, pre-treatment, energy conversion and distribution (e.g., Mafakheri and Nasiri 2014). In constrast to conventional supply chains, the focus is on the upstream activities like harvesting, collecting, pre-treatment, transport and storage. This strongly influences the modeling of those supply chains that rarely include distribution and other activities, e.g., demand planning after energy conversion. In general terms, the integrated parties in biomass-based supply chains are (Adams et al., 2011)

- the supplier of biomass,
- transportation and distribution entities,

Harvesting/ Collection	Pre-treatment	Energy conversion	Distribution	Further
Farmer/producer	Facility operator	Conversion operator	Distributor	Network designer
Independent supplier	Farmer/producer	Farmer/producer	Retail outlets	Policy makers
Landowner/counties	Conversion operator	Distributor	End-consumer	Investors
Import businesses	Owner of machinery	Transport companies	Transport companies	
Owner of machinery	Transport companies	Storage operator	Storage operator	
Transport companies	Storage operator			
Storage operator				

Fig. 4 Stakeholders in biomass-based supply chains

- energy production facility developers and operators,
- the government and utility firms who provide the incentives, and
- the end-users.

Figure 4 provides an overview of all the stakeholders involved in the different functional parts of a biomass-based supply chain in the considered literature. In each stage resp. functional area we position the stakeholders in the order of their importance. Some of the stakeholders (stakeholders' classes) occur more than once. We try to signal this fact using the same background color in Fig. 4. Please note, that e.g., the farmer during the harvest/collection does not necessarily have to be the same as during the pre-treatment phase or the energy conversion.

At the biomass acquisition stage, the respective biomass can be achieved in different ways depending on the type. Besides the typical supplying entities like farmers, producers and collectors, other stakeholders are involved in the harvesting and collection process. E.g., biomass can also be received in form of residues and waste accrued at mills or industrial companies (Freppaz et al., 2004). In Fig. 4 we address them as independent suppliers. Offshore farmers producing microalgae and macroalgae (Yue et al., 2014) represent a special type of farmers that should not be omitted. Some papers do not specify organizational units, but consider different counties or states (Kim et al., 2011). To ensure demand fulfillment in each period and avoid penalty cost, in some cases the import of biomass is permitted as an alternative (Dal-Mas et al., 2011). Another party that may influence harvesting or collection decisions is the owner of machinery since many farmers do not possess their own equipment but have to loan it from an independent company or from other farmers. Also transport companies may be important for the decisions to be made if the trucks should be loaded on-farm (Ba et al., 2016).

As pre-treatment can be operated not only at a special pre-treatment facility, but also at the sourcing site or the energy conversion plant, possible involved stakeholders are farmers, producers, collectors and the operator of the pre-treatment and energy conversion facilities. Some pre-treatment methods require special equipment provided by central companies, thus the availability of such machinery is also crucial for decision-making (Ba et al., 2016).

At the energy conversion stage, obviously the conversion operator handles the midstream segment. But also the farmers and distributors may influence the decisions, e.g., if the optimal conversion capacity has to be determined (Nasiri and Zaccour, 2009). Some authors examine the case that the conversion operator possesses his own transportation fleet and compare it with the situation in which independent transportation companies are employed (Zhu et al., 2011).

The parties in the downstream segment can be of different nature. Of cause the company responsible for distributing the end-product is the most important stakeholder. Inbetween those firms and the (private or industrial) end-consumer may be retail-outlets, e.g., gas stations. As already mentioned, many OR models in the literature consider just the upstream and midstream segment, i.e., so far research papers about the downstream structure are rare in the literature.

Sometimes storage operators accept the responsibility and need to be considered in the planning process. On a subordinate level, some other institutions may influence the biomass-based supply chains. To include all functional areas in the construction process of the whole supply chain, a network designer is needed (Zhu et al., 2011). Furthermore, for sustainability incentives and to protect the environment connected with biomass-based supply chains, third parties like governments and non-governmental institutions, i.e., policy makers, come into play. To enable the development of biofuels, investors need to provide capital, thus also having a voice during the planning process (Baños et al., 2011). During all stages of the supply chain, biomass, biofuels and interstage products can be stored in different modes (e.g., Yılmaz and Selim 2013).

4.2 *Structure of the Planning Tasks*

In the following, we present the different planning tasks we identified in the reviewed literature and classify them into the four determined functional areas (Fig. 3). Furthermore, we arrange them in the respective area according to the corresponding planning horizon. Please note, that in conventional supply chains the planning tasks are determined for just one stakeholder. E.g., an automotive manufacturer has to perform all activities from procurement to sales planning in order to maximize its profit. The different suppliers, manufacturers and customers can be distinguished clearly and easily. But along the biomass-based supply chain many different combinations of participating stakeholders are possible depending on the type of biomass, the way of pre-treatment and energy conversion and the type of the end-product. Also different stakeholders may execute different tasks like harvesting and pre-treatment or energy conversion and distribution. As we can not assign a stakeholder to just one area, we classify the planning tasks based on their relevant functional area/stage in the biomass-based supply chain.

4.2.1 Harvesting/Collection

Harvesting and collecting the biomass represents the first stage of biomass-based supply chains. The strategic decisions provide a basis for all the following determinations. First of all, the utilized **types of biomass** are selected and the **available lands resp. areas are identified and allocated** (Ba et al., 2016; Mafakheri and Nasiri, 2014). Meeting the demand regarding all uncertainties is the fundamental objective (Iakovou et al., 2010).

In the producer's perspective also the **selection of biomass suppliers** could be a possible decision (Yue et al., 2014). After having fixed the strategical decisions, further restrictions have to be taken into account, such as land availability and crop rotation. For example Murray (1999) introduced the "Unit Restriction Model" and the "Area Restriction Model", which both ensure that no adjacent blocks are harvested at the same time. The corresponding planning task is called **harvesting planning**. Obviously, this model can only be applied to certain types of biomass.

The counterpart of harvesting planning, the **collection planning**, can be used for wood residues in forests for example (Mafakheri and Nasiri, 2014). It has to be determined, at what time to harvest/collect which quantity of biomass in which area. Simultaneously, the **number and the type of equipment and personnel** for those operations need to be computed in the **personnel/equipment planning** (Sokhansanj et al., 2006). Both can be seen as a tactical decision for some months on an aggregated level and/or as an operational decision to fix exact harvesting/collection plans. Purely operational decisions are the **arrangement of working shifts**, the **allocation of operators to harvesting/collecting machines** as well as **vehicle routing** and **fleet scheduling** (Awudu and Zhang, 2012).

4.2.2 Pre-treatment

One major long-term decision in the pre-treatment section is the determination of the **type** of pre-treatment, which depends on the processed biomass. Different types reveal different advantages, i.e., reducing the moisture content, increasing the bulk density and the heating value or enabling a better handling for transportation (Hamelinck et al., 2005). Possible **locations** for the pre-treatment are the harvesting/collection sites, central collection points and the energy conversion facilities (Hamelinck et al., 2005). Carolan et al. (2007) analyzed a network of small decentralized pre-treatment units minimizing the storage and transportation cost. Also the **size** of the pre-processing plants can be optimized (Ba et al., 2016). To preprocess the biomass, a **production plan** has to be developed both on an aggregated level (tactical) and on a daily basis (operational level). Additionally, the associated **machinery** and **personnel** need to be assigned to the respective production plan on both levels.

4.2.3 Energy Conversion

The **location analysis** for energy conversion facilities has to incorporate uncertainties like the biomass supply, the energy demand and the transportation system (Sodhi and Tang, 2009). The **choice of the conversion technology** is restricted by the considered biomass and influenced by the way of pre-treatment or vice versa. Cameron et al. (2007) evaluated biomass gasification and direct combustion, which differ in their energy efficiency rates and the fixed capital. The **environmental impact** (McKendry, 2002b) and the **robustness** against energy prices (Mansoornejad et al., 2010) may also be considered for this strategic decision. Huang et al. (2010) built a model based on forecasts of the biomass sources and their availability to generate a schedule for the long-term decisions about the **capacity and the infrastructure** of energy conversion plants. Converting several types of biomass in one facility does not only influence the capacity, but also the **operational production plan**. While the optimal capacity can also be determined by a game theoretic approach evaluating the interactions of different stakeholders (Nasiri and Zaccour, 2009), the operational plan aims to meet energy demand in the most economical way (Rentizelas et al., 2009). Therefore, again a production schedule both on the tactical and operational level is necessary with the corresponding **machinery and personnel plans** to ensure the coordination of different material flows and the scheduling of multiple conversion tasks to provide the right amount of energy at the right time (Yue et al., 2014).

4.2.4 Distribution

As already mentioned, the **distribution planning** for biomass-based supply chains is not considerably explored in the literature. Many models just examine the upstream and midstream part of the supply chain, assuming a given distribution network and an a-priori known demand. The distribution sector has to ensure that the market and the consumers receive the product cost-effectively and sustainably (Yue et al., 2014). Hence, one strategic planning activity is the **assignment of customer-serving-areas to the different energy-conversion-facilities** (Yue et al., 2014). Problems occur if the existing pipeline systems are used for biomass-derived products because of different characteristics compared to conventional fuels (Bunting et al., 2011). So-called drop-in biofuels are compatible with the current infrastructure thus resulting in an easier and more widespread market acceptance (DOE/EERE, 2010b).

Dependent on the type of the final product, trains, barges, trucks and ships (DOE/EERE, 2010a) offer other opportunities for transportation. In the majority of cases **transportation and storage planning** are performed for the whole supply chain. Therefore, we explain them in the following subsection.

4.2.5 Integrated Planning Tasks

Since the different stages in the biomass-based supply chain can not be explicitly assigned to certain stakeholders, most of the time the optimization is executed for the whole network. Thus, we have to evaluate a complex system involving different market segments and actors (Elghali et al., 2007). The fundamental strategic decision affecting the whole supply chain is the **supply chain configuration**, responsible for the effective and efficient delivery of biomass (Awudu and Zhang, 2012). This includes the **scale of production** as one decisive parameter for bio-energy systems (Elghali et al., 2007). For many types of biomass several different possible pathways exist, enabling the network planner to consider a **cascading use of feedstocks** (United Nations Environment Programme, 2009). To settle the sometimes conflicting interests between stakeholders, **supply and demand contracts** are necessary with specific terms of delivery and fixed payments (Iakovou et al., 2010). Especially in biomass-based supply chains, **sustainability issues** play an important role in long-term decisions thus social, economic and environmental impacts of decisions need to be addressed (Awudu and Zhang, 2012).

On the tactical level integrated decisions are **inventory planning**, **fleet management** and **logistics management** (Iakovou et al., 2010). The way of storage is not only influenced by the type of feedstock, but also constrained by transportation options. Therefore, on-field storage can reduce the overall delivery-cost in many cases (Huisman et al., 1997). Also the desired safety stock level has to be determined (Ba et al., 2016). In general, a shorter harvest season leads to a higher number of storage facilities needed as a buffer capacity (Uslu et al., 2008). For transportation planning, the availability, the transportability and the energy demand impose further restrictions (Diekema et al., 2005). Hence, an important cost factor arises due to the low energy density of biomass (Mayfield et al., 2007). Not only the distance between two nodes is crucial, but also the speed set by road properties and infrastructure have to be incorporated in decision making (Möller and Nielsen, 2007). Additionally, the environmental consequences need to be examined like the emissions of CO_2 and other pollutants (Forsberg, 2000).

To put it in a nutshell, it has been shown, that an integrated planning of the aforementioned tasks could lead to a reduction of up to 20% in total cost of biomass-based supply chains (Dunnett et al., 2007).

5 Conclusion

Biomass-based supply chains incorporate fundamental differences compared to conventional ones. We illustrated those differences treating the challenges resulting of handling the biomass. Also different uncertainties appear due to the features of biomass like the moisture content and the seasonal harvesting.

We showed that it is not sufficient to simply adapt the Supply Chain Planning Matrix by Stadtler et al. (2015) to transfer it to biomass-based supply chains.

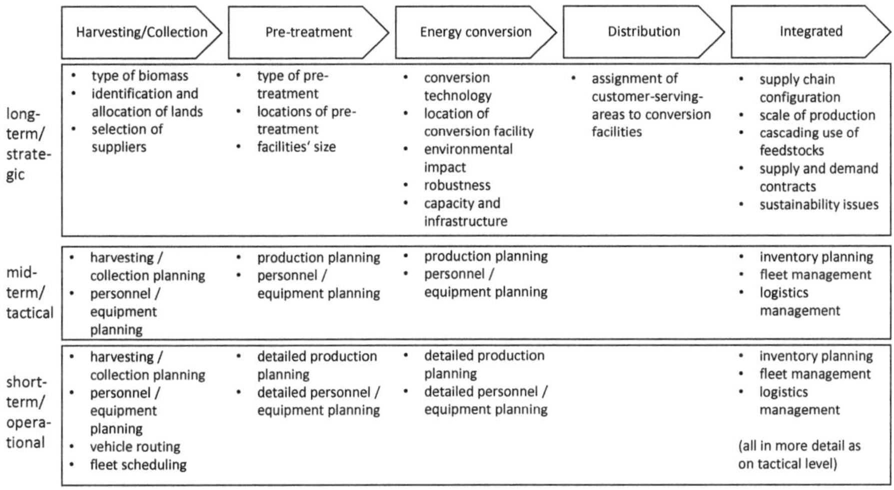

Fig. 5 Resulting assignment of planning tasks in the biomass-based supply chain

The different supply chain structure and stakeholders create a very complex network with several different possible configurations. Therefore, we classified the relevant planning tasks according to the supply chain's functional area and the planning horizon. However, no hierarchical order of the planning tasks is intended. Figure 5 summarizes our results.

The process of identifying the relevant planning tasks in the ever-expanding bio-energy sector revealed several topics for future work. As shown above, the stakeholders' assignment to the planning tasks is not yet clear. It depends on the type of the biomass considered and the supply chain configuration. Therefore, we will further investigate the stakeholders' integration in a similar construct as the Supply Chain Planning Matrix. Another very promising topic is working on incentives for a better cooperation of the stakeholders.

References

Adams, P., Hammond, G., McManus, M., & Mezzullo, W. (2011). Barriers to and drivers for UK bioenergy development. *Renewable and Sustainable Energy Reviews, 15*(2), 1217–1227.

Allen, J., Browne, M., Hunter, A., Boyd, J., & Palmer, H. (1998). Logistics management and costs of biomass fuel supply. *International Journal of Physical Distribution and Logistics Management, 28*(6), 463–477.

Altman, I., & Johnson, T. (2008). The choice of organizational form as a non-technical barrier to agro-bioenergy industry development. *Biomass and Bioenergy, 32*(1), 28–34.

An, H., Wilhelm, W. E., & Searcy, S. W. (2011). Biofuel and petroleum-based fuel supply chain research: A literature review. *Biomass and Bioenergy, 35*(9), 3763–3774.

Awudu, I., & Zhang, J. (2012). Uncertainties and sustainability concepts in biofuel supply chain management: A review. *Renewable and Sustainable Energy Reviews, 16*(2), 1359–1368.

Ba, B. H., Prins, C., & Prodhon, C. (2016). Models for optimization and performance evaluation of biomass supply chains: An operations research perspective. *Renewable Energy, 87*(2), 977–989.

Bals, B., Rogers, C., Jin, M., Balan, V., & Dale, B. (2010). Evaluation of ammonia fibre expansion (AFEX) pretreatment for enzymatic hydrolysis of switchgrass harvested in different seasons and locations. *Biotechnology for Biofuels, 3*, 1–11.

Baños, R., Manzano-Agugliaro, F., Montoya, F., Gil, C., Alcayde, A., & Gómez, J. (2011). Optimization methods applied to renewable and sustainable energy: A review. *Renewable and Sustainable Energy Reviews, 15*(4), 1753–1766.

Beamon, B. M. (1998). Supply chain design and analysis: Models and methods. *International Journal of Production Economics, 55*(3), 281–294.

Bunting, B. G., Bunce, M., Barone, T. L., & Storey, J. M. (2011). *Fungible and compatible biofuels: Literature search, summary, and recommendations* (No. ORNL/TM-2010/120). Oak Ridge National Laboratory (ORNL); Fuels, Engines and Emissions Research Center.

Cameron, J. B., Kumar, A., & Flynn, P. C. (2007). The impact of feedstock cost on technology selection and optimum size. *Biomass and Bioenergy, 31*(2), 137–144.

Carolan, J. E., Joshi, S. V., & Dale, B. E. (2007). Technical and financial feasibility analysis of distributed bioprocessing using regional biomass pre-processing center. *Journal of Agricultural and Food Chemistry, 5*(2), 1–29.

Charles, M. B., Ryan, R., Ryan, N., & Oloruntoba, R. (2007). Public policy and biofuels: The way forward? *Energy Policy, 35*(11), 5737–5746.

Christopher, M., & Towill, D. (2001). *An Integrated Model for the Design of Agile Supply Chains.* Bradford: MCB UP Ltd/Emerald Group Publishing Limited.

Costello, R., & Finnell, J. (1998). Institutional opportunities and constraints to biomass development. *Biomass and Bioenergy, 15*(3), 201–204.

Čuček, L., Klemeš, J., Varbanov, P., & Kravanja, Z. (2012). Reducing the dimensionality of criteria in multi-objective optimisation of biomass energy supply chains. *Chemical Engineering Transactions, 29*, 1231–1236.

Cundiff, J. S., Dias, N., & Sherali, H. D. (1997). A linear programming approach for designing a herbaceous biomass delivery system. *Bioresource Technology, 59*(1), 47–55.

Dal-Mas, M., Giarola, S., Zamboni, A., & Bezzo, F. (2011). Strategic design and investment capacity planning of the ethanol supply chain under price uncertainty. *Biomass and Bioenergy, 35*(5), 2059–2071.

Diamantopoulou, L., Karaoglanoglou, L., & Koukios, E. (2011). Biomass cost index: Mapping biomass-to-biohydrogen feedstock costs by a new approach. *Bioresource Technology, 102*(3), 2641–2650.

Diekema, W. H., De Mol, R. M., Annevelink, E., & Elbersen, H. W. (2005). Combining goals in the logistics bio-energy chains. In *14th European Biomass Conference, Paris, France* (pp. 495–498).

DOE/EERE. (2010a). *Biopower technical strategy workshop summary report.*

DOE/EERE. (2010b). *National algal biofuels technology roadmap.*

Dunnett, A., Adjiman, C., & Shah, N. (2007). Biomass to heat supply chains applications of process optimization. *Process Safety and Environmental Protection, 85*(5), 419–429.

Elghali, L., Clift, R., Sinclair, P., Panoutsou, C., & Bauen, A. (2007). Developing a sustainability framework for the assessment of bioenergy systems. *Energy Policy, 35*(12), 6075–6083.

EUBIA. (2007). *About biomass.*

European Union. (2003). Promotion of the use of biofuels or other renewable fuels for transport: Directive 2003/30/ec.

Forsberg, G. (2000). Biomass energy transport analysis of bioenergy transport chains using life cycle inventory method. *Biomass and Bioenergy, 19*(1), 17–30.

Freppaz, D., Minciardi, R., Robba, M., Rovatti, M., Sacile, R., & Taramasso, A. (2004). Optimizing forest biomass exploitation for energy supply at a regional level. *Biomass and Bioenergy, 26*(1), 15–25.

Gold, S., & Seuring, S. (2011). Supply chain and logistics issues of bio-energy production. *Journal of Cleaner Production, 19*(1), 32–42.

Grant, N., & Clarke, A. (2010). Biomass—A burning issue. *The Association for Environment*, 2–5.

Hamelinck, C. N., Suurs, R. A., & Faaij, A. P. (2005). International bioenergy transport costs and energy balance. *Biomass and Bioenergy, 29*(2), 114–134.

Hoogwijk, M., Faaij, A., van den Broek, R., Berndes, G., Gielen, D., & Turkenburg, W. (2003). Exploration of the ranges of the global potential of biomass for energy. *Biomass and Bioenergy, 25*(2), 119–133.

Huang, Y., Chen, C.-W., & Fan, Y. (2010). Multistage optimization of the supply chains of biofuels. *Transportation Research Part E: Logistics and Transportation Review, 46*(6), 820–830.

Huisman, W., Venturi, P., & Molenaar, J. (1997). Costs of supply chains of *Miscanthus giganteus*. *Industrial Crops and Products, 6*(3), 353–366.

Iakovou, E., Karagiannidis, A., Vlachos, D., Toka, A., & Malamakis, A. (2010). Waste biomass-to-energy supply chain management: A critical synthesis. *Waste management (New York, N.Y.), 30*(10), 1860–1870.

Ignaciuk, A., Vöhringer, F., Ruijs, A., & van Ierland, E. (2006). Competition between biomass and food production in the presence of energy policies: A partial equilibrium analysis. *Energy Policy, 34*(10), 1127–1138.

Kerckow, B. (2007). Competition between agricultural and renewable energy production. *Quarterly Journal of International Agriculture, 46*(4), 333–347.

Kim, J., Realff, M. J., & Lee, J. H. (2011). Optimal design and global sensitivity analysis of biomass supply chain networks for biofuels under uncertainty. *Computers and Chemical Engineering, 35*(9), 1738–1751.

Koh, L. P., & Ghazoul, J. (2008). Biofuels, biodiversity, and people: Understanding the conflicts and finding opportunities. *Biological Conservation, 141*(10), 2450–2460.

Krajnc, N., & Domac, J. (2007). How to model different socio-economic and environmental aspects of biomass utilisation: Case study in selected regions in slovenia and croatia. *Energy Policy, 35*(12), 6010–6020.

Kudakasseril Kurian, J., Raveendran Nair, G., Hussain, A., & Vijaya Raghavan, G. (2013). Feedstocks, logistics and pre-treatment processes for sustainable lignocellulosic biorefineries: A comprehensive review. *Renewable and Sustainable Energy Reviews, 25*, 205–219.

Kumar, A., & Sokhansanj, S. (2007). Switchgrass (*Panicum vigratum*, L.) delivery to a biorefinery using integrated biomass supply analysis and logistics (IBSAL) model. *Bioresource Technology, 98*(5), 1033–1044.

Kumar, A., Sokhansanj, S., & Flynn, P. C. (2006). Development of a multicriteria assessment model for ranking biomass feedstock collection and transportation systems. *Applied Biochemistry and Biotechnology, 129*(1–3), 71–87.

Mafakheri, F., & Nasiri, F. (2014). Modeling of biomass-to-energy supply chain operations: Applications, challenges and research directions. *Energy Policy, 67*, 116–126.

Mansoornejad, B., Chambost, V., & Stuart, P. (2010). Integrating product portfolio design and supply chain design for the forest biorefinery. *Computers and Chemical Engineering, 34*(9), 1497–1506.

Mayfield, C. A., Foster, C. D., Smith, C. T., Gan, J., & Fox, S. (2007). Opportunities, barriers, and strategies for forest bioenergy and bio-based product development in the Southern United States. *Biomass and Bioenergy, 31*(9), 631–637.

McKendry, P. (2002a). Energy production from biomass (part 1): Overview of biomass. *Bioresource Technology, 83*(1), 37–46.

McKendry, P. (2002b). Energy production from biomass (part 2): Conversion technologies. *Bioresource Technology, 83*(1), 47–54.

Mentzer, J. T., & DeWitt, W. (2001). Defining supply chain management. *Journal of Business Logistics, 22*(2), 1–25.

Möller, B., & Nielsen, P. S. (2007). Analysing transport costs of danish forest wood chip resources by means of continuous cost surfaces. *Biomass and Bioenergy, 31*(5), 291–298.

Mula, J., Peidro, D., Díaz-Madroñero, M., & Vicens, E. (2010). Mathematical programming models for supply chain production and transport planning. *European Journal of Operational Research, 204*(3), 377–390.

Murray, A. T. (1999). Spatial restrictions in harvest scheduling. *Forest Science, 45*(1), 45–52.

Nasiri, F., & Zaccour, G. (2009). An exploratory game-theoretic analysis of biomass electricity generation supply chain. *Energy Policy, 37*(11), 4514–4522.

OECD/FAO. (2007). *Agricultural outlook 2007–2016.*

Papapostolou, C., Kondili, E., & Kaldellis, J. K. (2011). Modelling biomass and biofuels supply chains. *Computer Aided Chemical Engineering, 29*, 1773–1777.

Peidro, D., Mula, J., Poler, R., & Lario, F.-C. (2009). Quantitative models for supply chain planning under uncertainty: A review. *The International Journal of Advanced Manufacturing Technology, 43*(3), 400–420.

Perry, M., & Rosillo-Calle, F. (2008). Recent trends and future opportunities in uk bioenergy: Maximising biomass penetration in a centralised energy system. *Biomass and Bioenergy, 32*(8), 688–701.

Pimentel, D., Marklein, A., Toth, M. A., Karpoff, M. N., Paul, G. S., McCormack, R., Kyriazis, J., & Krueger, T. (2009). Food versus biofuels: Environmental and economic costs. *Human Ecology, 37*(1), 1–12.

Rentizelas, A., Tatsiopoulos, I., & Tolis, A. (2009). An optimization model for multi-biomass tri-generation energy supply. *Biomass and Bioenergy, 33*(2), 223–233.

Rohde, J., Meyr, H., & Wagner, M. (2000). Die Supply Chain Planning Matrix. *PPS Management, 5*(1), 10–15.

Roos, A., Graham, R. L., Hektor, B., & Rakos, C. (1999). Critical factors to bioenergy implementation. *Biomass and Bioenergy, 17*(2), 113–126.

Sadaka, S. (2009). *Gasification, producer gas and syngas*. Agriculture and Natural Resources (Vol. 1051). Fayetteville, Ark.: Cooperative Extension Service; University of Arkansas, U.S. Dept. of Agriculture and county governments cooperating.

Sahinidis, N. V. (2004). Optimization under uncertainty: State-of-the-art and opportunities. *Computers and Chemical Engineering, 28*(6), 971–983.

Saidur, R., Abdelaziz, E., Demirbas, A., Hossain, M., & Mekhilef, S. (2011). A review on biomass as a fuel for boilers. *Renewable and Sustainable Energy Reviews, 15*(5), 2262–2289.

Scott-Kerr, C., Johnson, T., Johnson, B., & Kiviaho, J. (2009). Bioethanol-status report on bioethanol production from wood and other lignocellulosic feedstocks. In *63rd Appita Annual Conference and Exhibition, Melbourne, 19–22 April 2009* (p. 3). Appita Inc.

Searcy, E., Flynn, P., Ghafoori, E., & Kumar, A. (2007). The relative cost of biomass energy transport. *Applied Biochemistry and Biotechnology, 137–140*(1–12), 639–659.

Sharma, B., Ingalls, R., Jones, C., & Khanchi, A. (2013). Biomass supply chain design and analysis: Basis, overview, modeling, challenges, and future. *Renewable and Sustainable Energy Reviews, 24*, 608–627.

Simchi-Levi, D., Kaminsky, P., & Simchi-Levi, E. (2008). *Designing and managing the supply chain: Concepts, strategies, and case studies*. Mcgraw-Hill/Irwin series operations and decision sciences (3rd ed.). Boston: McGraw-Hill/Irwin.

Sodhi, M. S., & Tang, C. S. (2009). Modeling supply-chain planning under demand uncertainty using stochastic programming: A survey motivated by asset–liability management. *International Journal of Production Economics, 121*(2), 728–738.

Sokhansanj, S., Kumar, A., & Turhollow, A. F. (2006). Development and implementation of integrated biomass supply analysis and logistics model (IBSAL). *Biomass and Bioenergy, 30*(10), 838–847.

Stadtler, H., Kilger, C., & Meyr, H. (Eds.). (2015). *Supply chain management and advanced planning: Concepts, models, software, and case studies. Springer texts in business and economics* (5th ed.). Berlin, Heidelberg: Springer.

Thornley, P., Rogers, J., & Huang, Y. (2008). Quantification of employment from biomass power plants. *Renewable Energy, 33*(8), 1922–1927.

United Nations Environment Programme. (2009). Towards sustainable production and use of resources: Assessing biofuels.

Upreti, B. R. (2004). Conflict over biomass energy development in the United Kingdom: Some observations and lessons from England and Wales. *Energy Policy, 32*(6), 785–800.

Uslu, A., Faaij, A. P., & Bergman, P. (2008). Pre-treatment technologies, and their effect on international bioenergy supply chain logistics. Techno-economic evaluation of torrefaction, fast pyrolysis and pelletisation. *Energy, 33*(8), 1206–1223.

Yılmaz, S., & Selim, H. (2013). A review on the methods for biomass to energy conversion systems design. *Renewable and Sustainable Energy Reviews, 25*, 420–430.

Yue, D., You, F., & Snyder, S. W. (2014). Biomass-to-bioenergy and biofuel supply chain optimization: Overview, key issues and challenges. *Computers and Chemical Engineering, 66*, 36–56.

Zhu, X., Li, X., Yao, Q., & Chen, Y. (2011). Challenges and models in supporting logistics system design for dedicated-biomass-based bioenergy industry. *Bioresource Technology, 102*(2), 1344–1351.

The Use of Biomass for Energy Production and Organic Fertilizer for Mitigating Climate Change and Improving the Competitiveness of the Agricultural Enterprise: The Case of UPAP in Puriscal, Costa Rica

Olman Quirós Madrigal and Lady Arias Fallas

Abstract The farmer's association UPAP, in the Canton of Puriscal, Costa Rica, has been making efforts to transform its firm in order to adapt new technologies to reduce the production costs. This is done by reusing biomass (manure from cattle mainly) to generate energy and produce organic fertilizers which are distributed among farmers affiliated to the Association. The UPAP's main economic activity is livestock auction, which generates a large amount of biomass (cattle excreta). The guiding question to support the research was: how to manage the biomass produced in the cattle auction to mitigate the negative externalities and to get economic benefits? The research was developed taking into consideration the importance of the proposal for the climate change adaptation and allow the agribusiness to solve its negative impact on the environment while maintaining competitiveness. With the support of experts from the MAG an assessment of the activity to specify the type of biodigester that was to be built was prepared. This diagnosis included: number of animals, time spent in each auction, amount of water and manure and the vermicompost process. The preliminary results based on the gross profit indicator show a clear economic disadvantage in relation to investment in mitigating the negative externalities of agribusiness and show a

UPAP: Puriscal Farmers Association (Unión de Pequeños Agricultores de Puriscal, Costa Rica).
Project: "Red Iberoamericana de Bioeconomía y Cambio Climático". Coordinated by Universidad Nacional Autónoma, León, Nicaragua. Supported by "Programa Iberoamericano de Ciencia y Tecnología para el Desarrollo" (CYTED: http://www.cyted.org).

O. Quirós Madrigal (✉) • L. Arias Fallas (✉)
Agricultural Economics and Agribusiness Department, University of Costa Rica, 2060 San José, Costa Rica
e-mail: olman.quiros@ucr.ac.cr; lady.arias@ucr.ac.cr

S. Dabbert et al. (eds.), *Knowledge-Driven Developments in the Bioeconomy*, Economic Complexity and Evolution, DOI 10.1007/978-3-319-58374-7_16

shortfall of US$675.92 (₡ 361,621.00) per month. The inefficiency in the use of biogas as well as the time spent to produce vermicompost seem to be the most important factors that must be analyzed to make the project more efficient.

1 Introduction

Agriculture is at a crossroad: farmers should work to produce food for a growing population but also, they must reduce the negative environmental impacts resulting from the emission of greenhouse gases which directly influence the climate change that is facing the world community today. This also leads to a concern in the farming business: competitiveness. This dilemma shows that "business as usual" (Guy 2012) is no longer an option to face the challenges of the Millennium.

These impacts and adaptation needs are especially important for farmers in developing countries. Given this need, UPAP, in the Canton of Puriscal, Costa Rica, has been making efforts to transform its company in order to adapt new technologies to reduce production costs by reusing biomass (manure from cattle mainly) to generate energy and produce organic fertilizers which are distributed among the farmers affiliated to the Association. The UPAP's main economic activity is livestock auction (cattle bidding/auction), which generates a large amount of biomass (cattle excreta). This biomass must be transformed into other inputs in order to be recycled or reused. The goal is to improve competitiveness of the agri-food system in which different actors are involved: the UPAP Puriscal Canton and small farmers.

Considering the current laws (199th Zeledón 1999b; Decreto Nr. 31849-MINAE-S-MOPT-MAG-MEIC 2004; Zeledón 1999a), the UPAP requested support from MAG experts to plan and develop the "Project: Construction of a biodigester and production of vermicompost" in order to comply with the current demands regarding waste management in farm activities (Mora 2015). The guiding question to support the Project was: how to manage the biomass produced in the cattle auction to mitigate the negative externalities and also to get economic benefits from the necessary investments?

2 Objective

The main objective was to identify and assess the impacts generated by the construction and operation of the biodigester and the production of vermicompost at UPAP's farm, as an alternative method for treating livestock manure at the auction's facilities.

The main specific objectives were :

1. To identify the environmental importance of the recycling proposed, with the construction of a biodigester and vermicompost treatment.
2. To conduct a preliminary economic analysis of the implementation of the technologies identified in the project.

3 Methods

The research was developed in four phases, taking into consideration the importance of the issue and the need to organize information as a basis for disseminating proposals that allow agribusinesses to solve its negative impact on the environment while maintaining competitiveness. The requirements that the law of Health and Environment defined to allow these agribusiness activities were also considered. The stages of the research were:

1. Meetings with experts from the Ministry of Agriculture and Livestock (MAG) in the Canton of Puriscal, UPAP was identified as a representative organization for the small and medium producers of the Canton.
2. The regulatory measures (laws, norms) were reviewed.
3. Visits to the UPAP's farm and staff interviews of both the Association and MAG were conducted.
4. Review of secondary information based on UPAP's statistics of marketing, sales, costs, revenues.

The different activities were carried out during the period of December 2014 to April 2015.

4 Agribusiness Description and Main Activity

The UPAP is an organization made up of small and medium farmers in the Canton of Puriscal and it owns a farm in this Canton (Fig. 1). The principal activity of this farm of 14 ha, is the marketing of cattle in the region as a strategy to get the best prices for producers. This marketing is done through a "cattle auction Program" which is organized twice a week: Tuesdays and Thursdays (Mora 2015). The activity begins with the reception of the animals on Mondays and Wednesdays, from 4:00 p.m. to 8:00 p.m., and even the day of the auction from 6:00 a.m. to 10:00 a.m. The auction begins at 10 a.m. and takes about 2–3 h, with an average transaction of 180 animals per auction. This means that an average of 360 animals are sold per week. The permanence of this number of animals in the barnyard auction facilities produces about 72 m^3 of biomass (manure) weekly (Chavarría 2015; Fallas 2015).

Fig. 1 Location of UPAP's farm in Costa Rica. Source: Google-Earth 2015

Table 1 Different uses of biogas

Objectives	Result	Applications
Energy	Caloric energy: home kitchens, heating Electricity	Cooking food, heating facilities Cooling systems, mechanical milking systems
Compost	Compost and lombricompost	Organic fertilizer to be used in agricultural activities

5 Proposal for Handling the Biomass

The biomass produced can be used for energy production or composting in agricultural production. At the same time, the production of energy can be used directly as heat energy or can be used directly in home kitchens or it can be transformed into electrical energy to use it in agribusiness activities and others (Table 1).

The process that takes place in these systems is the same that takes place in the intestinal tract of animals, but this is done in a controlled manner. Its main objective is waste management, biogas production and biofertilization. The goal of

Table 2 Composition of biogas

Gas	Volume ratio (in %)
Methane (CH_4)	40–70
Carbon dioxide (CO_2)	30–60
Other gases	1–5
Hydrogen (H_2)	0–1
Hydrogen sulfide (H_2S)	0–3

Source: Gon (2008)

Table 3 Ration of manure and methane produced by animals

Type of manure	Live weight	Daily amount of manure (kg)	Methane (%)
Pigs	50	10	70
Dairy cattle	500	34	65
Horse	500	10	65
Sheep	20	1.5	70
Chickens	1.5	0.02–0.03	65–70

Source: Gon (2008)

this process is to stop producing organic wastes and use them as a raw material for domestic or commercial uses. This results into income or savings in the production process (Gon 2008). The composition by volume of biogas produced is shown as follows (Table 2).

The biogas production depends on the material that was fed into the digester and the amount of methane depends on the amount of manure. The produced methane from the biodigester depends at the same time on the type of animals (Table 3).

From the analysis of the available information, and based on discussions with experts from the MAG, the use of biogas for cooking was allowed. The justifications for reaching this decision were:

1. The UPAP has a restaurant at the cattle auction facilities. The restaurant can use this resource to provide food service to Auction's visitors.
2. The cost of electricity has become one of the factors that negatively affect the competitiveness of firms in Costa Rica.
3. It requires no investment in the purchase of equipment to transform biogas into energy.

The cost of labor was primarily considered to produce the compost. Therefore, making vermicompost using the Californian red worm (*Eisenia foetida*) was the best alternative identified. This activity allows the UPAP to use the underutilized infrastructure (Fig. 2).

Fig. 2 UPAP's cattle auction facilities

6 The Biodigester

According to different authors, using a biodigester is one of the best alternatives for the treatment of organic matter (manure) from animal farms. It can be defined as a bioreactor which under anaerobic conditions is designed to promote a suitable environment for bacteria that breaks down organic matter to produce biogas and an effluent to be used as agricultural fertilizer (Gon 2008; Irusta 2011).

With the support of experts from the MAG (Guerrero 2015; Elizondo 2015) an assessment of the activity to specify the type of biodigester that was to be built was prepared. This diagnosis included: number of animals, time spent in each auction, amount of water and manure and recognition of facilities. This information served as a basis for defining the type of digester that was to be built, through the following data:

1. Amount of manure per auction: 0.2 m^3
2. Excreta and water ratio is 1:4

The specifications of the type of biodigester were: tubular with 19 m long with a diameter of 2.5 m and a capacity of 30 m^3. The material used is permaflex geomembrane Amanco (PVC). Further notes were considered to determine the dimension of the biogestor such as the approximate retention time (TDR) to properly process the manure, which is a month. When released it is 85% free of contaminants (Guerrero 2015) (Fig. 3).

The process is as follows: once the auction is over, the facilities are washed only with water, a process that takes about 6–7 h with two workers. The material (manure) represents the raw material collected to feed the biodigester and to produce vermicompost. The manure is collected in an external collection tank where it

Fig. 3 The biodigester

Fig. 4 Separator tan of purines and biomass and the pump system

is stored and directed to four sedimentation tanks. External tank dimensions are 1.5 m by 4 m long and are filled on average 10–12 times a week. When this is filled, the process of separating liquids and solids begins. The solid manure is deposited in the sedimentation tanks for the production of vermicompost and liquid is conducted to the biodigester. The gas produced is used at the Restaurant located at the headquarters of the UPAP, which is opened on Tuesdays and Thursdays. The gas is also used in a house for the farm worker (Fig. 4 and Table 4).

Table 4 The biodigester components are summarized as follows

Parts of the biodigester	Description
Loading tank	For the collection of "raw materials" (manure)
Digestor	Receptor shaped tube for the organic matter placed in a pit, where wastes are decomposed (anaerobic process)
Unloading tank	Biomass from the digester is deposited
Plastic cover	Placed on the digester to prevent the air inlet and gas outlet
Pipes, valves and stopcocks	Accessory parts to connect the entire system of the digester

Fig. 5 Biomass collector for the vermicompost

7 The Vermicompost

The resulting solid material from four sedimentation tanks described above, is dried "in idle mode" for 8 days (Fig. 5). When dried, this material is passed to the vermicompost process. On average, 50–70 kg of organic material is extracted per week. This material is stored in these tanks for 6 months and it should be transferred tank by tank. In the last one is where the organic material is extracted for later packaging. The process starts with the first tank at ground level with 10 kg of red worm and about 200 kg of the dry manure, and it stays there for 2 months. The worms are fed every 10 days before being passed to the next tank. After 6 months 50% of the last tank is filled and the dry manure is transformed now in organic fertilizer and is separated then from the red worms. At the end of 6 months, the resulting compost or organic fertilizer is transported to a "galerón" where the drying

Fig. 6 Organic fertilizer

process is completed and bags of 20–25 kg are filled. This last process takes 22 days on average and 90 bags are obtained (Fig. 6). These are then sold at ₡4000.00 per bag (exchange rate: 1 US$ = ₡535.00).

8 Preliminary Results

8.1 The Economic Importance of Biogas and Vermicompost

To calculate the biogas production, it was assumed that each animal produces an estimated of 13 kg manure per day and the biogas conversion from this manure is 0.03 (Gon 2008). Manure production of 360 animals is 2283.42 kg/day. However, it should be taken into account that the average time of animals in the corrals is 2.5 days/week. Therefore, the ratio of 50% in the production of manure/week is accepted. Based on this ratio of the availability of manure production/week, the production of 85.62 m^3 biogas/week (Table 5) was estimated.

According to the objectives, the produced biogas must represent a saving for the agribusiness, since energy is produced and waste is reduced. Assuming that with 1 m^3 biogas (5500 Kcal/m^3) there is a replacement of 0.60 m^3 of natural gas (LPG) of 9300 Kcal/m^3 of calorific value. In the agribusiness-UPAP, the amount of biogas consumption is 30.72 m^3/week and is used in the livestock auction facilities and 26.88 m^3/week is used in the worker's house for a total consumption of 57.60 m^3/week (Table 6).

Regarding the economic value of biogas, the approximate value of the savings due to the replacement of liquefied petroleum gas (LPG) for biogas (GN) produced in the biodigester was calculated. Considering that 1 kg of LPG is equivalent to 1.28 m^3 of biogas and that the LPG market value is estimated at US$1.32

Table 5 Calculation of biogas production m^3/week

Type of biomass	Amount kg/day	Estimated length of stay	Biogas conversion m^3/kg	Total production of biogas m^3/día	Weekly production of biogas (2,5 days of occupancy)
Manure	2283.42	50%[a]	0.03	34.25	85.62

[a]Length of stay means the time which the animals remain in the auction because they are not all day

Table 6 Consumption calculation of biogas per week

	Gas use intensity	Number of days per week	Daily operating time hours/day	Total amount biogas/week in m^3	Total consumption m^3 biogas/week according% intensity
Kitchen facilities at UPAP' restaurant	100%[a]	2	12	30.72[b]	30.72
Stove of worker house kitchen	50%[a]	7	6	53.76[b]	26.88
Total (week)					57.60

[a]This relationship is done based on the information provided by workers of UPAP's farm
[b]Calculation based on a consumption ratio of 1.28 m^3 of biogas/hour according to Gon (2008)

Table 7 Estimated value of weekly savings using reference gas LPG

	Unit	m^3 biogas quantity consumed weekly	Amount of biogas equivalent to LPG	Cost LPG/kg	Weekly value in ₡ of biogas
Biogas: worker's house kitchen	Monthly	30.72[a]	24[b]	708,33[c]	16,999.92
Biogas: UPAP farm kitchen	Monthly	26.88[a]	21[b]	708,33[c]	14,874.93
Organic fertilizer	Biannual				
Total					₡31,874.85

[a]For conversions are estimated
LPG heating value of 11,800 Kcal/kg. Calorific value of GN 9300 Kcal/m^3
[b]The relationship between the two (1.28) indicates that 1 kg LPG is equivalent to 1.28 m^3 of GN
[c]Cost of 1 kg LPG

(₡708.33) per kg, the value of the biogas used is US$238.32 (₡127,499.40) per month (Table 7).

Moreover, the income from organic fertilizer was estimated since every 6 months on average 90 bags of 20–25 kg are sold at a price of ₡4000 per bag. The average value of this fertilizer is sold at a price of ₡60,000.00 per month. Therefore, the total income reaches US$350.47 (₡187,499.00) per month.

Table 8 Labor costs

Labors	Weekly hours	Amount of hours/week	Cost/hour	Total
General labors	Hours	56.00	₡1320.00	₡73,920.00
Permanent worker	Hours	48.00	₡1320.00	₡63,360.00
Total				₡137,280.00

Additionally, the labor cost (two workers) for the maintenance activities of the biodigester and vermicompost process requires a full-time worker and another part-time for washing the facilities (4 h per week). The total cost of labor amounts to US$256.60 (₡137,280.00) per week for a total of 1026.39 (₡549,120) per month (Table 8).

8.2 *Cost/Benefit Ratio*

The benefits of the biodigester and the production of organic fertilizer project are represented by savings in the cost of electricity due to the use of biogas and to the compost (organic fertilizer) sales. The main cost is due to labor because it is essentially a fixed cost. The cost of investing in the infrastructure of the biodigester is not included here since this is a preliminary analysis of gross profit. These preliminary results show a clear economic disadvantage in relation to the investment in mitigating the negative externalities of agribusiness and show a shortfall of US$675.92 (₡361,621) per month. The inefficiency in the use of biogas as well as the time used in the production of vermicompost seem to be the most important factors that must be analyzed to make the project more efficient.

9 Conclusions and Recommendations

1. It is important to consider that this is an initial study to evaluate the use of biogas and vermicompost. The purpose is to collect and systematize information that can demonstrate the economic and environmental viability of investments in technologies to reduce emissions of greenhouse gases to mitigate actions that cause climate change. More research is required under tropical conditions that can allow the development of appropriate and more efficient technology than the one adopted by the UPAP-project.
2. Optimal use of biogas is not done, this is affecting the results of the preliminary economic analysis. The efficient use of resources should be improved. In analyzing the relationship between the estimated biogas per week of 85.65 m^3/week where consumption barely reaches 57.60 m^3/week, it becomes clear that there is a significant underutilization of the resources produced. It is also necessary to

evaluate the process of production of organic fertilizer in addition to labor. This is a factor that also requires an analysis to improve efficiency due to the costs to agribusiness.

3. The socio-environmental benefits of investments for mitigation of greenhouse gases have not been analyzed in greater depth. It is necessary to evaluate the reduction of gases released into the environment. In addition, with the use of biogas as an energy source the use of firewood is avoided. This has a significant environmental benefit when considering that 1 m^3 of gas prevents deforestation of 0.33 ha of forest (MAG 2010).
4. Progress must be made in further studies that include long-term investments such as infrastructure in general. Likewise, more research is required to make new tools available for economic and environmental analysis.

References

Decreto Nr. 31849-MINAE-S-MOPT-MAG-MEIC. (2004). Reglamento General sobre los Procedimientos de Evaluación de Impacto Ambiental. La Gaceta Nr. 125 del 28 de junio del 2004. San José, Costa Rica.

Gon, L. M. (2008). *Guía para proyectos de biodigestión en proyectos agropecuarios.* Tesis de Maestría, Facultad de Ciencias de la Tierra y de Ambiente. Universidad Católica de Santa Fe. Argentina.

Guy, H. (2012). La bioeconomía y la agricultura baja en carbono. In: Agricultura y cambio climático: del diagnóstico a la práctica. Serie Seminarios y conferencias Nr. 71. Segundo seminario Regional Agricultura y Cambio Climático, Santiago de Chile, 23 y 24 noviembre 2011. CEPAL/NACIONES UNIDAS/FAO. Chile 2012.

Irusta, G. I. (2011). *Planificación y Evaluación Económica, Financiera y de Impacto Ambiental de la Instalación de un Biodigestor en un criadero de cerdos.* Tesis Grado de Ingeniero Agrónomo. Universidad Nacional de Río Cuarto-Córdoba. Argentina.

MAG (Ministerio de Agricultura y Ganadería). (2010). Guía técnica para la difusión de tecnologías de producción agropecuaria sostenible. Ficha técnica Nr. 23: Biodigestores. San José, Costa Rica.

Zeledón, R. (1999a). *Código Ambiental: Ley General de Salud* (2da edición, pp. 257–320). San José: Editorial Porvenir-Fondo Jurídico.

Zeledón, R. (1999b). *Código Ambiental: Ley Orgánica del Ambiente* (2da edición, pp. 4–28). San José: Editorial Porvenir-Fondo Jurídico.

Interviews

Chavarría, H. (2015). UPAP-worker.

Dagoberto, E. (2015). That was an interview.

Elizondo, D. (2015). MAG – expert. Puriscal, Costa Rica.

Fallas, G. (2015, Marzo de). UPAP-worker. Puriscal, Costa Rica.

Guerrero, L. (2015). MAG-expert. Puriscal, Costa Rica.

Mora, N. (2015). UPAP Manager. Puriscal, Costa Rica.

Bioethanol as the Sole Solvent for Vegetable Oil Extraction and Biodiesel Production

Marisa Aparecida Bismara Regitano-d'Arce, Naiane Sangaletti-Gerhard, and Larissa Braga Bueno-Borges

Abstract Economic and environmental repercussions of oil reserves depletion have led to the implementation of programs promoting the use of alternative fuels such as ethanol and biodiesel. Brazil has great potential for the production of these fuels, as sugarcane and soybean are national major commodities. Biodiesel is produced mainly by transesterification, a process in which oils or fats react with short-chained alcohols in the presence of a catalyst. Hexane is used worldwide in the industrial solvent extraction. This solvent has considerably higher flammability, explosiveness and toxicity compared to ethanol. Since the 1980s, the Laboratory of Oils and Fats at ESALQ-USP Agricultural College has developed a line of research on ethanol oil extraction as a way to also explore the regional importance and availability of this feedstock. The product of soybean oil extraction with ethanol is a miscella (oil + solvent) that, after cooled down to less than 30 °C, separates into three phases: rich-in-alcohol miscella (poor miscella), rich-in-oil miscella (rich miscella) and gum (crude lecithin). The poor miscella, composed of approximately 91% ethanol, can be used as solvent in subsequent extractions. With the natural phase separation, poor miscella carries the majority of polar substances such as phospholipids (0.4%), water and free fatty acids (0.7%). It can be said that the poor miscella promotes a partial refining of the rich miscella. The latter contains 90% oil and 7.8% ethanol and is suitable for biodiesel production without the need for desolventization and refining steps, contributing to the energy recovery of the process. In addition, miscella's oxidative stability in accelerated tests is three times higher than that of degummed oil. Meal produced from ethanol extraction also has a higher quality than hexane-extracted ones due to antinutritional compounds elimination. The economic and energy analyses of the ethanol process reveal that it requires adjustments to ensure higher efficiency. However, biodiesel from rich miscella via alkaline catalyst can be considered a promising alternative from several points of view, provided the whole process is executed in a single industrial plant using solely ethanol as the solvent extraction and acyl acceptor in the transesterification reaction.

M.A.B. Regitano-d'Arce (✉) • N. Sangaletti-Gerhard • L.B. Bueno-Borges
Department of Agroindustry, Food and Nutrition, Laboratory of Oils and Fats, College of Agriculture "Luiz de Queiroz" (ESALQ), University of São Paulo, Avenida Pádua Dias, 11, 13418-900 Piracicaba, São Paulo, Brazil
e-mail: marisadarce@usp.br; naiane@usp.br; larissabueno@usp.br

S. Dabbert et al. (eds.), *Knowledge-Driven Developments in the Bioeconomy*, Economic Complexity and Evolution, DOI 10.1007/978-3-319-58374-7_17

1 Introduction to the Biodiesel Production

Fossil fuels are widely applied in industry, agriculture and transportation, playing an essential role in the national economy. However, the high energy demanded by the industrial and domestic sectors, in addition to pollution problems caused by the use of such fuels has increased interest and need for the development of renewable sources of energy with lower environmental impact (Meher et al. 2006). Although petroleum is abundant on the planet, it is finite. In addition, many countries rely substantially on its importation and suffer from price fluctuations of feedstocks and derivatives, such as gas and diesel. Thus, the economic and environmental repercussions of oil reserves depletion have led to the implementation of programs promoting the use of alternative fuels. These fuels derive from many renewable sources such as plants, agro-industrial, forest and domestic biomasses, and are widely associated to their main representatives: ethanol and biodiesel (Angarita et al. 2008).

The use of biofuels in internal combustion engines has been known for more than 100 years (Shay 1993). In 1900, the World Exhibition in Paris has presented to more than 50 million people a diesel engine operating with crude peanut oil. The prototype won the highest prize of the exposition to the Société Française des Moteurs Diesel, bearer of the licenses to produce diesel engines at the time. Over the subsequent years, Rudolph Diesel continued his work culminating with the construction of the first ship powered by an internal combustion engine (Naegel and Althuser 2008). In 1912, during his lecture at the Institute of Mechanical Engineers in the United Kingdom, Diesel confirmed his conviction on the feasibility of the use of vegetable oils in diesel engines pointing them as a promising alternative energy source (Shay 1993; Knothe 2006).

In Brazil, since the 1920s, the National Institute of Technology (INT) has carried out tests with alternative and renewable fuels in engines (Angarita et al. 2008). As a matter of fact, Brazil, such a strong agricultural country, has great potential for biofuel production as sugarcane and soybean are national major commodities (Hausman 2012). In addition to originating from renewable sources, biodiesel from bioethanol and soybean oil has advantages over petroleum-based fuels such as the absence of sulfur and aromatics, high cetane number, a higher flash point and a lower emission rate of hydrocarbons, CO and CO_2 toxic particles, besides a less hazardous storage and handling (Knothe and Razon 2017).

Biodiesel comprises alkyl esters of long chain carboxylic acids (esters of fatty acids) mainly produced by transesterification where vegetable oil or animal fats (triacylglycerol) react with short-chained alcohols (acyl acceptor) such as methanol and ethanol in the presence of a catalyst. The triacylglycerol is converted stepwise into diacylglycerol, monoacylglycerol and finally glycerol (Fig. 1) (Ma and Hanna 1999; Meher et al. 2006). In this reaction, approximately 1 kg of biodiesel and 0.1 kg of glycerol are generated for every kilogram of oil.

Transesterification may be catalyzed by homogeneous or heterogeneous catalysts. Homogeneous catalysts include alkalis and acids. The most commonly used

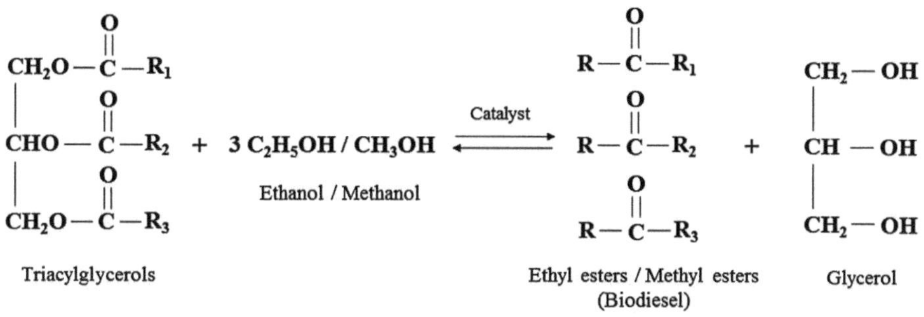

Fig. 1 General equation for transesterification of triacylglycerols

alkaline catalysts are sodium or potassium hydroxides and sodium methoxide or ethoxide; the most used acid catalysts are sulfuric, hydrochloric and sulfonic acids. Heterogeneous catalysts include ion exchange resins, metal complexes and organic polymers, and biocatalysts (enzymes) (Balat 2007). In addition, transesterification may be carried out, in the absence of catalysts, with alcohol in its supercritical state (Saka and Kusdiana 2001; Ranganathan et al. 2008), by using microwaves (Lam et al. 2010; Liao and Chung 2011), ultrasound (Ji et al. 2006; Deng et al. 2010; Lam et al. 2010) or hydrodynamic cavitation (Ji et al. 2006).

Esters can be generated from alcohols such as methanol, ethanol, propanol, isopropanol, butanol and isobutanol (Nimcevic et al. 2000), however only methanol and ethanol are used in the commercial transesterification reaction for biodiesel production. Methanol is produced from petroleum refining and is the worldwide choice of alcohol for fatty acid esters production for use as biodiesel (Haas and Foglia 2006) due to the lower oil/alcohol ratio required and lower cost in countries like Germany, in comparison to ethanol (Bozbas 2008). In Brazil, however, the use of ethanol is fully justified (Haas and Foglia 2006).

Ethanolysis, a transesterification reaction using ethanol as the acyl acceptor, is gathering more attention as this alcohol is known for its low toxicity and renewable character. Moreover, ethyl esters have lower Cloud and Pour Points than methyl esters avoiding fuel flowing problems in diesel engines' pipes and filters (Dunn 2006; Encinar et al. 2002). Methanol is more reactive in oil transesterification than ethanol as its carbon chain is shorter (Nimcevic et al. 2000); however methyl and ethyl esters yields from different feedstocks are very similar. Both alcohols may produce high esters conversion rates, in the range 96–98% (Kucek et al. 2007; Cernoch et al. 2010; Nimcevic et al. 2000; Rashid and Anwar 2008; Rashid et al. 2012).

The most widely used feedstocks for biodiesel production are soybeans, canola (rapeseed), palm, peanut, rapeseed, sunflowerseed, cottonseed, rice, camelina, coconut, pequi, macaúba, tucumã, babassu (Haas and Foglia 2006; Lima and Castro 2010; Soriano and Narani 2011). However, as the feedstock prices account for 70–80% of the total cost in the biodiesel production (Haas and Foglia 2006; Castro et al. 2010) alternative sources of inedible oils and fats have attracted researchers' attention, seeking for a feasible oil source and avoiding problems with the world food production. Some of the alternative feedstocks are: frying oils (Alcantara et al. 2000; Felizardo et al. 2006; Encinar et al. 2007; Lam et al. 2010); tallow (Ma et al. 1998; Alcantara et al. 2000; Kurzin et al. 2007) jatropha (*Jatropha curcas*) (Tiwari et al. 2007); neem oil (*Azadirachta indica*) (Ragit et al. 2011); India chestnut oil (*Sterculia foetida*) (Bindhu et al. 2011) microalgae, yeasts and fungi (Miao and Wu 2006; Chisti 2007; Singh and Singh 2010) chicken fat (Singh and Singh 2010); sewage sludge (Mondala et al. 2009; Sangaletti-Gerhard et al. 2015) among other materials cited by Singh and Singh (2010) and Azócar et al. (2010).

Vegetable oil can be extracted by pressing or solvent or a combination of both. The mechanism involves the extraction of oils from fruit pulps and seeds and comprises the following processes: lixiviating, washing, diffusion and dialysis. In order to improve oil extraction efficiency, the feedstock needs to be prepared by decreasing the grain size, reduction of grit width which results in area increase, and heating (industrial steps breaking, flaking and conditioning), exposing the spherosomes to the solvent action (Johnson and Lusas 1983). Pressing is the most economical and recommended process for oil-rich feedstocks while solvent extraction, besides being more expensive, is the efficient alternative to pressing, ensuring oil exhaustion in the meal.

Hexane is the most widely used solvent in the industrial vegetable oil extraction, dissolving the intracellular oil without reacting with other feedstock components. However, hexane has some disadvantages, such as higher flammability, explosiveness and toxicity compared to ethanol. Alternatively, solvents as n-heptane, n-propanol, iso-propyl alcohol and ethanol can be employed yielding as much oil as hexane (Gandhi et al. 2003) provided some adaptations to the process are made.

Ethanol was used for the first time in an industrial plant as solvent for soybean oil extraction, in 1934 in China. Since then, there have been studies and interest in the technical and energy feasibility of using ethanol as a substitute for hexane. This fact, plus the regional availability of bioethanol, has led the Laboratory of Oils and Fats at the Department of Agroindustry, Food and Nutrition of the Agricultural College ESALQ-USP to start research on this topic in the early 1980s, and since then, ethanol oil extraction has been improved.

2 Ethanol Oil Extraction and Products

Several solvents may be used in the extraction process and the choice will depend primarily on the main product desired (oil and meal). In hexane oil extraction, miscella (solvent + oil) should be distilled for separation of the crude oil for further refining and solvent recovery by condensers for subsequent use in extractions (Johnson 2002). The meal obtained in this process should be fully desolventized (solvent recovery) and toasted (in the case of soybeans) to be adequate for consumption as food or feed. However, desolventization followed by toasting is the most energy consuming industrial step in the oil extraction process due to the elevated temperatures necessary to remove the residual solvent and the longer heating times to remove undesirable and toxic substances (antinutritional compounds) from the meal (Sheehan et al. 1998). The use of ethanol as the solvent in the soybean oil extraction generates a liquid product that, after cooled down to less than 30 °C, results in three phases: rich-in-alcohol miscella (poor miscella), rich-in-oil miscella (rich miscella) and gums (crude lecithin). The poor miscella can be used as solvent in subsequent extractions. Up to 83% yield efficiency was reached with the use of the lean miscella, containing 91% ethanol, as solvent in our intermittent extractor for three cycles plus a last cycle with 99% ethanol (pure solvent). With the natural phase separation, poor miscella carries the majority of polar substances such as phospholipids (0.4%), water and free fatty acids (0.7%) due to the high concentration of ethyl alcohol in its composition. It can be said that the poor miscella promotes a partial refining of the rich miscella.

The rich miscella containing 90% oil and 7.8% ethanol is already suitable to produce biodiesel without the need for desolventization and refining steps, actively contributing to the energy recovery of biodiesel production from this feedstock. In the case oil should be meant for human consumption, classic refining would be much simpler and cheaper. Figure 2 shows the process of extraction using hexane and ethanol as solvents.

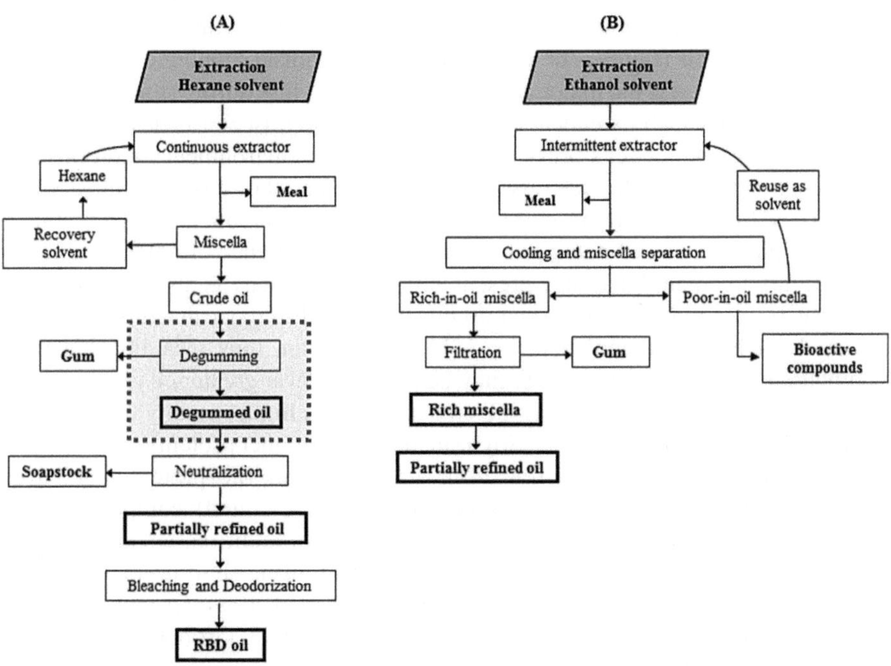

Fig. 2 The vegetable oil extraction process using hexane (**a**) and ethanol (**b**) as solvents. RBD oil = alkali refined, bleached and deodorized oil

Table 1 Soybean rich miscella and degummed oil composition and characteristics

Parameter	Rich-in-oil miscella[a]	Degummed oil[b]
Lipids (w/w%)	90.0 ± 1.0	98
Alcohol (w/w%)	7.6 ± 0.1	–
Non-volatile matter (w/w%)	91.7 ± 0.4	–
Phospholipids (w/w%)	0.6 ± 0.0	0.02
Unsaponifiable matter (w/w%)	1.0 ± 0.2	1.5
Peroxide value (meq/kg miscella)	10.9 ± 0.2	10.0
Acid value (% oleic acid)	0.4 ± 0.1	0.7
Water (w/w%)	0.3 ± 0.2	–
Oxidative stability at 110 °C (h)	22.0 ± 2.0	7.1[c]

[a]Sangaletti-Gerhard et al. (2014a)
[b]Sheehan et al. (1998)
[c]Unpublished data

The rich miscella has very similar characteristics to the degummed soybean oil (Table 1), representing an economical advantage over the hexane extracted crude oil, which requires further steps to preserve its quality after storage (degumming, alkali refining, bleaching and deodorization) (Fig. 2). Furthermore, although the storage requirements of soybean rich miscella are identical to hexane-extracted oils, miscella's oxidative stability in accelerated tests is three times higher than the degummed oil. This feature is of utmost importance from an economic point of view, since lipid oxidation is the major cause of vegetable oil loss during storage (Sherwin 1976). A product with extended shelf life such as the miscella is desirable for its ability to withstand longer storage periods and harsher environment conditions than hexane-extracted oils. The rich miscella is a lipid material of more complexity than oils derived from conventional extraction; besides the oil, ethanol, small amounts of phospholipids, water and non-polar antioxidants, the presence of polar compounds with affinity to ethanol that may present antioxidant capacity is also expected. The same can be inferred about the poor miscella, with the difference that this co-product is largely richer in ethanol than its "sibling" miscella, thus meaning that polar compounds of antioxidant capacity may be present in ever greater concentrations.

Meal produced with ethanol as the oil solvent has a higher quality than the ones generated from hexane extractors. Soybean meal has a higher protein content (48%), is lighter in color, as well as free from antinutritionals (no protease inhibitors, lectins, phytates, saponins and oligosaccharides were found) compared to the hexane extracted (Sangaletti-Gerhard et al. 2014a). In addition, ethanol removes undesirable compounds, such as chlorogenic acid from sunflowers (Regitano-d'Arce et al. 1994), gossypol from cottonseed (Hron et al. 1994; Groppo 2015) and aflatoxin from peanuts (Fonseca and Regitano-d'Arce 1993; Groppo 2015), adding value to the meal (Fig. 3).

Although, as mentioned above, rich-in-ethanol miscella (poor miscella) may have a high concentration of undesired compounds, it is also rich in antioxidants and

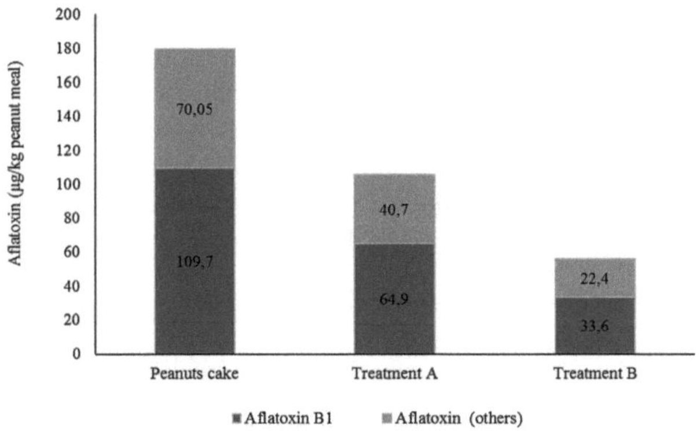

Fig. 3 Reduction of aflatoxin content after oil extraction using ethanol 99% (Treatment A) and ethanol 90 % (Treatment B) (Groppo 2015)

oligosaccharides (sucrose, raffinose and stachyose), which adds value to the final product.

3 Biodiesel Production from the Rich-in-Oil Miscella

The lipid feedstock, alcohol (known as acyl acceptor) and catalyst used in the transesterification reaction are chemical components that can interfere with the yield of biodiesel.

The chemical characteristics and quality of the feedstock determine the type of catalyst required. Catalysts are enzymatic or chemical substances that facilitate the reaction, reducing reaction time and promoting high yields. The most used and common homogeneous catalysts are alkalis (NaOH and KOH) and alkylated metal (sodium methoxide or ethoxide); a group of affordable chemicals that are easily applied. However, in order to achieve high yields of esters, these catalysts require a purer lipid source, mostly free of water and free fatty acids (D'Ippolito et al. 2007; Canakci and Van Gerpen 1999). On the other side, heterogeneous catalysts include ion exchange resins, metal complexes, organic polymers and biocatalysts (enzymes). These catalysts are advantageous over homogeneous ones because they are easily separated from the product, significantly reducing the number of purification steps and making their reuse possible. Methanol is the most commonly used alcohol in biodiesel production for its fast conversion and associated high yields, however it is fossil-derived and highly toxic compared to ethanol. Since ethanol is easier to handle, comes from a renewable source and its use is fully encouraged in countries such as Brazil, transesterification with ethanol as the acyl acceptor and alkaline catalyst is increasingly being investigated and optimized, with results comparable to those obtained from the use of methanol.

Feedstocks like crude soybean oil cannot be directly used for biodiesel production. There is a need for degumming and alkali refining for phospholipids and free fatty acids removal, respectively. A hypothetical substitution of the degummed hexane-extracted oil for the ethanol-extracted rich miscella would minimize the exposure of the lipid material to oxidation conditions during refining, eliminating costs associated to refining, transportation and storage (including oil losses), since the proposed model integrates feedstock and biodiesel production in the same industrial plant.

Reaction-wise, traces of water and oil acidity adversely affect the yield of the esterification. Oils that present more than 0.5% FFA and 0.3% water face the formation of soaps when alkaline catalysts are used, reducing the yield of biodiesel (Freedman et al. 1984). Furthermore, when phospholipids are present, their emulsifying activity hinders biodiesel purification step when alkaline catalyst is applied and they present inhibitory activity towards enzymatic catalysts (Watanabe et al. 2002; Sangaletti-Gerhard et al. 2014a).

Our rich soybean oil ethanolic miscella has consistently presented low concentrations of water (0.3%), free fatty acids (0.4%) and phospholipids (0.6%), which allows direct transesterification between the rich miscella and ethanol (acyl acceptor) using alkali (sodium hydroxide) and enzymatic catalysts to happen without the need for any oil refining steps (Sangaletti et al. 2013; Sangaletti-Gerhard et al. 2014a).

In our work (Sangaletti-Gerhard et al. 2014a), rich soybean oil miscella and ethanol have undergone transesterification under a 1:12 molar ratio (rich miscella to ethanol) and 0.6% NaOH (catalyst). After 60 min of reaction at 30 °C, 97.2% ethyl esters were produced, showing a large advantage over the conventional industrial process that utilizes temperatures higher than 60 °C (Sangaletti-Gerhard et al. 2014b). Besides this high performance, biodiesel produced from the rich-in-oil miscella presented all characteristics within the standardized limits set by the Brazilian National Petroleum, Natural Gas and Biodiesel Agency (ANP). The schematic representation of the biodiesel production chain directly using soybean ethanolic miscella is shown in Fig. 4.

Although alkyl esters are the main product of the transesterification reaction, recovery of glycerin is desirable due to the large number of industrial uses of this component. This tri-alcohol has been widely used as enzyme stabilizer and its protective capacity derives from humectant characteristics and viscosity. The glycerin layer concentrates most of the excess alcohol used in the reaction, plus catalyst residues and water. It has to be purified prior to commercialization, especially when produced via methyl route, because methanol is a compound of high toxicity that must be completely removed by distillation (Filippis et al. 2005). Transesterification by the ethyl route produces glycerin of high commercial value (US\$1.78–2.22 kg^{-1}) (ICIS 2014) for its absence of solubilized methanol. Both the rich and poor miscella generated from ethanol extraction have shown strong evidence of the presence of antioxidant compounds of polar nature. These, in turn, are transferred to the biodiesel and, consequently, the produced glycerin. Thus, the residual glycerin derived from ethyl route may further contain a high content

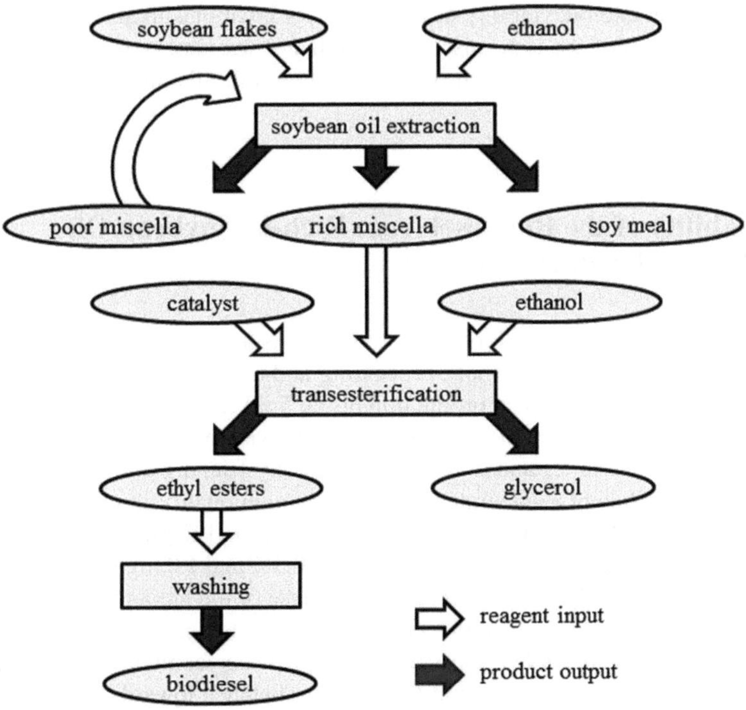

Fig. 4 Reagents and products (and co-products) involved in direct transesterification of ethanolic soybean miscella

of phenolic compounds extracted directly from the feedstock by the action of the ethanol, adding even more value to a product that can be readily absorbed by the market due to its non-toxic origin.

Novozym®435 is the trade name of lipase B from *Candida antarctica* immobilized on acrylic resin (polymethylmethacrylate). It is widely studied as a lipase enzyme catalyst for its robustness, specificity, and excellent conversion to esters, reaching 99% yield in the production of biodiesel. However, we found 85.4% ethyl esters yield in the direct transesterification using rich miscella and ethanol (molar ratio of 1:4, rich miscela:99% ethanol) and 9.5% catalyst at 40 °C for 24 h. We also observed that phospholipids and glycerol were adsorbed to the surface of the resin that supports the lipase, hindering the contact of the enzyme with the substrate. Furthermore, phosphorus and ethanol can cause catalyst poisoning and enzyme denaturation, respectively, resulting in deactivation of the catalyst. In our research (Sangaletti-Gerhard et al. 2014b) in collaboration with the University de la Frontera, in Chile, we observed that the addition of 5% tert-butanol as a co-solvent, prolonged Novozym®435 lipase activity for three more reaction cycles and increased ethyl ester yield to 94%. This showed that the tert-butanol kept glycerol and phospholipids solubilized in the substrate, delaying their contact with the Novozym®435. The energetic and economic viability of the processes involving alkaline and enzymatic transesterification of the rich miscella and ethanol was assessed in comparison to the conventional reaction, which uses degummed oil, alkaline catalyst, methanol and ethanol.

4 Viability of the Biodiesel from Rich-in-Soybean Oil Miscella Productive Chain

The main economic factor in biodiesel production is the cost associated to feedstock production (oil or fat and alcohol), processing and logistics, comprising up to 75% of the final product cost. Ideal production of the lipid feedstock requires, primarily, its extraction from the animal or plant matrix followed by further purification using as few steps as possible, under an environmentally friendly condition (low waste generation), producing a substrate with good enough quality in order to guarantee high yields of biodiesel (Ma and Hanna 1999; Haas and Foglia 2006). This means that water contents and acid values greater than 0.3% and 1.0 mg KOH/g (0.5% free fatty acid), respectively, are not acceptable because of the incomplete conversion to esters in conventional transesterification reaction due to soap formation (Freedman et al. 1984; Ma et al. 1998). Furthermore, high water contents dilute ethanol or methanol (acyl acceptors) affecting the reaction balance towards the formation of the product (Cernoch et al. 2010). Thus, in order to achieve high yields an excess amount of acyl acceptor is required, leading to increased costs of the process also due to solvent recovery at the end of the reaction. Given the favorable chemical characteristics of the rich miscella (reduced water content, phospholipids and acid values) (Table 1), the direct transesterification (non-refined rich miscella as the substrate) was possible using ethanol as the acyl acceptor and two different types of catalysts: (i) the enzyme, having the strong environmental claim of replacing chemical compounds for natural reactants and; (ii) alkaline, being the most used worldwide. Processes with both catalysts have achieved high conversions of fatty acids ethyl esters (FAEE).

Energetic (Table 2) and economic viability (Table 3) of the biodiesel production chain using the soybean rich miscella, ethanol and alkaline catalyst (as the alternative process) was determined and compared with the conventional biodiesel production using oil extracted with hexane and methanol as the acyl acceptor. Further details on the energy flow of each step of the conventional process of biodiesel and processes using the rich miscella can be found in Sangaletti-Gerhard et al. (2014b).

Figure 5 illustrates the similarity in the total amount of energy demanded by the crop production step (around 20 MJ) and soybean flaking (around 14 MJ) in both cases. However, the extraction step using ethanol required three times more energy than the conventional process, so a reduction of 60% would be necessary in order to make the ethanol process competitive. This energy consumption is later

Table 2 Energy flow inventory of the process steps and energy balance to produce 1 kg of biodiesel (Sangaletti-Gerhard et al. 2014b)

Process stages	Conventional process		Alternative process	
	MJ	%	MJ	%
Input energy				
Soybean crop	19.6	35.4	20.3	29.5
Flaking	14.1	25.6	14.6	21.2
Extraction	10.4	18.8	31.6	46.0
Refining	1.7	3.1		
Transesterification	9.5	17.1	2.3	3.3
Total	55.3		68.8	
Soybean hull	7.2	6.5	7.4	6.5
Meal	63.8	57.7	66.0	58.4
Biodiesel	37.0	33.4	37.0	32.7
Glycerine	2.7	2.4	2.7	2.4
Total	110.7		113.1	
EBAO[a]	17.9		22.8	
EROI[b]	2.00		1.64	

[a]EBAO: energy balance of biodiesel production from energy input allocated to soybean oil 18%
[b]EROI: energy return on investment

compensated in the biodiesel production stage, which demanded four times less energy (2.3 MJ) than to the conventional transesterification process (9.5 MJ).

The energy return on investment (EROI), which is determined by the amount of energy output (OP) divided by the energy invested (IP) by the economic system, was 2.0 and 1.64 for conventional and alternative process, respectively, which means that the system produces more energy than it consumes (Table 2).

Evidently, the energy balance (Σ output energy - Σ input energy) is positive for both processes, the conventional (55.4 MJ) and alternative (44.3 MJ), demonstrating that the process with ethanol has a solid proposal, but also shows great potential for improvement. Moreover, the method using ethanol as solvent saves 1.7 MJ of energy compared to the conventional extraction refining step (degumming and alkali refining) and generates less waste (gums and soapstock), since not always the gum is used or meant for the production of lecithin.

The economic analysis confirmed the results obtained in energy analysis in which the ethanol extraction process requires adjustments, especially regarding reduction of the consumption of anhydrous ethanol employed in the final extraction cycle. According to Table 3, the cost of producing 1 kg of rich miscella (US$7.32) was three times higher than to produce refined soybean oil (US$2.12), meaning that 70% of the volume of ethanol consumed should be reduced in the extraction step. Moreover, different and new extractor models can be designed to attain higher yields of oil extraction. As a promising option, the rich-in-ethanol miscella contains high levels of phospholipids, antioxidants and possibly other compounds of polar affinity.

Table 3 Cost of soybean oil extraction process with ethanol and hexane and biodiesel production

	Material			Cost		
		Hexane extraction	Ethanol extraction	USD	Hexane extraction	Ethanol extraction
	Unit	Quantity	Quantity	(US$ Unit^{-1})	US$	US$
Input						
Extraction						
Soybean	kg	5.955	5.926	0.34[a]	2.02	2.02
Electricity	kWh	0.131[b]	0.016	0.16[c]	0.02	0.002
Steam	kg	1.644[b]		0.03[d]	0.05	
Hexane	L	0.0153[b]		0.31[e]	0.01	
Ethanol	L		9.996	0.53[f]		5.30
Refining						
Degummed oil	kg	1.002[g]		2.1[h]	2.1	
Electricity	kWh	0.013[g]		0.16[c]	0.002	
Steam	kg	0.209[g]		0.03[d]	0.01	
Diesel	L	0.005[g]		0.53[c]	0.003	
NaOH	kg	0.005[g]		0.65[e]	0.003	
Total					2.12	7.32
Transesterification						
Refined oil	kg	1.058[i]		2.12[j]	2.24	
Rich miscella	kg		1.088	7.32[k]		7.97
Methanol	L	0.242[i]		0.53[e]	0.13	
Ethanol	kg		0.145	0.53[f]		0.08
NaOH	kg	0.011[i]	0.007	0.65[e]	0.007	0.005
Total	kg				2.38	8.06
Output						
Meal	kg	3.919[l]	3.983[m]	0.29[a]	1.14	1.16
Glycerine	kg	0.105	0.105	0.02[e]	0.02	0.02
Total					1.16	1.18

[a] Index Mundi commodities—(2016)
[b] Sheehan et al. (1998)
[c] Electricity price for industry in Germany in 2011 (International Energy Agency 2012)
[d] Steam was calculated by using the equation $C_e = 100 \cdot C_f / H_f . \eta$ and $C_s = C_e(h_s - h_{fw})$, where, Ce is energy unit cost (US$/1000 kJ), C_f is diesel unit cost (US$/ton), H_f is calorific value of diesel (low value) kJ/kg, C_e is energy unit cost (US$/1000 kJ), h_s is specific enthalpy of steam at boiler pressure (kJ/kg) and h_{fw} is specific enthalpy of feed water (kJ/kg)
[e] International Construction Information Society—ICIS (2014)
[f] Anhydrous ethanol price industry of Brazil in 2016 (Center for Advanced Studies on Applied Economics 2016)
[g] Dorsa (2004)
[h] Cost input of extraction oil process using hexane for 1 kg refined oil
[i] Keera et al. (2011)
[j] Cost input of extraction oil and refining process using hexane for 1 kg biodiesel
[k] Cost input of extraction oil and refining process using ethanol for 1 kg biodiesel
[l] Toasted meal
[m] Not toasted meal

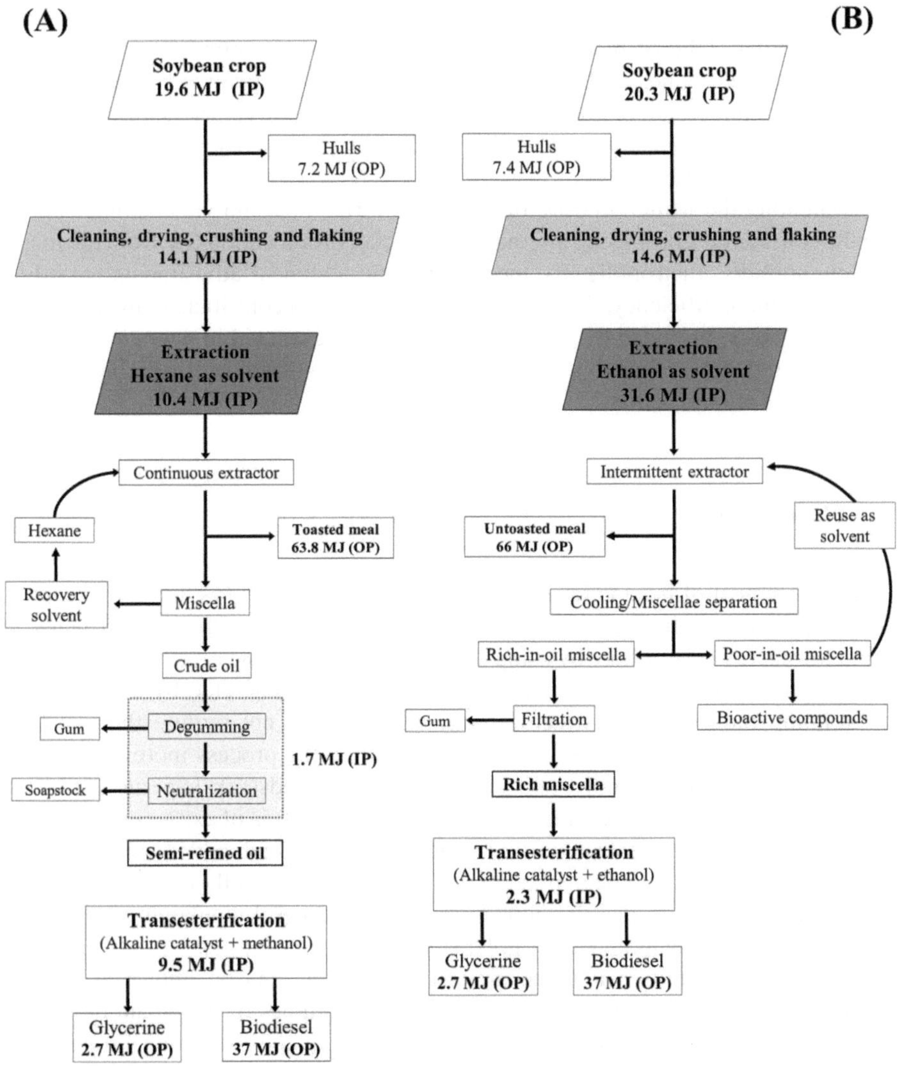

Fig. 5 Biodiesel production chain by the conventional method (**a**) and the alternative method using ethanol as a substitute of hexane and methanol (**b**). *IP* input, *OP* output (Sangaletti-Gerhard et al. 2014b)

These features can be of interest to the pharmaceutical and food industry, adding value to the alternative process.

Assessing the biodiesel production chain of both processes, biodiesel produced by the conventional way (US$2.38) would cost half as much as the biodiesel from alternative process (US$8.06), which is justified by the high cost of feedstock (rich miscella).

Considering that the cost of biodiesel is the difference between input and output process values, 1 kg of biodiesel produced from the conventional and alternative process attained values of US$1.22 and US$6.88, respectively. Comparing these values to commercial biodiesel prices (US$0.83 kg^{-1}) (ICIS 2014), technical adjustments are needed to make the extraction process using ethanol solvent economically feasible.

Evaluating the whole process, biodiesel from rich miscella via alkaline catalyst can be considered a promising alternative, especially for locations far from the major diesel suppliers' metropolis. By means of some technical adjustments regarding improvement of efficiency, the process should be able to contribute to the reduction of energy costs, provided the whole process can be executed in a single industrial plant using solely ethanol as the solvent extraction and acyl acceptor in the transesterification reaction.

5 Conclusions

The replacement of hexane (as the extractant) and methanol (as the acyl acceptor) for ethanol in the biodiesel production chain is promising as it contributes to reducing energy expenditure by using only ethanol for both the oil extraction and the transesterification reaction, reducing waste production generated by the vegetable oil refining industry. Oil extraction with ethanol does not require the usual oil refining steps prior to biodiesel production, making the process more economical and environmentally friendly, and produces a lipid feedstock less susceptible to deterioration and loss by oxidation reactions. The integration of extraction and direct transesterification, generates non-toxic high quality and high added value products and co-products that can be applied by other industries as well. The latter includes the detoxified meal, gums, the poor miscella and glycerol, containing substantial amounts of antioxidants and other polar substances.

The efficiency and cost of the process as a whole can be improved through intensification of research efforts into optimization studies of the ethanolic extraction and the further use of co-products generated in the extraction process and transesterification.

References

Alcantara, R., Amores, J., Canoira, L., Fidalgo, E., Franco, M. J., & Navarro, A. (2000). Catalytic production of biodiesel from soy-bean oil, used frying oil and tallow. *Biomass and Bioenergy, 18*, 515–527.

Angarita, E. E. Y., Lora, E. E. S., Costa, R. E., Lamônica, H. M., Finguerut, J., Ramirez, G. A. S., & Nascimento, M. A. R. (2008). Produção de óleos vegetais e biodiesel: tecnologia e análise de ciclo de vida. In L. A. B. Cortez, E. E. S. Lora, & E. O. Gomez (Eds.), *Biomassa para energia* (732 p). Campinas: Editora Unicamp.

Azócar, L., Ciudad, G., Heipieper, H. J., Muñoz, R., & Navia, R. (2010). Improving fatty acid methyl ester production yield in a lipase-catalyzed process using waste frying oils as feedstock. *Journal of Bioscience and Bioengineering, 109*(6), 609–614.

Balat, M. (2007). Production of biodiesel from vegetable oils: A survey. *Energy Sources, Part A, 29*, 895–913.

Bindhu, C. H., Reddy, J. R. C., Rao, B. V. S. K., Ravinder, T., Chakrabarti, P. P., Karuna, M. S. L., & Prasad, R. B. N. (2011). Preparation and evaluation of biodiesel from *Sterculia foetida* seed oil. *Journal of the American Oil Chemists' Society, 89*, 891–896.

Bozbas, K. (2008). Biodiesel as an alternative motor fuel: Production and policies in the European Union. *Renewable and Sustainable Energy Reviews, 12*, 542–552.

Canakci, M., & Van Gerpen, J. (1999). Biodiesel production via acid catalysis. *Transactions of the ASAE, 42*(1984), 1203–1210.

Castro, A. M. G., Lima, S. M. V., & Pinheiro, N. P. (2010). A indústria de produção de biodiesel no Brasil. In A. M. G. Castro, S. M. V. Lima, & J. F. V. Silva (Eds.), *Complexo Agroindustrial de Biodiesel no Brasil: Competitividade das cadeias produtivas de matérias-primas* (712 p). Brasília, DF: Embrapa Agroenergia.

Center for Advanced Studies on Applied Economics—CEPEA. (2016). *Anhydrous ethanol index, São Paulo state*. Accessed February 17, 2016, from http://cepea.esalq.usp.br/english/ethanol/

Cernoch, M., Hajék, M., & Skopal, F. (2010). Ethanolysis of rapeseed oil—Distribution of ethyl esters, glycerides and glycerol between ester and glycerol phases. *Bioresource Technology, 101*, 2071–2075.

Chisti, Y. (2007). Biodiesel from microalgae. *Biotechnology Advances, 25*, 294–306.

D'Ippolito, S., Yori, J., Iturria, M., Pieck, C., & Vera, C. (2007). Analysis of a two-step, noncatalytic, supercritical biodiesel production process with heat recovery. *Energy and Fuels, 21*(4), 339–346.

Deng, X., Fang, Z., & Liu, Y. (2010). Ultrasonic transesterification of *Jatropha curcas* L. oil to biodiesel by a two-step process. *Energy Conversion and Management, 51*, 2802–2807.

Dorsa, R. (2004). *Tecnologia de óleos vegetais*. Campinas: Ideal.

Dunn, R. O. (2006). Propriedades a baixas temperaturas e desempenho do biodiesel. In G. Knothe, J. V. Gerpen, J. Frahl, & L. P. Ramos (Eds.), *Manual de biodiesel* (pp. 91–132). São Paulo: Edgard Blucher.

Encinar, J. M., González, J. F., Rodríguez, J. J., & Tejedor, A. (2002). Biodiesel fuels from vegetable oils: Transesterification of *Cynara cardunculus* l. oils with ethanol. *Energy and Fuels, 16*, 443–450.

Encinar, J. M., González, J. F., & Rodríguez-Reinares, A. (2007). Ethanolysis of used frying oil. Biodiesel preparation and characterization. *Fuel Processing Technology, 88*, 513–522.

Felizardo, P., Correia, M. J. N., Raposo, I., Mendes, J. F., Berkemeier, R., & Bordado, J. M. (2006). Production of biodiesel from waste frying oils. *Waste Management, 26*, 487–494.

Filippis, P. D. E., Borgianni, C., & Paolucci, M. (2005). Rapeseed oil transesterification catalyzed by sodium phosphates. *Energy and Fuels, 19*(10), 2225–2228.

Fonseca, H., & Regitano-d'Arce, M. A. B. (1993). Aflatoxin removal of peanut meals with aqueous ethanol. *Scientia Agricola, 50*(1), 154–156.

Freedman, B., Pryde, E., & Mounts, T. (1984). Variables affecting the yields of fatty esters from transesterified vegetable oils. *Journal of the American Oil Chemists Society, 61*(10), 1638–1643.

Gandhi, A. P., Joshi, K. C., Jha, K., Parihar, V. S., Srivastav, D. C., Raghunadh, P., Kawalkar, J., Jain, S. K., & Tripathi, R. N. (2003). Studies on alternative solvents for the extraction of oil soybean. *International Journal of Food Science and Technology, 38*, 369–375.

Groppo, S. S. (2015). *Tortas de algodão e amendoim como matérias-primas para produção de miscela etanólicas, farelo e biodiesel* (109 p). Piracicaba: Dissertação (Mestrado)—Centro de Energia Nuclear na Agricultura.

Haas, M. J., & Foglia, T. A. (2006). Matérias-primas alternativas e tecnologia para a produção de biodiesel. In G. Knothe, J. V. Gerpen, J. Frahl, & L. P. Ramos (Eds.), *Manual de Biodiesel* (340p). São Paulo: Editora Blucher.

Hausman, C. (2012). Biofuels and land use change: Sugarcane and soybean acreage response in Brazil. *Environmental and Resource Economics, 51*, 163–187.

Hron, R. J., Kuk, M. S., Abraham, G., & Wan, P. J. (1994). Ethanol extraction of oil, gossypol and aflatoxin from cottonseed. *Journal of the American Oil Chemists' Society, 71*(4), 417–421.

Index Mundi Commodities. (2016). *Commodity prices—Vegetables oils and protein meal*. Accessed February 17, 2016, from http://www.indexmundi.com/commodities/?commodity=soybeans

International Construction Information Society—ICIS. (2014). *Indicative chemical prices AeZ*. Accessed February 17, 2016, from http://www.icis.com/chemicals/channel-info-chemicals-a-z

International Energy Agency—IEA. (2012). *IEA statistics: Electricity information*. Accessed February 17, 2016, from http://www.iea.org/media/training/presentations/statisticsmarch/electricityinformation.pdf

Ji, J., Wang, J., Li, Y., Yu, Y., & Xu, Z. (2006). Preparation of biodiesel with the help of ultrasonic and hydrodynamic cavitation. *Ultrasonics, 44*, e411–e414.

Johnson, L. (2002). Recovery, refining, converting, and stabilizing edible fats and oils. In C. C. Akohn & D. B. Min (Eds.), *Food lipids: Chemistry, nutrition, and biotechnology* (pp. 241–292). New York: Marcel Dekker.

Johnson, L. A., & Lusas, E. W. (1983). Comparison of alternative solvents for oils extraction. *Journal of the American Oil Chemists' Society, 60*(2), 229–242.

Keera, S. T., El Sabagh, S. M., & Taman, A. R. (2011). Transesterification of vegetable oil to biodiesel fuel using alkaline catalyst. *Fuel, 90*, 42–47.

Knothe, G. (2006). Analyzing biodiesel: Standards and other methods. *Journal American Oil Chemists' Society, 83*(10), 823–833.

Knothe, G., & Razon, L. F. (2017). Biodiesel fuels. *Progress in Energy and Combustion Science, 58*, 36–59.

Kucek, K. T., Oliveira, M. A. F. C., Wilhelm, H. M., & Ramos, L. P. (2007). Ethanolysis of refined soybean oil assisted by sodium and potassium hydroxides. *Journal American Oil Chemists' Society, 84*(4), 385–392.

Kurzin, A. V., Evdokimov, A. N., Pavlova, O. S., & Antipina, V. B. (2007). Synthesis and characterization of biodiesel fuel based of esters of tall oil fatty acids. *Russian Journal of Applied Chemistry, 80*(5), 842–845.

Lam, M. K., Lee, K. T., & Mohamed, A. R. (2010). Homogeneous, heterogeneous and enzymatic catalysis for transesterification of high free fatty acid oil (waste cooking oil) to biodiesel: A review. *Biotechnology Advances, 28*, 500–518.

Liao, C. C., & Chung, T. W. (2011). Analysis of parameters and interaction between parameters of the microwave-assisted continuous transesterification process of *Jatropha* oil using response surface methodology. *Chemical Engineering Research and Design, 89*, 2575–2581.

Lima, S. M. V., & Castro, A. M. G. (2010). A indústria A agroindústria de óleo vegetal para a produção de biodiesel. In A. M. G. Castro, S. M. V. Lima, & J. F. V. Silva (Eds.), *Complexo agroindustrial de biodiesel no Brasil: Competitividade das cadeias produtivas de matérias-primas*. Brasília, DF: Embrapa Agroenergia.

Ma, F., Clements, L. D., & Hanna, M. A. (1998). The effects of catalyst, free fatty acids, and water on transesterification of beef tallow. *American Society of Agricultural Engineers, 41*(5), 1261–1264.

Ma, F., & Hanna, M. A. (1999). Biodiesel production: A review. *Bioresource Technology, 70*, 1–15.

Meher, L. C., Sagar, D. V., & Naik, S. N. (2006). Technical aspects of biodiesel production by transesterification—A review. *Renewable and Sustainable Energy Reviews, 10*, 248–268.

Miao, X., & Wu, Q. (2006). Biodiesel production from heterotrophic microalgal oil. *Bioresource Technology, 97*, 841–846.

Mondala, A., Liang, K., Toghiani, H., Hernandez, R., & French, T. (2009). Biodiesel production by in situ transesterification of municipal primary and secondary sludges. *Bioresource Technology, 100*, 1203–1210.

Naegel, P., & Althuser, J. (2008). Les débuts du moteur Diesel en France (pp. 2–7). Connaissance de la Meuse.

Nimcevic, D., Puntigam, R., Wörgetter, M., & Gapes, J. R. (2000). Preparation of rapeseed oil esters of lower aliphatic alcohols. *Journal of the American Oil Chemists' Society, 77*(3), 275–280.

Ragit, S. S., Mohapatra, S. K., Kundu, K., & Gill, P. (2011). Optimization of neem methyl ester from transesterification process and fuseel characterization as a diesel substitute. *Biomass and Bioenergy, 35*, 1138–1144.

Ranganathan, S. V., Narasimhan, S. L., & Muthukumar, K. (2008). An overview of enzymatic production of biodiesel. *Bioresourse Technology, 99*, 3975–3981.

Rashid, U., & Anwar, F. (2008). Production of biodiesel through optimized alkaline-catalyzed transesterification of rapeseed oil. *Fuel, 87*, 265–273.

Rashid, U., Ibrahim, M., Ali, S., Adil, M., Hina, S., Bukhari, I.H., & Yunus, R. (2012). Comparative study of the methanolysis and ethanolysis of maize oils using alkaline catalysts. *Grasas y Aceites, 63*(1), 35–43 (Enero-Marzo).

Regitano-d'Arce, M. A. B., Gutierrez, E. M. R., & Gunther, P. A. S. (1994). Estudo da cinética de extração do ácido clorogênico de tortas de girassol com etanol. *Ciência e Tecnologia de Alimentos, 14*(2), 149–159.

Saka, S., & Kusdiana, D. (2001). Biodiesel fuel from rapeseed oil as prepared in supercritical methanol. *Fuel, 80*, 225–231.

Sangaletti, N., Cea, M., Regitano-d'Arce, M. A. B., Vieira, T. M. F. S., & Navia, R. (2013). Enzymatic transesterification of soybean ethanolic miscella for biodiesel production. *Journal of Chemical Technology and Biotechnology, 88*, 2098–2106.

Sangaletti-Gerhard, N., Cea, M., Risco, V., & Navia, R. (2015). In situ biodiesel production from greasy sewage sludge using acid and enzymatic catalysts. *Bioresource Technology, 179*, 63–70.

Sangaletti-Gerhard, N., Romanelli, T. L., Vieira, T. M. F. S., Navia, R., & Regitano-d'Arce, M. A. B. (2014b). Energy flow in the soybean biodiesel production chain using ethanol as solvent extraction of oil from soybeans. *Biomass and Bioenergy, 66*, 39–48.

Sangaletti-Gerhard, N., Vieira, T. M. F. S., Groppo, S. S., Rodrigues, J. R., & Regitano-d'Arce, M. A. B. (2014a). Alkaline-catalyzed ethanolysis of soybean oil ethanolic miscella. *Fuel, 116*, 415–420.

Shay, E. G. (1993). Diesel fuel from vegetable oils: Status and opportunities. *Biomass and Bioenergy, 4*, 227–242.

Sheehan, J., Camobreco, V., Duffield, J., Graboski, M., & Shapouri, H. (1998, May). *Life-cycle inventory of biodiesel and petroleum diesel for use in an urban bus* (Final Report No.: NREL/SR-580e24089. Contract No.: DEAC0283CH10093). Golden, CO: National Renewable Energy Laboratory.

Sherwin, E. (1976). Antioxidants for vegetable oils. *Journal of the American Oil Chemists' Society, 53*, 430–436.

Singh, S. P., & Singh, D. (2010). Biodiesel production through the use of different sources and characterization of oils and their esters as the substitute of diesel: A review. *Renewable and Sustainable Energy Reviews, 14*, 200–216.

Soriano Jr., N. U., & Narani, A. (2011). Evaluation of biodiesel derived from *Camelina sativa* oil. *Journal of the American Oil Chemists' Society, 88*, 917–923.

Tiwari, A. K., Kumar, A., & Raheman, H. (2007). Biodiesel production from Jatropha oil (*Jatropha curcas*) with high free fatty acids: An optimized process. *Biomass and Bioenergy, 31*, 569–575.

Watanabe, Y., Shimada, Y., Sugihara, A., & Tominaga, Y. (2002). Conversion of degummed soybean oil to biodiesel fuel with immobilized *Candida antarctica* lipase. *Journal of Molecular Catalysis—B Enzymatic, 17*(3–5), 151–155.

MIX
Papier aus verantwortungsvollen Quellen
Paper from responsible sources
FSC® C105338

If you have any concerns about our products,
you can contact us on
ProductSafety@springernature.com

In case Publisher is established outside the EU,
the EU authorized representative is:
Springer Nature Customer Service Center GmbH
Europaplatz 3, 69115 Heidelberg, Germany

Printed by Libri Plureos GmbH
in Hamburg, Germany